# THE EXTRACTION STATE

# THE
# EXTRACTION STATE

## A HISTORY
### OF
## NATURAL GAS
### IN
## AMERICA

# CHARLES BLANCHARD

UNIVERSITY OF PITTSBURGH PRESS

Published by the University of Pittsburgh Press, Pittsburgh, Pa., 15260
Copyright © 2020, University of Pittsburgh Press
This paperback edition, 2021
All rights reserved
Manufactured in the United States of America
Printed on acid-free paper

Cataloging-in-Publication data is available from the Library of Congress

ISBN 13: 978-0-8229-6676-0
ISBN 10: 0-8229-6676-X

Jacket design by Melissa Dias-Mandoly

# THE EXTRACTION STATE

# PART I

## MAKING USE OF A USELESS
## BYPRODUCT, 1878–1954

If you were to ask a crowd which state is the historic home of the oil and gas industry, surely the overwhelming and emphatic response would be Texas. You might hear Oklahoma or Louisiana as well—what I'll refer to collectively as the Southwest. And while it is true that those states became the most important in North American oil and gas, in its early days the young industry was centered in Pennsylvania, West Virginia, Ohio, and western New York—Appalachia. It was in Appalachia, in the second half of the 1800s, where the first oil and gas wells were drilled, the first gas pipelines were laid, and the first consumers switched from coal-derived manufactured gas to natural gas.

I begin the story of the natural gas industry in the late 1870s. At this point the Industrial Revolution had made coal an important and widely used fuel in industry, though most people still used wood to heat their homes and businesses. The world's first oil boom began in northwest Pennsylvania in 1859, and thousands of prospectors had rushed to the region in search of black gold, much in the same way they had rushed to California a decade earlier in search of actual gold. The lucky struck oil. The unlucky struck nothing. The truly unlucky struck an uncontrollable, unstorable, and entirely unusable compound: natural gas.

# CHAPTER 1

## THE SMOKY CITY

Picture yourself standing on the sidewalk, looking out on a broad cobblestone street. A row of brick townhomes stands behind you. A man is walking across the street, toward you, wearing a denim shirt buttoned all the way up to his neck. With each step he takes, the rim of his porkpie hat flaps slightly, softened by years of wear. The two sunken breast pockets on his shirt are both stained black with soot from where his fingers reach for tobacco to fill the pipe hanging from his mouth. Soot-strewn suspenders hold up a pair of black-creased dungarees, which hide a pair of tattered leather boots. He steps from the street to the sidewalk, nods to you, and makes his way further into town.

Your intent focus on this man's attire is not due to sartorial interest but rather because the city that surrounds you is cloaked in smoke, as if you are staring at it through goggles made of sea glass.

The view is not entirely obscured, though. As the man walks away, you can easily make out the outlines of a large factory, with a forest of chimneys rising forty or fifty feet in the air, the ones closest to you producing smoke of a light gray, while the ones farther away seem to belch pure black. If the wind kicks up and clears the air, you will see a five-hundred-foot-tall wall of dirt on the other side of the Monongahela River, now known as Mount Washington, but that the old-timers still call Coal Hill.

This is Pittsburgh in 1878. The "Smoky City," as it was known colloquially, was industrializing at a rapid clip. As such, it was the newfound home of many thousands of immigrants, chiefly from Central and Eastern Europe. Germans and Poles would lend Pittsburgh much of its unique linguistic and culinary character; it was just two years earlier, in 1876, when Henry John Heinz sold his first jar of Pittsburgh-made ketchup, after going bankrupt trying to sell horseradish, pickles, and sauerkraut. A less numerous, but equally important, group of immigrants came from England—many from Sheffield—which had, since the time of Chaucer, been the steel capital of the world. If vinegar defined Pittsburgh's culinary character, steel defined its economy.

By 1878 Pittsburgh was home to 150,000 people and hundreds of factories that accounted for roughly 50 percent of the United States' production of glass, 40 percent of its iron, and 15 percent of its steel (but almost 70 percent of its higher-quality crucible steel—the variety in which Sheffield had long held a monopoly).[1] Indeed, 1870 to 1910 is considered Pittsburgh's golden age, as its population swelled from 86,000 to more than 500,000— more than double the national average rate of growth, and almost totally on the back of heavy industry.

This was a frenetic period of technological advancement and economic growth. Just twenty years prior, Pittsburgh's first blast furnace for making pig iron was completed,[2] but now the Bessemer process had been commercialized and Andrew Carnegie, the Star-Spangled Scotsman, was turning Pittsburgh into the world's preeminent steel city. Carnegie's first Bessemer mill, the Edgar Thomson Steel Works, produced its first steel in 1875 in a town called Braddock, just outside Pittsburgh.

Pittsburgh was a natural home for the emergent US steel industry, both because of its geography and its geology. The Allegheny and Monongahela Rivers join in downtown Pittsburgh to form the Ohio, the main tributary of the Mississippi. This made it the easternmost city with access to the Mississippi, and earned it the moniker "Gateway to the West" by the early 1800s. Pork, corn, and whiskey flowed to Pittsburgh from the east, where it was loaded onto barges bound for markets to the south and west, and often as far away as the Caribbean. Indeed, Pittsburgh's earliest industry was shipbuilding, and it was from Pittsburgh docks that the first steamship, Robert Fulton's *New Orleans*, made the journey to its namesake city.

Probably more important even than the Ohio River to Pittsburgh's eventual status as an industrial powerhouse, though, was the Pittsburgh Coal Seam. The summit of Mount Washington looms only about five hundred feet above downtown Pittsburgh, so it is really more of a large hill than a small mountain. It would, of course, have been unthinkable to affix George Washington's great name to a mere hill. But before the area began to be developed as a residential community in the 1860s, it had a more befitting name: Coal Hill.

Today, the neighborhood atop Mount Washington is affluent, encircled by green spaces and home to many restaurants and bars that capitalize on the beautiful views. Beginning around 1830, though, and continuing until city officials began allowing the trees lining Mount Washington's slopes to grow back in the 1920s, the area was so devoid of life that there would hardly have been a difference if the following black-and-white photograph had been taken in color.

Settlers began exploiting Coal Hill as early as the 1760s, and by the early 1800s coal production had grown so much that Pittsburgh had become

**FIGURE 1.1:** Mount Washington (aka Coal Hill) in the 1920s

the only US city of any note where coal, rather than wood, served as the primary fuel source. The coal bed that gave Coal Hill its name was part of a much larger formation called the Pittsburgh Coal Seam, which actually underlies over eleven thousand square miles in Pennsylvania, West Virginia, and Ohio.

G. Follansbee, who was superintendent of the Pittsburgh Chamber of Commerce in 1882, wrote quite lyrically about the city's bounty:

> Coal is the sunshine of long past ages stored in an ebon[y] casket awaiting man's unlocking to serve his purposes. It is the lever of most potent energy; driving the wheels of manufactures and of commerce; rescuing the metals of the world from the baser mold which surrounds and contaminates them: dispelling the gloom of night and melting the severity of winter. Its possession by State or community is, then, of primary and inestimable importance.
>
> The wonderful prosperity of Pittsburgh as manufacturing center is chiefly due to its possession of a vast vein of bituminous coal, unexcelled in quality for gas, steam, and domestic purposes.[3]

Not only was coal from the Pittsburgh Coal Seam ubiquitous and cheap to mine, with much of it lying practically on the surface, it was also well suited for producing something called coke. It was this, more than anything else, that earned Pittsburgh the moniker "Steel City."

## COAL, COKE, IRON, STEEL, AND BEER

A few sentences must be devoted to the process by which iron ore and coal are made into steel. We will start first with coal—more precisely, the high-quality metallurgical coal needed for producing iron and steel, which was found throughout the Pittsburgh Coal Seam.

Coal itself is incidental to ironmaking, except that it can be used as a heat source. Coke, however, is essential. Producing coke is very simple—if coal is heated to around 1,100 degrees Celsius in the absence of oxygen, you get two things: coke and coke oven gas. For our intents and purposes, coke is basically pure carbon, and coke oven gas is a by-product that can also be burned for heat.

Iron ore is mined from the earth. The "iron" in the ore is not pure, metallic iron but rather an oxide of iron. To create metallic iron, the oxygen must be removed, and that is what coke does. During the smelting process, iron ore and coke are combined in a blast furnace, air is blown in to introduce oxygen, and the coke combusts. The coke's carbon combines with oxygen from the air to form carbon monoxide, which in turn steals oxygen away from the iron oxide, yielding pig iron. To make steel via the Bessemer process, air is basically just blasted through molten pig iron. Easy.

However, not all coals can be used to make coke. The conditions inside a blast furnace are extraordinarily violent, and coke needs to withstand the enormous stresses without breaking. If it disintegrates, the ash blocks the flow of oxygen that drives the entire reaction. Coals that are too soft can function effectively as a heat source, but are structurally worthless. But coal from the Pittsburgh Coal Seam was very hard, and thus well suited for ironmaking. The abundance of this hard coal made Pittsburgh a natural home for iron and steel manufacture.

While coke's most essential role is certainly in ironmaking, where it is irreplaceable, it also boasts certain advantages over coal as a fuel for heat. It is cleaner burning, as it has been stripped of the volatile components that boil off as coke oven gas during the coking process. This meant less smoke inside the Pittsburgh homes and businesses that burned coke instead of wood. It is also what allowed coke to replace wood-derived charcoal in a brewery in Derbyshire, England, in 1642. The brewmasters had experimented with burning coal to roast the malt, but this imparted in the beer foul, sulfurous taste. Coke, however, burned cleanly, and, unlike charcoal, it imparted no brownness to the malt, producing instead a golden, pale brew. Coke is what gave the world pale ale.[4]

Back to 1878, and our man with the dirty shirt and pork pie hat. On the other side of the steel mill he is walking toward are its rail facilities, which bring in iron ore from Michigan and Wisconsin. Indeed, Pittsburgh actually imports most of its iron ore. Its competitive advantage in iron and steelmaking stems wholly from its great seams of hard coal.

The locomotive turns around and carries away finished steel, which is turned into ships, bridges, railroad tracks, and more locomotives. Only after the 1890s does steel-framed building construction become common, and the first Ford Model T does not roll off the assembly line until 1908. The steel that the trains do not take are loaded onto barges and steamed down the Ohio River, to markets foreign and domestic.

## THE HAYMAKER WELL

While Pittsburgh's economic growth had been impressive throughout the 1870s, and would continue to be for several more decades, a highly industrial economy fueled exclusively by coal suffers on many quality-of-life measures. The Smoky City was not so named for an innocuous morning fog as it was for the black soot that occupied every cupboard corner, floorboard crack, and fingernail bed. In 1846 the *Pittsburgh Gazette* wrote of Coal Hill, clearly visible from much of downtown Pittsburgh:

> In the days of its glory, which covered with trees from summit down to the edge of the water, it was the fairest portion of our surrounding scenery. But, now how changed! At its base vast furnaces belch forth dense clouds of flame and smoke, its steep side has been cut down by large quarries, and all along near its top a dozen yawning throats pour down a dozen railroads its rich treasures. Tree and shrub have been reft from their fast hold, and the old hill now stands before us with scarred sides and almost shaven crown.[5]

In 1866 a travel essayist for the *Atlantic Monthly*, James Parton, ventured to the top of Coal Hill, which had only just been recently rechristened Mount Washington, and famously described the city as "hell with the lid taken off." Records show that the smoke had only gotten worse in the twelve years between Parton's essay and 1878—not surprising, as the population had doubled, pig iron production had more than doubled, and steel output had risen by a whopping 50 percent per year.[6] The man with the sooty shirt pockets, then, is not unsanitary—he's merely a typical late nineteenth-century working-class Pittsburgher.

The year 1878 was going by much like 1877, as far as most Pittsburghers were concerned—industry continued to grow rapidly, with coal consumption in lockstep. But in the hills about twenty miles east of the city, two brothers who had been tooling around with what they hoped would be an oil well were about to bring in the biggest "gasser" the world had yet seen.

Obadiah and Michael Haymaker might have looked and dressed similarly to our man outside the factory, but they would have been covered in brown dirt rather than black soot, for they had spent months trying to drill a well in the still-forested hills east of Pittsburgh, in Murrysville. They were using outdated equipment, and it took them a full year to reach a depth of just 400 feet. (For context, today's modern rigs drill thousands of feet each day.) By 500 feet, the Haymakers had run out of money, and began selling interest in the well to speculators and promoters. This gave them the capital that they needed to keep drilling, which they did until they hit 1,400 feet, when the well gave "a terrific roar and rumble that was heard fifteen miles away."[7]

The Haymaker brothers would probably have felt a mix of terror and triumph as the well came to life, after more than a year of dogged effort. A good oil well would make them rich. But their jubilation was cut short when, instead of the gentle gurgle of oil, they heard only the deafening hiss of gas.

After the blowout of November 3, 1878, the Haymaker well began producing an estimated thirty to forty million cubic feet of natural gas per day. This well, on its own, would have supplied the entire East Coast with gas. But instead, lacking a pipeline and a market, it vented its gas straight into the sky.

It was, perhaps, inevitable that such a spectacle would attract curious onlookers, even in the evening hours, when lanterns were needed to light the way through the woods. One such lantern-wielding onlooker inadvertently caught the gas stream on fire, leading to a conflagration of biblical proportions. The well burned for four years straight, a lighthouse in the Alleghenies. A historian wrote, some years later: "Its flaming fire, issuing from the earth could be seen at night at a distance of eight or ten miles, and its roaring sound was distinctly heard for five or six miles."[8]

The Haymaker and other wells to be drilled in the Murrysville gas field had the potential to clean up the smog that had become Pittsburgh's trademark, inasmuch as they could displace coal with natural gas, which is free from soot. But the Haymaker well burned uncontrollably in the woods, without a pipeline through which to market its production.

Eventually, the well was brought under control. Then came the fight over who owned its bounties. In 1882 one of the Haymakers's outside investors, H. J. Brunot, invited Joseph Pew to come to Murrysville to see the still-burning well. A deal was negotiated to sell the well to Pew and his business partner, Edward Octavius Emerson, who was a cousin of Ralph Waldo Emerson (the family apparently had a penchant for outlandish middle names). Pew and Emerson wanted to build a pipeline to connect the well to Pittsburgh, where they would sell its gas to factories that burnt coal for heat. Complicating matters, though, was the well's contested ownership. A "promoter" from Chicago had offered to purchase the well for $20,000 earlier that same year, and sent a down payment of $1,000. While initially keen, Brunot and the Haymakers had given up on him after no more payments arrived. After making more solid arrangements with Pew and Emerson, Brunot tried to refund the $1,000 to the Chicago promoter, but the promoter would not cash the check and give up his supposed ownership of the well. Brunot then deposited the money directly in the promoter's account, and thought the matter resolved.

The events of November 26, 1883, proved that it was not. That day, fifty men armed with rifles and bayonets, representing the Chicago promoter, came to take the well by force. The Haymakers then arrived with a posse of ten. Cooler heads, in this case, did not prevail, and when the battle was over, Obadiah Haymaker had been bayoneted four times and died before reaching home. His brother Michael escaped unharmed. He moved to San Antonio, Texas, where he died an old man in his nineties. The Chicago promoter served time in jail. Pew and Emerson became the uncontested owners of the Haymaker well.[9]

## NATURAL GAS IS KING IN PITTSBURGH

Pew and Emerson's company, which they called Penn Fuel Gas Company, delivered the first molecule of natural gas to Pittsburgh's Sixteenth Street station in January 1883. Penn Fuel had bought the Haymaker well and additional acreage in Murrysville and constructed a 5 5/8 inch iron pipeline into the city. When Pittsburgh received its first natural gas, the *New York Times* heralded its arrival as panacea to Pittsburgh's pollution problems, writing that "the place will lose its world-renowned title of 'Smoky City.' The inhabitants are rejoicing at the prospect."[10]

While Pittsburgh was not the first American city to receive natural gas, it was the first that mattered. (The writer apologizes to the good people of Fredonia and Barcelona, New York, and Titusville, Pennsylvania.) This was so not only because it boasted a larger population than the small towns that had used limited volumes of natural gas before, but because it had such a ready displacement market. Enter the story of manufactured gas.

Recall that heating coal in the absence of oxygen yields two products: coke and coke oven gas. Coke is very nearly pure carbon. Coke oven gas is about one-half hydrogen and one-third methane, with the remainder consisting of inert gases with no heating value. Until the 1870s, coke oven gas and manufactured gas were virtually synonymous. Then, a major technological improvement added oil to increase the heating value and somewhat purify the coke oven gas. The resultant product—still known as manufactured gas[11]—quickly became ubiquitous, while natural gas remained rare. Natural gas burned more cleanly, and boasted more energy per cubic foot, but was only used near to where it happened to be found.

The natural gas industry was in its infancy in the 1880s, but manufactured gas had been around since the early 1800s. American engineers brought the technology over from England, who would argue that they had invented it and not the French, which is open to debate. Squabbling over the genesis aside, manufactured gas got its start in the United States when a portrait painter fittingly named Rembrandt Peale nearly went bankrupt after opening a museum in Baltimore.

The Peale family were artists of note: both Rembrandt and his father, Charles, had painted portraits of George Washington, who was a close acquaintance of the family. Charles diversified his earnings when he opened a museum in Philadelphia around 1790, which featured not portraits nor even paintings, as one might expect, but animal specimens. The first two exhibits were of a dried paddlefish from the Allegheny River and a badly preserved Angora cat.[12] Eventually, more collections were added, and the museum grew successful enough to support the entire Peale family. Aging Charles turned the reins over to one of his sons, Rubens, in 1810.

Rubens was the only Peale who did *not* wish to become an artist, striving instead to deftly manage the family's museum business. Impressed with the gas streetlights he had seen on a trip to Europe some years earlier, he decided to install gaslights in the Philadelphia museum in 1814 in order to draw larger crowds. The gimmick worked, and museum attendance skyrocketed, with the gaslights becoming as much of an attraction as the exhibits.

Rubens's brother, Rembrandt, was an artist, and a talented one at that. He completed more than six hundred paintings in his lifetime, which are still displayed at museums across the United States, and was just as prolific a father as a painter, siring nine children with his wife. His large family required more cash than his art could provide, so Rembrandt took a cue from the Peale playbook and opened a museum in Baltimore in 1814.

Rembrandt's superior credentials notwithstanding, business at the museum was less than brisk. Much of this was due to poor timing: the British had attacked Baltimore in September 1814 as part of the War of 1812, keeping many Baltimoreans otherwise preoccupied. Having spent a princely sum to get the museum up and running, Rembrandt began looking for a way to drum up business. Once again, Rembrandt chose to mimic in Baltimore what his father and brother had done in Philadelphia, and installed gaslights in his museum. Once again, the gimmick worked—the museum began to prosper.

Where Rembrandt's path diverged from his father and brother's was in his realization that gaslight was more than a novelty. He wanted to light the streets and homes of Baltimore, not just the hallways of his museum. To this end, in 1816 Rembrandt and some associates founded the Gas Light Company of Baltimore, and in 1817 the company lit its first municipal streetlamp.[13]

They thought that gaslights would be adopted quickly, but the uptake was painfully slow. The incumbent whale oil and tallow industries mounted a hostile public relations campaign, charging that running gas lines through the streets was sure to lead to explosions and deaths, and the company laid

less than two miles of pipelines over its first twenty years.[14] Despite his
museum's popularity, Rembrandt could not afford to wait even a few years;
he had always had a poor mind when it came to his finances, and held considerable debt. Rembrandt quickly began bickering with his associates, who
grew weary of his histrionics and used his precarious financial position to
force him out of the business after just a few years. Rembrandt ceded control of the Baltimore museum to Rubens several years later, which let him
focus fully on his true passion: portrait painting.

Even with Rembrandt Peale gone, the Gas Light Company of Baltimore did not add customers at a rapid clip for some time. The gasification
drive, though, spread across the country regardless, not from a home base
in Baltimore but rather from London, where the technology had a firm seat.
The next major city to install municipal gas lighting was New York City
in 1823, where a group of "men of standing and means" formed the New
York Gas Light Company and beat back negative publicity from the whale
oil and tallow men. Unlike in Baltimore, gaslight was adopted rapidly in
New York, perhaps because of its greater wealth and larger population, and
Broadway quickly became the brightest street in the country. The other major US cities fell like dominoes over the years: Chicago got manufactured
gas in 1850, San Francisco in 1854, and by the 1870s manufactured gas
illuminated virtually every thoroughfare worthy of mention.[15]

Across the United States, manufactured gas was already an entrenched
industry, and by the time that Pittsburgh received natural gas in 1883, an
early incarnation of a business model that prevails to this day had already
been established. This business model is referred to by academics and prac-
titioners alike as the *utility compact*.

Utilities—in this case gas utilities—would receive a franchise from
the city granting them the right to dig up streets and otherwise use public
spaces in return for providing service at a "just and reasonable" rate. The
franchise agreement was a contract, and terms varied, especially in the early
days of the industry. Most importantly, some cities granted utilities an exclusive franchise—monopoly power—within their service territory, while
others did not, leaving markets open for competition.

It is thus unsurprising that executives from a Pittsburgh-based manufactured gas utility called Fuel Gas Company, which had been granted an
exclusive franchise under Pennsylvania's Manufactured Gas Act to supply
the city with "heat from gas," felt betrayed when in 1883 Pew and Emerson's competing Penn Fuel began to pipe natural gas from the Haymaker
well into Pittsburgh. Fuel Gas Company had raised capital to build a facility to turn coal into coke and coke oven gas, lay gas distribution lines,
and sign up customers based on the guarantee that their exclusive franchise
would shield them from competition. They immediately sued, but Penn

Fuel argued that the former's exclusive franchise only applied to sales of manufactured gas, while Penn Fuel sought to distribute natural gas. The protection afforded to utilities under exclusive franchise agreements, Penn Fuel argued, was only from lookalike competitors.

Ultimately, the battle went to the Pennsylvania Supreme Court, which agreed with Penn Fuel that Fuel Gas Company's charter granted them a monopoly only in the distribution of manufactured gas, and that Penn Fuel was marketing a wholly different commodity. This, in turn, led to Pennsylvania passing the Natural Gas Companies Act in 1885, which expressly allowed natural gas firms to compete for business with manufactured gas firms and also with each other—natural gas companies were not initially given a monopoly within their service territory. Penn Fuel was the first to file for a charter, and Fuel Gas Company was not far behind.[16]

In fact, Fuel Gas Company had tried to buy the Haymaker well some years before, but was beaten to the punch by Pew and Emerson. Fuel Gas Company executives realized that natural gas was a superior alternative to manufactured gas; they did not want to block its introduction to Pittsburgh, they just wanted to be the ones who introduced it, and to retain their monopoly.

## THE UTILITY COMPACT AND REGULATED MONOPOLY

The theme of entrenched interests, like Fuel Gas Company, pressuring lawmakers, administrators, and the courts to block potential competitors from entering the market is repeated ad infinitum in the history of regulated industries. Often justifiably, as regulatory barriers to entry are an essential part of the utility compact.

The basic logic of the utility compact is as follows: Certain industries are inherently suited to be best operated as monopolies. These "natural monopolies" are generally industries characterized by high upfront costs and low variable costs, which offer increasing returns to scale. The market is most efficiently satisfied by one, and only one, firm—to the naturally monopolistic victor goes the spoils. This is in contrast with markets where competition improves efficiency. The ice cream truck business, for example, is not a natural monopoly. Each truck is fairly inexpensive, but requires a high amount of labor and fuel to operate. Nothing would be gained by dispensing ice cream cones from an eighteen-wheeled tractor-trailer, and children would probably be quite scared. The gas pipeline business, however, is a natural monopoly. After a large initial outlay, the pipeline is cheap to operate, and larger pipelines have lower costs per unit of gas transported. This means that building a second pipeline where only one is needed is wasteful.

If competition is unworkable in a naturally monopolistic industry, then how to control for a monopoly's tendency to exploit its dominance? How to, and whether to, for in a capitalist economy like the United States, some monopolies are allowed to exploit their dominance without intervention.

The term *affected with a public interest* is a phrase that has been applied to certain industries since it was coined in medieval England, except that it was then spelled *publick*. When an industry is deemed public, the government protects consumers by regulating the rates that the monopoly is allowed to charge. This protectional regulation against exploitation is the more visible side of public utility regulation. Equally, however, the government is interested in the long-term health of public industries and in preventing duplicative wastefulness. Natural monopolies are subject to outside "raids." Wasteful though it may be to build a second pipeline parallel to the first, the raiding company may be willing to incur losses while it bankrupts the incumbent. Left alone, unprotected by any regulation, the risks of competition would be too high to raise the large amount of upfront capital required, so no industry would develop, or would develop chaotically. This, too, is an unacceptable outcome for a public industry.

The government thus offers public industries a quid pro quo: protective regulation, designed to ensure just and reasonable prices for consumers, alongside promotional regulation, designed to spur investment and shield against competition.

In the utility world, rate setting serves both promotional and protective purposes. The regulator authorizes the utility to charge a rate high enough and for a period long enough to fully amortize its substantial upfront investments, plus a reasonable margin, but no more. This is known as a "cost-plus" rate structure. An additional dimension of promotional regulation is achieved by granting franchise exclusivity, or partial exclusivity—entry control shields the natural monopoly from competition.

This model both encourages the formation and stability of the industry and, when properly implemented, ensures that consumers receive the lowest cost of service possible. Not only is duplication avoided, promotional regulations translate into low risk, meaning that capital is raised at low rates. And that cheap, ever-ready capital translates into low and stable costs for consumers. This is what makes the utility compact a win-win for both the company and the regulator.

Over the past two centuries, entrenched utility interests have scored major regulatory victories: cases where the government acted as a savior to their interests. They have also suffered major losses: cases where the government has acted, willingly or unwillingly, against them. By and large, though, the history of the utilities' political capitalism has been defined by minor losses: cases where the government did not agree to set policy to protect them fully from competitive forces but also did not expose them to "ruinous competition" from would-be raiders. In these minor losses, such as the one Fuel Gas Company suffered in 1884, utilities are spared from lookalike competitors but are not insulated from broader market dynamics.

Since 1884 gas market regulators in the United States have hewed to a mostly pragmatic application of the tenets underlying the utility compact. This has spared us from the underlying logic's most important flaw—namely, that complete monopoly protection would stifle the competition that leads to "disruptive" improvements from nonlookalike competitors, and ultimately into lower rates for consumers.

# CHAPTER 2

## GOING BACK TO SMOKE

The number of smoky days in Pittsburgh declined precipitously through the 1880s after natural gas began supplanting coal and manufactured gas. Coal consumption within Pittsburgh plummeted from three million tons in 1884 to just one million tons a few years later. Interestingly, these figures came from a study written in 1913 called *The Economic Cost of the Smoke Nuisance to Pittsburgh*, which also pointed out that Pittsburgh's sooty air had earned it the interesting title of "the greatest laundry town in the country."[1]

Andrew Carnegie was so enamored with the nonpolluting qualities of natural gas that he wrote a paper documenting how its use had improved both the cleanliness and profitability of his facilities: "In our steel rail mills we have not used a pound of coal for more than a year, nor in our iron mills for nearly the same period. The change is a startling one. The houses being whitewashed, not a sign of the dirty fuel of former days is to be seen, nor do the stacks emit smoke."[2]

Sadly, the clear air would not last. By 1890 coal use was rising again, as production from the Murrysville field had begun to decline. While it may have been unknown to the engineers of the day, this was due to the fact that the gas-producing formations of Appalachia comprise low-permeability rock ("tight" gas). This meant that the gas was flowing to the wellbore through natural fractures in the rock. Wells that produce through fracture networks, which include the hydraulically fractured wells common today, inherently exhibit high rates of initial production followed by steep rates of decline. Think of it like this: if you pour water through a sieve, most will come straight out the other side, and only a small amount will continue dripping down after you are through pouring; whereas, if you pour water onto a rag, it will drip out the other side more slowly, but for longer, at a more constant rate. Appalachian gas wells were mostly sieves.

The first gas shortages cropped up just a few years after natural gas service began, and, combined with growth in overall industrial output, led to coal consumption in Pittsburgh beginning to grow rapidly once again.

In 1892 a speaker at a meeting of the Engineers' Society lamented: "We are going back to smoke! We had four or five years of wonderful cleanliness in Pittsburgh, and we have all had a taste of knowing what it is to be clean."[3]

Carnegie echoed the threat in an 1898 address to the Pittsburgh Chamber of Commerce, declaring: "The man who abolishes the smoke nuisance in Pittsburgh is foremost of us all."[4]

Although several smoke-control ordinances were passed in the 1890s and 1900s, they were often weakly enforced and anyway did not get to the root of the problem, which was that natural gas was in short supply. New supplies were needed, as the local fields were insufficient. This required a major change in the structure of the still-young natural gas industry—an interconnected, regional network was needed.

## STANDARD OIL AND THE FIRST REGIONAL GAS NETWORKS

Joseph Pew and Edward Emerson had renamed their Penn Fuel Company as Peoples Natural Gas in 1885, not long after connecting their Murrysville wells. At first, the Murrysville field was Pittsburgh's sole source of supply, and gas flowed to the city under its own pressure. Peoples and other companies did buy acreage farther afield, sometimes as far as fifty miles outside the city, and two technological innovations helped alleviate the supply shortage.

It could be said that George Westinghouse, whom history remembers for prevailing over Thomas Edison in the "War of the Currents" (Westinghouse's preferred alternating current was widely adopted, while Edison's direct current was abandoned), got into electricity distribution via gas distribution. Already a rich man by his mid-twenties from inventing the railway air brake, which used compressed air to control the deceleration of trains, at thirty-seven Westinghouse was anxious to use his hydraulics know-how to control the flow of natural gas, so he had a team of men drill a well in the backyard of his Pittsburgh mansion, which he had named Solitude. The well came in "with the infernal violence of a volcano" on May 29, 1884, and caused a great stir in the genteel neighborhood.[5] Westinghouse then bought a nonfunctioning company solely for the fact that it held a valid charter to distribute gas in Pennsylvania. Somewhat ironically, since it served only Pittsburgh, it was called the Philadelphia Company, and was for a time the largest gas producer in the United States, through its main production subsidiary, Equitable Gas. (In 2017, 130 years later, Pittsburgh-based Equitable Gas—now called EQT Corporation—would once again become the largest gas producer in the nation, surpassing ExxonMobil.)

Over the next few years, and, remarkably, at the same time as Westinghouse Electric was building scores of alternating current electric generating stations, George Westinghouse patented almost thirty gas-related inventions, including new meters and leak prevention and detection equipment. One of his devices automatically shut off the gas supply if pressures dropped below where a pilot light could hold a flame, which drastically reduced the chances of flammable gas building up in homes and businesses.[6] Westinghouse's inventions helped increase the adoption rate of natural gas appliances in the early days of the industry and laid waste to claims from oil men that the fuel posed too much danger, undetectable and flammable as it was.

Westinghouse was also the first to build telescoping gas pipelines, which were thinner and operated at higher pressure near the well, and grew wider and lower pressured as they approached consumers. This improved the flow of gas, and was also the basis for alternating current transmission, which uses transformers to step up voltage for long-distance transport and then step it down for use in distribution to customers.

For his part, Joseph Pew installed a compressor station that used a rotary engine to send the gas down the pipe at higher pressure.[7] It achieved a similar effect to Westinghouse's telescoping pipe of coaxing more gas out of the wells, but ultimately both innovations were Band-Aids on a battle wound. What was needed for the industry to scale was a system of long-distance main lines connecting many different sources of supply with the franchise territories of several local distributors.

This change could not have been expected to be led by the distributors themselves, though. Companies like Pew's Peoples and Westinghouse's Philadelphia were, by their very nature, just as parochial as the larger manufactured gas companies around the United States: their business was tied wholly and necessarily to a single city. Protection under the utility compact, after all, did not extend beyond the borders of the municipal franchise territory, and certainly not across state lines.

Moreover, in the early 1880s, there were practical constraints to how far gas could be transported. Heavy, wrought-iron sections of pipe still had to be screwed together individually by groups of men wielding absurdly large calipers, who came to be known as "tong gangs." While it was possible for crews to thread these pipes over long stretches, the process was extremely labor-intensive and expensive. Still, threaded wrought iron represented substantial progress from the very first pipes, if they can even be called that, which were made of Canadian pine with channels bored through them and tar sealing their joints, making them as effective at distributing gas into the dirt as they were into cities.

Another reason that a regional solution was needed was that Pittsburgh was not the only city going back to smoke. The reader can be forgiven for not being immediately familiar with the town of Gas City, Indiana, which was at the center of the Indiana Gas Belt, but its name tells of a brief period where Indiana was an emergent natural gas giant.

Gas City was incorporated in 1892 with a population of 150 people. Two years later, that number was 25,000. It may have had the most obvious name of the Gas Belt towns, and it did sit atop the best acreage, but dozens of other communities were also home to wells, which any visitor could easily tell, especially if they visited at night. East Indiana was a poster child for waste: owners kept a flare burning constantly atop each well just to show onlookers that gas was flowing—these were called flambeaux. It is even said that Indiana farmers harvested at night, by gaslight. The gas was so abundant that many town officials offered free gas supply to manufacturers if they relocated to Indiana, which many did. Officials especially targeted producers of glass, whose manufacture requires enormous amounts of heat, and by 1900 Indiana was the second-largest glass-producing state in the nation, having occupied only a minor role a decade earlier.[8]

In 1891 east Indiana wells were connected to Chicago, over 120 miles away, in what was the longest pipelines of its day.[9] Unfortunately, the wanton wastefulness led to a predictable and tragic end: pressures dropped, and, ultimately, the pipelines ran empty. In 1907 the last molecules of Indiana gas were transported to Chicago.[10] After that, the city was forced to switch back wholly to manufactured gas.

The historian David Waples wrote: "In the early twentieth century, any community in the Appalachian region with a natural gas well and a supply of pipe was home to its own gas company."[11] This decidedly local arrangement was fine, so long as the wells were able to keep the lines full of gas at sufficient pressure. But as the wells and the fields they produced from were depleted, supply from farther afield was necessary.

The shortages of the late nineteenth century did terrible damage to the reputation of the emergent natural gas industry, with many government officials and investors decrying it as comprised of fly-by-night hucksters. This brought into focus a chicken-and-egg problem—that, in order to scale to become a strategic commodity of national importance, such as coal, oil, or even manufactured gas, which would merit serious capital investment, natural gas service would need to be proven reliable. Reliability, meanwhile, would require the type of redundancies offered by a large, regional network of gas wells and pipelines, which would require serious capital investment.

It would take an extremely well-capitalized company with a grand strategic vision, as well as advances in pipeline technology, to overcome these challenges and create a regional gas network.

John D. Rockefeller's Standard Oil needs little introduction. Already by 1880 it controlled more than 90 percent of US oil refinery production and was flush with cash. Standard conducted its day-to-day operations through a complicated and cascading network of subsidiary companies, all of which, though, followed to the letter orders from above. Although Standard

| Oil/gas window | Depth (km) | Temp (C) |
|---|---|---|

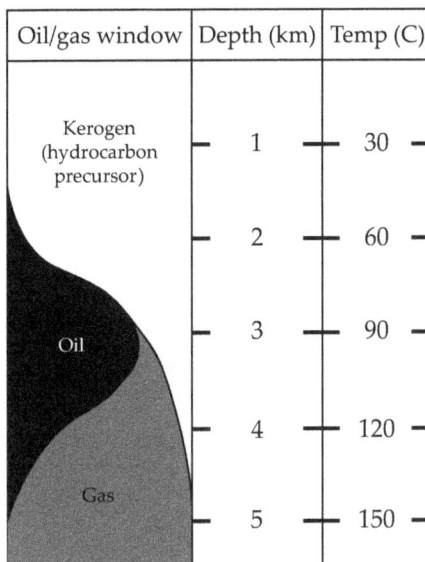

**FIGURE 2.1:** Hydrocarbon generation

focused on controlling the transportation, refining, and marketing of oil and was less interested in joining the cacophony of independent producers who actually brought the stuff out of the ground, it did own several exploration and production (E&P) subsidiaries. One such E&P was South Penn Oil. In its search for oil, South Penn also discovered tremendous reserves of natural gas. This was perfectly typical. Most natural gas was found by accident in the search for oil, as had been the case with the Haymaker well. Westinghouse was a bit of an oddball for purposefully targeting gas.

Oil and natural gas are generated from the same material, plankton (not dead dinosaurs), by the same geological processes. The dead plankton fall to the seafloor, and are then buried in a basin. As sediment piles on top of the plankton, heat and pressure build, slowly cooking the organic material and breaking its long, complex molecules into shorter, simpler ones. Hydrogen and carbon share a strong chemical bond, so the cooking process naturally forms hydrocarbons—chains of carbon atoms surrounded by hydrogen atoms.

Oil and gas, then, share many similarities, and are mostly found together. But they occupy two different locations on the hydrocarbon continuum. Oil comprises longer chains, with each molecule having a greater number of carbon atoms, and is thus liquid. Gas starts as oil, but is left in the oven for longer, and thus had its hydrocarbon chains broken down further. Because there is a continuum of hydrocarbons, the definitions of what is oil and what is gas are somewhat arbitrary, but generally hydrocarbons with more than five carbon atoms are liquid at normal conditions and thus qualify as oil. Gas is mostly methane—one carbon and four hydrogens—with some slightly heavier molecules also present.

Gas is the end result of the hydrocarbon generation process: all oil will break down into gas if it is left to cook long enough, and there is thus naturally more gas in the earth than oil. Therefore, while it is common to find gas without oil—called "dry" or "non-associated" gas—it is less common to find oil without gas—"black" oil, or "dead" oil. They are usually found together, in varying proportions.

Until the 1950s, and certainly at the turn of the twentieth century, natural gas was still, in most circumstances, a nuisance byproduct, and in the worst case, a disaster. As such, there are dozens of quotes attesting to the worthlessness of natural gas without pipeline transport and a ready market. Among the best comes from Jack Tankersley, former CEO of Consolidated Natural Gas, one of Appalachia's largest pipeline companies and a direct corporate descendant of Standard Oil: "During those days, the first prayer of the roughneck was to find oil. His second prayer was not to find natural gas. Gas was not only useless, it was downright dangerous, capable of igniting a good oil well literally at the drop of a match."[12]

All of which is to say that gas did not occupy a position of strategic importance for any producer, including Standard. But it is also true that Rockefeller was a notorious penny-pincher, and sought out cost efficiencies even in areas that were profitable, and where most men would have contended themselves with those profits. Natural gas could be burned, and that heat gave it value, so Standard set up another operating company called Flaggy Meadow Gas Company, whose purpose at first was simply to pipe gas from South Penn's oil wells to nearby drilling rigs, where it was used as fuel. The gas was only useful inasmuch as it allowed South Penn to sell more oil: South Penn was the diamond, Flaggy Meadow the setting, and a utilitarian one at that.

Flaggy Meadow added local distribution to homes and businesses in the immediate vicinity, but it remained a small, parochial enterprise whose purpose was merely to "deal with" gas produced by South Penn, so that it could continue its profitable oil exploration. Until, that is, a key lieutenant of Rockefeller's had that grander vision for the gas business.

Calvin Payne set off from the family farm in Pennsylvania at the age of fifteen to work in the original oil fields near Titusville. While his formative years were spent in oil exploration, his legacy is in his contribution to the gas industry. He built one of the first long-distance pipelines from northwest Pennsylvania to Buffalo, New York, in 1886, a year after joining Standard, then assumed a leading role at yet another operating company that Standard had formed in 1881: National Transit Company.

National Transit was set up to manage all of Standard's pipeline interests, both gas and oil, and would soon also control its "upstream" natural gas interests (i.e., its production). Payne laid pipe in northwest Pennsylvania and western New York and organized a distribution company on the Ohio/West Virginia border, near Marietta, Ohio. By the turn of the century, Payne convinced Standard's senior management, Rockefeller in particular, to take bolder action to grow the natural gas business.

In 1898 Hope Natural Gas was organized to produce, gather, and transport all of Standard's West Virginia gas to customers in Ohio, where it would be distributed by the newly formed East Ohio Gas Company. East Ohio Gas was a utility, just like Pew and Emerson's Peoples and Westinghouse's Philadelphia Company. Hope, though, was something a bit new: a producer and transporter of wholesale natural gas. Hope did not serve homes and businesses directly, and hence did not fall under the auspices of the utility compact. It sold wholesale to utilities, who then conducted retail business.

Aided by a major technological advancement patented in 1885—Solomon Dresser's Dresser Coupling, which allowed gas to be moved at long distances with minimal leaks and without each string of pipe being screwed together by the surrealist tong gangs—Hope built a ten-inch transmission line from its fields in West Virginia through mountains and over the Ohio River, whereupon East Ohio Gas continued the line to Akron. This line, the Akron Ten Inch, was completed in 1899.

Three years later, National Transit was told to continue the line northward, to Cleveland, which was by then home to more than 350,000 people—a bit larger than Pittsburgh—and was being served by manufactured gas. Cleveland's sitting mayor, Tom Johnson, was at first skeptical, having little knowledge of the gas industry and unconvinced that East Ohio Gas—whose reputation and ownership were unknown to him—had the financial resources to complete such an undertaking. After the representative from National Transit told the mayor that East Ohio Gas was owned by Standard, Johnson changed his tone: "Well, that's different. Go to it and I'll help you."[13]

The year 1903 can be said to mark the true beginnings of Appalachia's integrated, regional gas network. This was the year that East Ohio Gas began delivering gas from West Virginia to Cleveland. It was also the year when National Transit bought Peoples from Joseph Pew, interconnecting its pipeline system and bringing it into the Standard fold. Now Rockefeller's juggernaut owned major reserves in both West Virginia and Pennsylvania and delivered to two major cities, Pittsburgh and Cleveland, and a host of smaller ones.

## DARK AGES IN THE NORTH,
## BIRTH OF AN INDUSTRY IN THE SOUTHWEST

Despite the grand vision and deep pockets of Standard and its chief competitor in Appalachia, Columbia Gas & Electric, the natural gas industry in Appalachia was once again in trouble by the mid-1910s. Production began to outpace reserves additions, as there was simply not enough gas in the ground to supply the rapidly industrializing region. Given the high rates of production decline from each well, producers were sprinting just to stay in the same place.

Chicago ceased importing gas from Indiana in 1907, which necessitated a full return to manufactured gas. Pittsburgh coal use had begun to tick up even as natural gas production had been rising, as iron, steel, and aluminum output skyrocketed; now that it was falling, coal was king once again in the Smoky City. Even Akron, which Standard had connected in 1899 as the terminus of the Akron Ten Inch, was running out: East Ohio Gas began building a manufactured gas facility there in 1919 because it could not guarantee long-term natural gas supplies.[14]

This period, from roughly 1907 through 1940 might be thought of as the dark ages of the natural gas industry in Appalachia. Keep in mind that natural gas had not yet been widely adopted outside of this region: the large East Coast cities—New York, Philadelphia, Boston, Baltimore, and Washington—were completely fueled by manufactured gas. So were Detroit, St. Louis, Milwaukee, and Minneapolis. Chicago, having received gas from Indiana, was a bit of an exception, and even there, natural gas was mixed with manufactured gas, as there had never been enough to supply the city's needs fully.

Geologists, businessmen, and policymakers alike believed that natural gas's days were numbered. The general feeling was that it was a superior fuel, when available, but that there was simply not enough of it to supply any meaningful quotient of the country's energy mix. Utilities reinvested in newer, larger manufactured gas production facilities—manufactured gas was back in vogue. It must have made some sense, as history is full of examples where society has replaced exhaustible, natural materials with synthetics (think rubber, sponges, and goose down).

That said, despite the fact that competition from natural gas was diminishing quickly, the real belle of the energy ball during the 1910s was not manufactured gas but electricity, which quickly and nearly completely usurped one of manufactured gas's main markets: lighting. In the north, gas—both manufactured and natural—became predominantly a fuel for heat.

Things looked bleak for natural gas in Appalachia in the 1910s. But just as the Dark Ages in Europe coincided with the Islamic Golden Age, the dark ages of gas in Appalachia coincided with the golden age of gas (and oil) in the Southwest.

The first oilfield in Texas was named after the town in which it is located, Corsicana, about fifty miles south of Dallas. It was discovered in June 1894 by the Corsicana Water Development Company, who were quite annoyed when they struck oil. The drillers cemented the walls of the well to prevent the oil from seeping in and continued drilling for another 1,400 feet downward, until they hit water.[15] So began the storied Texas petroleum industry: not with a bang, but with a whimper.

A few months later, local businessmen saw the promise of at least meager profits in Corsicana, and drilled the Lone Star State's first true oil well. In October 1895 the well came on, flowing two and a half barrels per day.[16] Many more were sunk, and the field's larger wells produced up to twenty-five barrels per day. At its peak, the field produced just over two thousand barrels per day.

These figures paled in comparison to analogues in Appalachia, where individual wells would often attain initial production of up to several hundred barrels per day, and the record stood at a superlative four thousand.[17] Appalachia was the powerhouse of US oil and gas production, even if production had begun to decline.

That all changed on January 10, 1901, when a well drilled on Spindletop Hill in Beaumont, Texas, about ninety miles east of Houston, blew out with a fury never before encountered. Spindletop's driller, Captain Anthony Lucas, had been hopeful that the well would produce five barrels per day. Instead, it shot one hundred thousand barrels per day 150 feet into the air, more than doubling total US oil production in an instant. Spindletop signaled the arrival of the Southwest, with its much larger and more productive fields, as the new center of the American petroleum industry.

The boom was on, and quickly. And while the ambitious businessmen chased crude oil, just as they had in Appalachia, and just as they continue to do to this day, several companies were formed to pipe the gas associated with oil production to markets.

One of the first large Southwest gas systems connected the newly discovered Caddo field in northwest Louisiana, near the Texas and Arkansas borders—the so-called Texarkana region—with Shreveport in 1906.[18] Little Rock, Arkansas, was connected in 1911, and a pipeline system making deliveries throughout the Texarkana region was eventually consolidated into the Arkansas-Louisiana Gas Company, or Arkla. Farther west, the Lone Star Gas Company was incorporated in Dallas in 1909 and built a sixteen-inch line from the Petrolia field on the Oklahoma/Texas border to Dallas, expanding its system to new fields in later years and eventually becoming one of the largest transporters of gas in the nation.

The Southwest initially gasified similarly to Appalachia, with pipes being built to connect "stranded" gas fields to cities, most of which were already being served by manufactured gas. There were some important differences, though. There were slightly fewer mom-and-pop gas companies that needed to be consolidated, partly because the technology existed immediately to use Dresser coupling to lay longer-distance pipeline fairly easily, and partly because there were simply fewer communities in the less-populated Southwest than there were in Appalachia. Another important distinction was that most natural gas companies in the Southwest were pipeline companies that did not distribute the gas themselves, as manufactured gas distribution companies had generally been set up in the years prior to Spindletop. The new gas companies served these distributors. In this way, companies such as Lone Star were similar to Standard's Hope Natural Gas, rather than East Ohio Gas or Peoples—transporters of wholesale gas, not distributors of retail gas.

Last, the new companies really were new, at least at first. The large oil and gas companies and utility combines of the day, Standard chief among them, were conspicuously absent from much of the Southwest. Standard, which controlled between 80 and 90 percent of the petroleum industry in Appalachia at the turn of the twentieth century, was wary of anti-monopoly sentiment in Texas, and was also unimpressed by the quality of Texas crude, which yielded a lower share of the refined product that Standard most desired: kerosene (keep in mind, this is before the advent of the automobile, so gasoline and diesel were not sought after).[19] Other large gas companies in the north—whether natural gas transporters and distributors like Columbia, or manufactured gas distributors such as Gas Light Company of Baltimore—were largely uninvolved in the development of southwestern gas reserves or pipeline systems. The utility mind-set, after all, is local in nature—what economies of scale would be gained from a northern utility essentially duplicating its business in the Southwest?

Had things kept going the way they had been, with new oil and gas fields being connected to neighboring cities, replacing or supplanting manufactured gas, these next few pages would essentially be a very boring repetition of the expansion of the industry in Appalachia, albeit with different places and names. But, perhaps for the first time, geology will serve to keep the story interesting: in their quest for the next Spindletop, Southwest producers began finding gas fields so large that it was simply inconceivable that local markets could absorb all the new supply.

The first of the gas giants was the Monroe field, discovered in 1916 on the outskirts of the town of Monroe in northern Louisiana.[20] It was the largest gas field yet discovered, and would become the main source of supply for four new pipeline systems serving Atlanta, St. Louis, New Orleans, and a dozen or so other southern cities.

Indeed, Monroe was so giant, with reserves initially estimated at 6.5 trillion square feet—more than five times as large as the next known largest field—that it was virtually the only supply needed for points south and east of St. Louis, especially since this region was not as heavily industrialized as Appalachia and the Midwest and experienced milder winters. Monroe's chokehold on this sizable market meant that the region was now saturated, and that any future giant fields would have to find other markets for their gas.

In 1917 natural gas production in West Virginia and greater Appalachia peaked at 845 and 1,430 million cubic feet per day,[21] respectively. Production from Pennsylvania had peaked in 1906.[22] Standard's Hope Natural Gas was adding horsepower to its compressors in a desperate attempt to try to suck more out of its wells, and buying gas from erstwhile competitors.[23] Gas utilities across Appalachia and the Midwest were busy building massive new manufactured gas facilities. And World War I was raging in Europe.

The Appalachian industrial complex faced gas shortages during the war, which earned the natural gas industry the federal government's attention and ire. Measures were passed outlawing flambeaux, the open flames atop wellheads, and daytime gas lighting. In 1920 Secretary of the Interior Franklin K. Lane gave a long address to policymakers and members of the gas industry, scathing what he saw as blatant wastefulness.

> As I have gone through the States where natural gas is, or has been abundant, I have felt a sense of outrage at the manner in which it was burned all day and all night in the street lamps, wastefully used in industries as a fuel, wastefully used in stoves for cooking. I think we can honestly say to ourselves that we have not applied ordinary thrift, or good sense, to the method of using natural gas. . . . We have been wasteful with natural gas—partly because it was a sort of by-product like the straw in the wheat field. We have been wasteful because we wanted to get rid of it and get the real thing that we were after, the oil; wasteful because we did not appreciate its value.[24]

Lane's speech also contained extremely thinly veiled threats to the industry that, if it did not take steps to conserve natural gas, the US government might compel it to do so. "I do not know what a governmental policy might be with regard to natural gas. It might be that all industry would be denied the use of natural gas and home use alone allowed. You know how facts are gathered—it might be shown by a statistical study satisfactory to a board that that would be the wise thing to do; and you know what the result upon your industry would be if such a policy were adopted."[25] Lane's words were eerily prescient of what would become of natural gas policy sixty years hence, during the Carter administration and after the Arab oil embargo. Nevertheless, they must have sounded ridiculous in the Texas

Volumes (million cubic feet per day)                                    % of production wasted

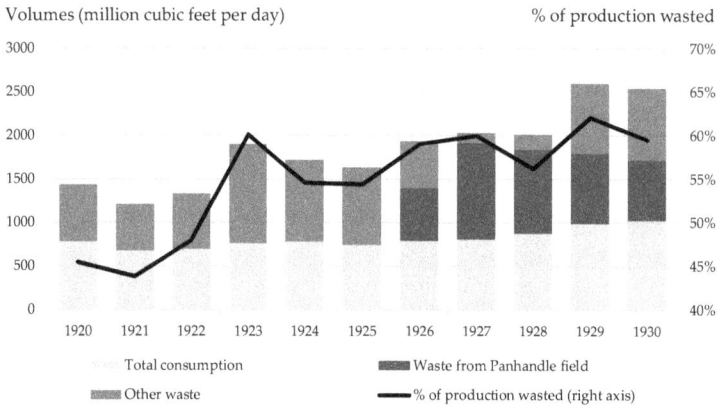

**FIGURE 2.2:** US gas production and volumes wasted

Panhandle, where a small company called Amarillo Oil had brought on the first well in the Panhandle field, about thirty miles north of Amarillo, two years earlier, in 1918.

This first well flowed 15 million cubic feet a day[26]—a large amount of gas, to be sure, but not a record-setting amount (recall that the Haymaker and the well George Westinghouse drilled in his backyard both produced over 30 million cubic feet a day). The next two wells were also very encouraging, but similarly did not break any records. But the fourth well that Amarillo drilled flowed an eye-popping 104 million cubic feet per day, proving that the Panhandle field was something spectacular indeed.[27] Contemporary estimates pegged the Panhandle field's gas reserves at 16 trillion cubic feet—almost three times the size of the massive Monroe field.[28]

Of course, there was no market for this staggering amount natural gas, so Amarillo and other producers continued their search for oil.

More wells were drilled, but oil proved relatively elusive in the gassy Panhandle for several years. During this search, producers sank the first well into the neighboring Hugoton field near Liberal, Kansas, in 1922.[29] Once again, drillers uncovered a giant. The Hugoton would go on to produce over thirty trillion cubic feet of gas, and continues to produce to this day. And although drilling did not begin in earnest until 1927, and contemporary reserves estimates had it looking slightly smaller than Panhandle, it was still known to be at least double the size of Monroe. Unfortunately, at least for the time being, the Hugoton was also nearly bereft of oil, which stymied development until the gas could find a market worthy of the enormous cost of transporting it from what was, essentially, the middle of nowhere.

It is worth repeating these reserves figures. Monroe, at six and a half trillion cubic feet, was five times larger than anything that had been discovered before. Then Panhandle clocked in at sixteen trillion. Then Hugoton at over ten trillion.

Thus, already prior to World War I, and to an almost absurd extent by the mid-1920s, there was a regional imbalance in natural gas supply and demand in the United States. Most producers in the Southwest who were lucky enough to have struck oil were flaring or simply venting into the atmosphere hundreds of millions of cubic feet per day of natural gas. By the mid-1920s, more than half of all natural gas being produced nationwide was being wasted.[30] The situation was worse for those with gassy acreage, who found themselves in the possession of trillions of cubic feet of unmarketable, essentially worthless, resources.

Meanwhile, *conservation* was the word in Appalachia, which was, almost in its entirety, going back to smoke.

# CHAPTER 3

## THE RISE OF THE POWER TRUST

While the natural gas market in Appalachia was still highly balkanized, before Standard began building large pipelines linking fields to cities hundreds of miles away, the electric industry was going through its own growing pains. Thomas Edison invented the incandescent lightbulb in 1878, and though electricity would totally displace gas as an energy source for lighting, it took more than a decade to reach consensus on how the electricity would be transmitted from the generating station to the bulb.

In hindsight, the central station model seems obvious. But Edison himself advocated for a decentralized system, where electricity was generated locally and transmitted only a short distance. His competitors, and eventually even his allies, recognized the superior economies of scale in building large power plants on the outskirts of cities, then sending the electricity to homes and businesses over long-distance lines. But that required a technology that Edison despised: alternating current.

The "War of the Currents" revealed Edison's hubris and cost him control of one of his namesake companies, Edison General Electric (which we know today simply as GE). But his key lieutenant, an Englishman named Samuel Insull, adopted the central station model with great zeal, and went on to become probably the single most influential person in the history of the utility industry. It was Insull, not Edison, who managed to electrify not just the homes of the well-heeled, but virtually the entire US population.

As Insull and his peers electrified the country, they stitched together local utilities and turned their holding companies into empires. By the onset of the Great Depression, Insull-controlled companies alone produced 11 percent of electricity in the United States. To defend this empire from competition—"raids," in the parlance of the times—Insull also controversially, but successfully, lobbied for government regulation. In exchange for a guarantee of local monopoly, Insull proposed forming state-level public utility commissions to oversee the rates that utility companies charged consumers.

In the 1910s, such commissions were created in almost every state; they prevented upstarts from encroaching on incumbents' service territory and were an important bromide for a public leery of monopoly. The holding companies were also seen as good shepherds of capital, as they brought affordable electricity, and later gas, to the masses. Thus, throughout the Roaring Twenties, the so-called Power Trust was regarded generally as a good thing.

But as the country entered the Depression, the companies were exposed as extremely financially weak. They had saddled their local utilities with debt, and were unable to operate in anything but a rapid-growth market. In the end, Insull declared bankruptcy, was charged with financial crimes, and fled the country to avoid prosecution. He died broke in a Paris subway station.

The Power Trust was foundational in creating a national energy network. They built hundreds of massive power plants, some of which are still in operation, and the nation's first long-distance natural gas pipelines, which linked the giant fields of the Southwest with population centers in the Midwest. As importantly, they helped create the state-regulated monopoly utility framework that prevails to this day.

But they also stymied competition, blocked the development of several important infrastructure projects, and engaged in deceptive financial practices to enrich their shareholders at the expense of consumers. Backlash against the Power Trust's more nefarious practices shaped many New Deal regulations, which created a federal regulatory apparatus that steered many of the US economy's largest industries for the next fifty years, until the Carter and Reagan administrations. Gas and electric utilities became subject to a litany of regulations, alongside airlines, banks, and the trucking and telecommunications sectors.

## THE WAR OF THE CURRENTS

When the gaslights on Broadway were illuminated in 1825, New York became the second major city to receive manufactured gas service, after Baltimore. Over the following decades, gaslights became common in homes, first among the rich, then among the merely well-off. But just as manufactured gas had displaced whale oil and tallow as lighting fuels, beginning in the 1870s, electricity began displacing gas. It was in 1880 when the gaslights on Broadway were replaced by super-bright electric arc lamps, which earned the thoroughfare the nickname "the Great White Way."

Arc lamps were wondrously luminous, but wholly impractical for anything but large public or industrial spaces. Indeed, they were so bright that actors performing in early motion-picture studios that used arc lamps wore sunglasses when off-camera. For a short while, it seemed that manufactured gas would retain its status as the go-to fuel for residential and commercial lighting.

Thomas Edison saw things differently. He had invented an improved telegraph device in 1874; used the funds from the sale of that patent to build a research laboratory in Menlo Park, New Jersey; and then invented the phonograph in 1877, to much fanfare. But even as Edison was on the road demonstrating his phonograph, his team of engineers in Menlo Park was developing what would become the incandescent lightbulb. The lightbulb has come to serve as shorthand for genius, and electricity became ubiquitous over the following years, relegating gas to a heating and cooking fuel. But the transformation was not immediate, and Edison did almost as much to slow the rollout of electricity to the masses during the last few years of 1880s as he did to encourage during the first few.

Edison's genius was to target the residential market, creating a product that would mimic the incumbent gaslight. The voltage required to do so was much lower than for arc lamps: 110 volts (still the standard in the United States). His hubris was his insistence that the most practical—indeed, the only practical—way to serve homes with electricity was via direct current. This meant that the electricity had to be transmitted at the same low voltage at which it was used, which was inefficient over distance and thus required the generators to be located quite close to the load. The thought was thus that the electricity industry would develop in a decentralized manner, with each city block, and potentially even each building, housing its own power plant.

At first, this was uncontroversial. Pearl Street Station in downtown Manhattan, the first commercial power plant, began operations in September 1882—incidentally, less than a year before the Haymaker well began supplying Pittsburgh with natural gas—and served the New York Stock Exchange as well as the private home of the most important financier of the era, J. P. Morgan. Morgan was an early investor in Edison's electric ventures and continued to purchase shares as the business grew, eventually becoming its largest single investor, with a larger stake than Edison himself. In 1884, though, a group of scientists in Europe made a great leap in transformer design, allowing the voltage of an alternating current to be increased or decreased, with only minor losses. This made it possible to transmit electricity over great distance at high voltage, and then "step down" the voltage before sending it into homes, and gave alternating current (AC) a notable advantage over direct current (DC). George Westinghouse, who began in the electric business in 1884 using a DC system, quickly saw the merits of AC and switched over. Edison's other main competitor, the Thomson-Houston Electric Company, did the same, leaving Edison as the only major company using DC and beginning the War of the Currents.

It was not immediately obvious who would win. Electric motors had been developed to work with direct current, not alternating current, and there was at first no way to meter an alternating current. But by far the most contentious argument between AC and DC revolved around safety. In 1888 Edison and a few allies began claiming that AC power posed a genuine threat to public safety, given the extremely high voltages being used. While Edison's claims certainly supported his own business, they were not false: there were several deaths that year in New York City from electric shock. In reality, the danger was from lax regulation and roughshod development, rather than anything inherent in AC technology. In New York, where the War of the Currents was being fought, dozens of wires hung on street corners with scarce protection from the elements, whereas other cities with AC systems required that lines be buried underground and had better standards for line insulation.

Things quickly took on a circus-like character, with an Edison-affiliated engineer making public demonstrations of killing dogs by running AC current through them, and even challenging George Westinghouse to an "electricity duel" in late 1888, suggesting that he would be willing to run any amount of direct current through his own body if Westinghouse would do the same with his preferred alternating current. Westinghouse declined. A few months later, when New York rendered its first sentence of death by electric chair, alternating current was used. The *New York Times* eventually settled on the new word *electrocution* to describe death by electric shock; a close runner-up was "Westinghoused."[1]

Westinghouse and AC were losing in the court of public opinion, and Edison was lobbying to have a law passed limiting high-voltage transmission, which would take away the crucial and only advantage AC held over DC. Within a few months, however, the war ended as quickly as it had begun. For all the grandstanding that Edison and his associates did, Westinghouse and Thomson-Houston had grown much more profitable. AC systems simply had better economics, and after the invention of an AC meter and AC motor (by former Edison employee Nikola Tesla), there was nothing direct current could do that alternating current could not.

In April 1889 various Edison companies merged to form Edison General Electric, and Thomas Edison received the biggest paycheck of his life. His young but trusted lieutenant, Samuel Insull, began assuming day-to-day management responsibilities, and began the process of converting their systems to AC. Almost exactly three years later, Thomas Edison was forced out, in both name and person. J. P. Morgan did not much care for Edison's ceaseless tinkering and side projects, cognizant though he was of the tendency for brilliant minds constantly to wander. (Edison had spent the prior two years trying to perfect a method to coax iron ore out of the ground via magnetism. It worked, but was horribly uneconomical, proving the axiom: they can't all be winners.) In April 1892 Morgan merged Edison General

Electric with Thomson-Houston and forced Edison himself out. *Edison* was dropped from the company's name, and GE was born.

## INSULL'S RISE TO PREEMINENCE

A technological marvel such as electricity surely can sell itself. But a large chunk of credit should be given to Samuel Insull, who rose from essentially an errand boy to become the man most responsible for electrifying the United States.

On February 28, 1881, Insull arrived on a ship in New York Harbor. Two years earlier he had taken a job with a banker representing Edison's interests in Europe, where the Edison Telephone Company was engaged in fierce competition with the Bell Telephone Company to introduce telephony to London. By virtue of hard work, and ingratiating himself with one of Edison's top engineers, Insull became acquainted with the great inventor himself. After Edison and Bell merged their UK telephone businesses in 1880, Insull was offered a job as Edison's private secretary. He left London, where he had a high-paying job and a strong network, and moved to the United States, where he knew no one.

This was more than a clerical post. For all his genius and resolve, Edison had become disorganized in his business affairs, whereas Insull, though only twenty-two years old, was deliberate and exacting. Edison wrote of Insull, whom he regarded very fondly and called Sammy: "His mind was much older than his years would indicate."[2]

After the merger of Edison General Electric and Thomson-Houston and the ouster of his mentor, Insull could have remained at GE, where he had been appointed vice president—in a time when that title was not given out nearly as often as it is today—making what in today's dollars would be roughly $1 million per year.[3] But J. P. Morgan had put former Thomson-Houston chief executive Charles Coffin in the top seat at GE, and Insull did not want to play second fiddle. Anyway, he thought that there was more money and power leading a vertically integrated utility, which owned power plants, high-voltage transmission lines, and low-voltage distribution lines, rather than a manufacturer like GE.

In July 1892 Insull left GE to become president of Chicago's second-largest electric utility, Chicago Edison, and was seen off with an elaborate steak dinner at Delmonico's in New York City. Once again, Insull moved from a high-paying job where he had a bright future to a place where he knew no one and his fortunes were far from guaranteed. This time he took a two-thirds pay cut, to boot.[4]

On arriving in Chicago, Insull immediately, and at a furious pace, began consolidating the city's utility industry by purchasing his competitors, integrating horizontally, just as Standard and Columbia had done in the

Appalachian gas industry. This was not limited to the classic big-fish-eats-little-fish acquisition process, either. On one occasion, the president of Chicago Arc Power and Light, a larger company, invited Insull to lunch to ask whether Chicago Edison was interested in purchasing some shares. Insull informed him that he was not interested in investing in a competitor, and that he actually intended to buy the larger company outright, which he did a year later, via issuing debt.

Insull's debt-fueled growth program at Chicago Edison made him his former employer's largest customer, as he built what were at the time the largest power plants in the world. So convinced was Insull in the superlative economics of large, central station economics that "often in defiance of his own engineers and almost always over the protests of those at GE, Insull would demand and receive generators two, three and four times as large as any others in existence."[5] Embracing the central station model—and with it AC transmission—was the first step toward bringing electricity to the masses. The next step was a bit more technical, but equally important.

Chicago Edison's electrical output, and hence revenues, were skyrocketing, but growing revenues do not always lead to growing profits (indeed, they often lead the other way). The company struggled to find the right price point that would allow it both to add customers and make money. The solution to this problem, which is one of Insull's great contributions to the utility industry, was not something he learned while working for Thomas Edison, or even thought of himself. Rather, it was something he observed while on vacation in the English seaside resort town of Brighton, in 1894.

Insull noticed that Brighton was very bright indeed, by virtue of the great many electric lights that lit its homes and businesses, and sought out the chief of the local utility. A meeting was quickly arranged, and Insull asked with characteristic purposefulness whether and how the company turned a profit. He was told that, in addition to measuring the total amount of power that customers used, the Brighton utility also measured when they used it.

For all intents and purposes, electricity cannot be stored. In the 1890s, as in the 2010s, batteries were far too expensive to serve as a demand-balancing mechanism. The "storability" of a commodity has tremendous implications on pricing, because demand is hardly ever constant. This is true for electricity, which in Insull's day was used mostly in the evening, for lighting; for gas, whose use does not vary so much on an hourly basis as it does seasonally, with most used during winter for heat; and even for oil, more of which is consumed during the summer driving season. Oil can be stored easily, in tanks, and one seldom hears about balancing supply with demand on a short-term basis. Gas is more troublesome, but can be stored underground in old, depleted fields or salt deposits that have been turned into underground caverns. Electricity is the most troublesome.

Because it cannot be stored, utilities must build enough generating capacity to serve the highest load they expect in a year, plus some extra in case something breaks. There are 8,760 hours in a year, and the electric utility must build its last watt of capacity knowing that it will sit idle for 8,759 of them. So, whereas oil and gas producers need only to drill enough wells to produce the annual average amount of demand, electric utilities need to build enough capacity to meet peak load.

Chicago Edison had been billing its customers based solely on volume—how much electricity they used. But this simple rate design benefits intermittent users of electricity at the expense of around-the-clock, or baseload, users. It is the intermittent users that cause the swings in demand, which forces the utility to build more capacity. To recover its investment, the utility must pass these costs back onto consumers. Charging on a purely volumetric basis means the baseload users end up subsidizing power plants that are idle most of the time, even though their needs are constant.

A two-part rate design solves this problem and introduces an economic definition that is important in every aspect of the energy industry: fixed versus variable costs.

In two-part rates, fixed (or sunk) costs are separated from variables. Fixed costs are determined based on peak load—that is, what capacity the utility must build in order to serve each user—while variable costs are paid based on the amount of electricity actually used. While both baseload and intermittent users must pay fixed costs, for baseload users, they represent a much smaller proportion of the total. Essentially, a fixed charge decreases the price paid per unit consumed, which benefits large users.

A fixed/variable rate structure properly accounts for the stress that intermittent users place on the system, and solved Insull's quandary of how to price electricity to both add users and turn a profit. The meter technology that the Brighton utility used was brought over to Chicago, allowing Chicago Edison both to move to a two-part rate structure and to measure total system load, which was invaluable in planning for the future.

Beginning in 1897, Insull used lower rates to target off-peak users, such as streetcar companies, which served daytime commuters; ice houses, which had their refrigeration loads peak during the heat of the day; and other industries whose electricity needs did not coincide with the evening lighting peak. The new rate structure made Chicago Edison immensely profitable and its president, Samuel Insull, one of the most powerful men in America. So when he began advocating for government regulation of utility franchises—his next great contribution to the industry—politicians and businessmen took note, even if they were horrified.

## THE CALL FOR REGULATION

In 1898 Insull gave a speech at the National Electric Light Association calling for utilities to be regulated as monopolies: "The best service at the lowest possible price can only be obtained . . . by exclusive control of a given territory being placed in the hands of one undertaking. . . . Charges for services [should be] fixed by public bodies to be based on cost, plus a reasonable profit. . . . The more certain [franchise] protection is made, the lower the rate of interest and the lower the total cost of operation will be, and, consequently, the lower the price of the service to public and private users."[6]

While this may not seem entirely dissimilar from the idea of the utility compact that had existed for some time in the manufactured gas industry, Insull was calling for monopoly power to be instituted universally in utility franchise agreements. He was also calling for a substantial deepening of regulatory oversight, and wanted to create state-level public utility commissions made up of specialists who would fix service rates, rather than entrust city councils to review rates as they deemed necessary.

At first, this alarmed businessmen and investors who wanted nothing but for the government to stay out of their affairs. Regulation was anathema to profit. Insull himself said as much, declaring in 1908: "The one absolutely desirable thing to do is to be able to conduct your affairs without coming in contact with the government."[7]

Insull's inconsistent views might make him seem schizophrenic—in the sense of the word's original meaning, "split mind"—but given the historical context, his call for regulation was, if anything, surprising only in that it had not come earlier. Though he was an electricity man through and through, Insull's changing worldview must have been informed to a large degree by what he had seen play out in Chicago's gas market. Peoples Gas Light and Coke Company was the city's largest and oldest gas utility, having been established and granted a perpetual, but nonexclusive, franchise in 1849. It coexisted with two other gas utilities in the relatively slow growth era, before a new technology introduced in the late 1870s spurred more rapid adoption of gaslights.[8] During the 1880s and 1890s, new gas companies sprang up and competed with Peoples on its own service territory, often laying lines directly besides Peoples', often in violation of municipal ordinances. (While Peoples had never been granted an exclusive franchise, these ordinances sometimes granted it an exclusive path on a given street.) The intense competition drove gas rates down from $2.38 per thousand cubic feet in 1881 to just $1.25 three years later, and just $1.00 by 1897, which bankrupted one of the gas companies, put two others dangerously close to receivership, and ate into the profitability of all.[9]

In response to intense lobbying by the gas firms, in 1897 the Illinois legislature passed the Gas Acts, which had two key provisions. The first effectively (and backhandedly) erected a steep barrier to entry by requiring the consent of the majority of property owners along a street before a new gas main could be laid (or electric wires strung), and by giving veto power to any individual property owner. For the incumbent utility to block entry was as simple as bribing a single property owner to object to construction. The second provision allowed gas companies to merge, which had previously been impossible.[10]

The results were predictable: Peoples acquired almost all of its competitors in 1897 and 1898 (it acquired the holdouts in 1907), becoming an unregulated monopoly. Its finances quickly improved, and it began to link the disparate systems into a coordinated, citywide network, setting it up to benefit from economies of larger scale.[11] This, and similar experiences in other parts of the country, must have proved to Insull the value of franchise exclusivity.

Insull had also, by the time of his pro-regulation speech, encountered more than his fair share of corrupt city officials, which convinced him that rate setting should be left to state officials. The wisdom of this preference was proven out in 1900, when Chicago passed an ordinance requiring gas companies to reduce their rates from $1.00 to just 75 cents—below cost. Peoples refused to comply, and one of its shareholders filed suit against the city. In 1904 Peoples won the suit, with the court ruling that "the regulation of the prices to charge consumers by gas companies is not one of the powers essential to municipal government . . . such power cannot be exercised by a city unless it has been delegated by the state."[12] But the city of Chicago was not about to give up so easily.

City politicians realized that utilities—both gas and electric—were easily demonized, and that they could garner votes by promising to slash rates. So they lobbied the Illinois legislature until it passed the Enabling Act in 1905, which granted the Chicago City Council the express right to set gas and electric rates. In 1906 the city and the gas companies came to an uneasy agreement on gas rates: eighty-five cents for five years. If this was not proof enough that local politicians might not be the best group to set gas and electric rates, what happened after the agreement expired in 1911 surely was.

Carter Harrison was running for mayor of Chicago in 1911, and one of his campaign promises was to reduce gas rates from eighty-five cents to just seventy cents. On assuming office, he acted quickly to make good on this promise, and enlisted an expert to conduct a thorough investigation of Peoples' finances, and to suggest a new, lower rate. The expert concluded that any rate below seventy-seven cents would not allow the company a fair return on capital. Disappointed with this expert, Harrison fired him and hired another, whom he paid five times as much, and who arrived at the desired number of seventy cents.[13]

By this time, several states had already come to the conclusion that municipal politicians, in seeking short-term political gain, might not always act in the long-term public interest, which required utilities to remain in sufficient financial health so as to attract capital at reasonable rates. Massachusetts was way ahead in this regard, having established a state public utilities commission, although with quite limited powers, in 1887. Several other states followed suit in the 1900s, and, indeed, the first expert that Mayor Harrison hired to determine a fair gas rate for Chicago—the one who had recommended seventy-seven cents and was then fired—was a member of Wisconsin's Public Utilities Commission, formed in 1907.[14] The Illinois Public Utilities Commission was formed in 1913, and by 1920, almost all states had transitioned to Insull's preferred method of regulation: state-regulated monopoly.

While the threat of capricious rate setting and ruinous competition was real, the actions of Insull and other utility captains across the country cannot be said to have been taken purely in the long-term public interest. It was certainly true that state-regulated, monopoly utilities could more efficiently raise capital and provide service within their franchise territory. It was equally true that, having obtained monopoly status, the captains had effectively built a moat around their castles.

Men like Insull, Rockefeller, Carnegie, and scores of others who built industrial empires all spoke at one time or another of ruinous competition, which would undo the order they had established. They also, at one time or another, practiced and benefited from ruinous competition. Before their empires were empires, they were companies that grew by buying amenable competitors, like Joseph Pew's Peoples (the Pittsburgh-based natural gas company, not the Chicago manufactured gas utility) or Chicago Arc Light. Failing that, they were more than happy to bankrupt, then buy, less amenable ones after what Rockefeller termed "a good sweating." During a sweating, the incumbent slashes rates to below costs, forcing the competition to do the same, lest they lose market share. Both companies lose money in the short run, but the well-capitalized incumbent is able to stand the heat for longer than the upstart competitor, who either surrenders or falls under receivership.

Monopoly by law allowed incumbent utilities to continue to earn profits without fear of would-be "raiders" instigating rate wars. This term, *raid*, was actually used by the utility captains, both verbally and often in written correspondence, to describe independents trying to enter their territory.

The economic historian Gabriel Kolko is regarded as having coined the term *political capitalism*, which he defines as "the utilization of political

outlets to attain conditions of stability, predictability, and security—to attain rationalization—in the economy."[15] Members of the Power Trust were
clearly shrewd economic and political capitalists, having advocated for and
received franchise exclusivity in both gas and electric distribution. As energy markets grew, so too did the extent of government involvement. But
as we examine how the government's role in the gas and electric industries
grew over the next several decades, it is useful to keep in mind that regulators were, more often than not, responding to requests from the private
sector, rather than imposing their will on business. A simplistic view, then,
holding that all regulation is antibusiness, while freedom from regulation
is pro-business, is not supported by the historical record (either ancient or
modern). Kolko puts it well: "It was never a question of regulation or no
regulation, of state control or laissez faire; there were, rather, the questions
of what kind of regulation and by whom."[16]

## UTILITY CAPTAINS AND THEIR HOLDING COMPANIES

Perhaps unsurprisingly, Peoples Gas Light and Coke appointed Samuel Insull as its chairman in 1913.[17] The great man of electricity was now a great
man of gas as well, and the preeminent utility captain in the United States.
These were the days of the Power Trust.

Insull extended the territory of his empire in lockstep first with the expansion of greater Chicago, then with the Midwest, and finally with that of
the United States as a whole. He formed utilities to serve some of the city's
larger suburbs, which were separate municipalities with separate governments, and bought others. He even extended service to some of the more
distant rural communities. But it seemed that even Insull could not expand
fast enough to keep pace with demand. And whereas Chicago Edison, now
called Commonwealth Edison, had to contend with only one municipal
government (as complicated and corrupt as it may have been), Insull now
had to deal with dozens of smaller governments.

Again, this was not unique to Chicago. In frantic search for electric and
gas service, and suspicious of business in general (keep in mind, this was the
Progressive Era that trust-busting Theodore Roosevelt had helped to usher
in), a drive toward municipal ownership began to build across the country.
In municipalization, new utilities were sometimes formed. Other times,
private companies were expropriated to create public ones, with shareholders being replaced by citizens. To the extent that Insull's support for state
regulation had not already caught on, the threat of municipalization galva-
nized the remaining utility captains behind the idea. By 1907 direct state
oversight of rates in exchange for monopoly protection had become the official preferred policy stance of the National Electric Light Association, the
group that Insull had addressed in his 1898 speech calling for regulation.[18]

Another consequence of skyrocketing demand and expansion beyond the urban core was the near-constant need for financing these increasingly disparate entities. This was thought to be most efficiently achieved by setting up a central holding company, which could funnel development dollars to where management thought they were best suited. Raising capital for a small Indiana utility with no assets and no track record might be difficult, but the holding companies were large, well-known enterprises, and so could raise capital at low rates and afford to take on a great deal of debt without spooking investors.

In 1912, the year before he took over as chairman of Peoples Gas Light and Coke, Insull formed Middle West Utilities to finance his expansion plans. Upon its formation, Middle West served 140 towns in southern Indiana. Five years later, it served 400 towns in thirteen states. The holding company proved to be an extremely effective model for building a (nearly) nationwide electric and gas infrastructure. In an account of the era, William Henderson wrote: "By placing huge amounts of capital at the disposal of a few highly skilled, technologically sophisticated entrepreneurs, the holding company device seemed to have the potential for drawing upon the benefits of central planning while stepping around the pitfalls of bureaucratic inefficiency and sloth."[19]

As it went in electricity, so it went with gas, both manufactured and natural. In the 1920s, a few large public utility holding companies grew to dominate the US natural gas market; by 1930 the largest four controlled nearly two-thirds of transmission and distribution.[20] (Notably, however, they controlled only 16 percent of natural gas production,[21] whereas the production of electricity was much more concentrated among the largest electric utilities: Insull-controlled companies alone produced 11 percent of US electricity in 1928.[22]) And while the combined Insull interests constituted by far the largest utility overall, they were not included in this list of natural gas heavyweights, as, to this point, they were engaged solely in the production and distribution of manufactured gas.

Who were the four natural gas behemoths, then? It probably does not come as a shock that Standard Oil made the cut; it may be more surprising to learn that it was actually the smallest on the list. The head of Standard's utility holding company, Standard Gas & Electric, was Christy Payne. If that name sounds familiar, that is because Christy was the son of Calvin Payne, who had run National Transit and been so instrumental in getting Standard into the natural gas business years earlier.

Columbia Gas & Electric ranked third largest, and largest by overall pipeline mileage. Columbia's rise has not been discussed in great detail, and that is because it was a product of dozens of mergers and acquisitions, which makes for very tedious reading. Suffice to say that Columbia's

**FIGURE 3.1:** Map of Standard and Columbia Gas & Electric's gas networks in Appalachia, c. 1930

network largely overlapped Standard's, except that it extended farther west, into the heart of Ohio, and south into Kentucky. Together, the two firms accounted for more than half of all natural gas distributed in by far the largest market in the United States: Appalachia. Philip Gossler, who began his career at Edison General Electric, was the chairman of Columbia, having held the top seat since he had assumed presidency of the company in 1909.

Coming in at number two was Electric Bond & Share Company (EBASCO), formed by GE in 1905, but having its true corporate origins in the 1890s. In many ways, EBASCO was the first incarnation of GE Capital, as its original purpose was to help small utilities to purchase GE–manufactured electric generation equipment by buying their securities and paying the parent company in cash. GE chose as president of EBASCO a man named Sidney Zollicoffer Mitchell, who had worked for GE as a young man and after that was instrumental in bringing electricity to the Pacific Northwest. In 1885 Mitchell and his associates built the first central generation station west of the Rockies, bringing incandescent light to Seattle, which was then home to just five thousand people.[23]

Through 1930 EBASCO was almost solely an electric utility, with its operations centered in Pennsylvania. The reason it clocked in at number two on the list of largest natural gas companies in 1930 was because it had only just—in March of that year—acquired the by-then very large United Gas system, which owned production, pipelines, and utilities in Louisiana and Texas. Indeed, the acquisition of United not only made EBASCO the second-largest gas company by transmission and distribution, but by far the largest in terms of owned production, since the heart of the United Gas system was the Monroe field in Louisiana.

An article in *Time* magazine written just after the EBASCO/United merger had been announced summarized the national situation concisely: "Significance of this consolidation lies in the extended influence of Electric Bond & Share; in the drift of the natural gas business away from its petroleum and toward its public utility affiliations; in the probable status of the newly formed company as the first of many far-reaching consolidations which should ultimately create a super-gas situation comparable to the already existing super-power systems in the electric field."[24]

EBASCO had purchased United from two slightly mysterious gentlemen named William Moody III and Odie Seagraves, whose businesses are referred to collectively as the Moody-Seagraves interests, and in addition to gas included hotels, railroads, and even cosmetics. Moody-Seagraves, like Columbia and so many others, had built their empire by acquiring smaller companies, starting in the Texas Gulf Coast and then linking the system with the Monroe field and extending service as far north as Dallas and as far south as Monterrey, Mexico. Where they were unique was that they had, from the beginning, focused on the development of gas resources, rather than oil.[25]

Moody and Seagraves retained a large stake in the consolidated company, and so remained quite influential in future decision-making. And though they and EBASCO did not play as large a part in the story of bringing the Southwest's huge reserves of gas to northern markets as Insull, Gossler, and Payne—at least, not to begin with—even their presence influenced the course of events, as we will see shortly.

And the largest natural gas company in the country? Where else, but Kansas. Today Kansas is not much associated with oil and gas, but it is home to two gas fields that were, in their time, some of the largest in North America. The Hugoton, in southwestern Kansas, was discovered in 1922 in the search for oil. However, a few decades before that giant was discovered, in the state's unpopulated west, eastern Kansas closely resembled a miniature version of Appalachia, with hand-dug wells delivering gas to small, nearby towns. The field supplying most of these wells was located in Iola, roughly halfway between Wichita and Kansas City, where drilling began in the late 1880s and gained pace in the 1890s, after several large zinc smelters, cement and brick plants were built to take advantage of the supply of natural gas. In 1904 pipelines were built to connect the Iola field to Kansas City, and by 1906, Wichita had natural gas.[26]

The situation in eastern Kansas was remarkably similar to the situation in the Indiana Gas Belt. In both cases, a surfeit of gas led to profligate waste. Municipalities attracted manufacturers by promising free gas supplies: the glass industry in Indiana grew to rival that of Pittsburgh,[27] and Kansas rapidly overtook Illinois to become the largest zinc-producing state in the United States.[28] Neither region's coal supplies were suitable for making iron-grade coke, so neither had any real prospects of giving the Pittsburgh metals industry a run for its money.

In a repeat of what happened in Indiana, production from eastern Kansas peaked in 1909 and dropped steadily thereafter.[29] Several companies sought out supplies farther afield, including from the colorfully named Hogshooter field in northern Oklahoma, but again, this only delayed the inevitable, and several eastern Kansas companies went into receivership between 1912 and 1914. They were bought by a gas company engineer from Columbus, Ohio, named Henry Doherty, who controlled a holding company called Cities Services.[30] Cities Services would come to almost completely control the gas markets of the mid-continent, which encompassed Kansas City, Wichita, Topeka, and Hutchinson, in Kansas; Kansas City, Joplin, and Springfield, in Missouri; and several towns in Oklahoma.

It was Henry Doherty who presided over the largest natural gas company in the United States in 1930. Although his home turf was the environs around Kansas City and Wichita, the company had become quite geographically expansive: Cities Services had brought together 120 companies in twenty-three states, plus Canada and Mexico, eventually serving over three hundred thousand customers.[31] In addition to his status as a business magnate, Doherty was a central figure in marketing gas to the masses, spending heavily on advertising and developing new uses for the fuel, and was among the first of the gas men to recognize the genius of the two-part rate structure that Insull had brought to the electric industry in 1897.

While Doherty is remembered as a lifelong student, who always thought of himself first and foremost as an engineer, he was just as much a member of the Power Trust as were Samuel Insull, Philip Gossler, Christy Payne, and William Moody and Odie Seagraves. These utility captains all conducted their business ruthlessly, but never deviated from the unwritten but universally acknowledged gentlemen's agreement: do not conduct raids into one another's territory.

The Power Trust had untold millions in financial resources, as well as political connections at almost every level, in almost every geography. But recall the natural gas market situation in the 1920s: Appalachia was running out, and the federal government was threatening to cut off all gas sales to nonresidential users; meanwhile, the Panhandle and Hugoton fields had trillions of cubic feet ready to be tapped. Even the Power Trust could not stop this tidal wave.

## A RAID TO REMEMBER

"If I would take a map of the Central United States and a pencil and a ruler, I could not draw a line to raise more hell than that one."[32] It was June 1930, and Columbia chief Philip Gossler was saying this to a man whom he had made rich the year before, and who had only months prior joined a new company that was planning to build a pipeline from the Southwest straight into the heart of the Power Trust's territory. This was a raid, pure and simple.

The man to whom Gossler was speaking was William Maguire, who got his start in railroads before moving over to the coking coal business, where he helped pioneer the so-called Roberts process of making coke from previously "uncokable" coal.[33] The process made St. Louis a major steel manufacturer in the 1920s. Had Maguire and the man who invented it, Arthur Roberts, been around forty years earlier, the advantage that Coal Hill conferred on Pittsburgh would not have counted for very much, and St. Louis, more directly downriver from the ore sites, might have been America's preeminent steel city.[34]

In 1928 Maguire started his own brokerage, arranging the acquisition of small midwestern utilities by larger firms. His largest transaction was a sale of natural gas properties to Philip Gossler's Columbia, which made him rich enough to contemplate retirement, in his mid-forties. But then Maguire, a short and blunt man, met a tall and brash Frank Parish, who just a few years earlier had established an upstart company that was competing in Kansas with Henry Doherty's Cities Services.[35]

Competing, that is, as a vulture might with a lion, over scraps. Cities Services had a stranglehold on the large markets of Kansas City and Wichita, and Parish's young company, Missouri-Kansas Pipeline Company, or Mo-Kan, was originally formed purely to act as a supplier of gas to Cities Services, not to serve customers of its own. Doherty was happy for the additional supply and to do business with Mo-Kan. In fact, the timing could not have been better, as far as Doherty was concerned, because public opinion had been building against the monopoly that Cities Services enjoyed in the region, so much so that he felt compelled to purchase a large stake in the local newspaper just to put an end to the negative editorials. Cities Services suggested that Parish join forces with a local wildcatter and an engineer who were developing a gas field just south of Kansas City; Doherty's lieutenants even introduced Parish to the underwriters that would help Mo-Kan raise capital.[36] The size and scope of Cities Services business in its home market was such that the threat of competition must not have even occurred to Doherty or his underlings, but Parish had grander plans for Mo-Kan than he let Cities Services know.

Two of the most powerful utility captains had unwittingly helped finance future competitors—Gossler by making Maguire wealthy enough to take a chance on a risky upstart, Doherty by nurturing Parish and Mo-Kan. Between 1928 and 1930, prior to Maguire's arrival, Mo-Kan gobbled up several distribution companies, seeking to become a legitimate challenger to Cities Services and the other large utility holding companies.[37] Mo-Kan's biggest threat to the Power Trust, however, came not from its ambitions on the distribution side, but from that hell-raising line that it hoped to build to transport natural gas from the Hugoton field into Power Trust territory in the Midwest and Appalachia.

The line would be the longest ever built, eventually spanning 1,250 miles and entering the territory first of Cities Services, then of North American Light & Power Company (whose customer base was in Illinois and Missouri, and of which Samuel Insull owned half), then of Middle West (of which Insull owned all), then of Columbia.

It is worth stepping back to 1883, when Joseph Pew and Edward Emerson's Penn Fuel first connected the Haymaker well to Pittsburgh's Sixteenth Street Station. The city's incumbent manufactured gas utility, Fuel Gas Company, cried foul and filed suit. But remember that Fuel Gas itself had tried to buy the Haymaker well and pipe its gas into the city. Penn Fuel had just beaten it to the punch. Not long afterward, Fuel Gas bought natural gas properties adjacent to the Haymaker and began transporting its own natural gas to Pittsburgh. It wasn't that the incumbent did not want natural gas in its territory, it was just that it wanted to control it itself.

This behavior cannot be ascribed purely to personal pride, although there is no doubt that pride plays a role; there is also rational economic motivation. In a capital-intensive industry, such as gas and electricity distribution, large sums are spent upfront on assets like pipelines, wires, power plants, and manufactured gas facilities, and amortized slowly but steadily over the asset's economic lifespan, which often exceeds twenty years. Fuel Gas had only begun serving Pittsburgh with manufactured gas in 1874, and its assets were not yet fully amortized when Penn Fuel began stealing its market in 1883. Moreover, even if a new entrant were economically inferior and posed no serious threat to the incumbent, any perceived breach of the protective regulation offered to the incumbent under the utility compact could spook investors, making capital more scarce and more expensive. Utility investors are buying a castle with a moat around it—fill in the moat, and the value of the castle drops substantially.

As it was in 1883, so it was 1930, only now the castles were much more grand. Samuel Insull, king of the Power Trust, presided over a large manufactured gas business into which he had poured millions of dollars building new capacity to keep pace with demand. He was far from being opposed to introducing natural gas in his territory; in fact, he had plans to do just that, but on his own terms and on his own timeline. As it was, landmen—the men who drive from farm to farm to negotiate leases and drilling terms with landowners on behalf of producers—were bumping up against each other in the Hugoton, Parish's against Insull's.[38] The utility captains were planning a pipeline of their own to take natural gas from the Southwest into Chicago. Upon arriving, though, the plan was to mix the natural gas with manufactured gas and allow Insull to recoup his investments before making a full-scale conversion.

This project was controlled jointly by Insull, Doherty, and Payne, as well as certain concerns from Texas who supplied it with gas, and was called the Natural Gas Pipeline Company of America, or NGPL (it was also commonly referred to as the Chicago Line). NGPL was to be a monster: a twenty-four-inch diameter system that would move gas 980 miles through nine compressors that kept the line packed with gas at a record-setting seven hundred pounds per square inch of pressure. This would allow 210 million cubic feet per day to move from the fields in Texas and Oklahoma into Chicago.[39] Compare that to the earlier natural gas line to Chicago, built in 1891 from the now largely dormant Indiana Gas Belt, which comprised two, parallel eight-inch lines running 120 miles that could move just 20 million cubic feet per day at three hundred pounds per square inch.[40]

Before detailing the Mo-Kan kerfuffle—which is high drama that includes corporate espionage, stock raids, indictments, congressional testimony, and even an old-fashioned fistfight—a few pages must address how rapidly pipeline technology had advanced in the latter half of the 1920s, and how the broader natural gas market was developing alongside it.

## THE GRAND DAMES OF LONG-DISTANCE, INTERSTATE PIPES

Recall the tong gangs—the groups of bearded woodsmen who, with their enormous calipers, manually screwed together twenty-foot lengths of wrought iron pipe. Clearly, efficiencies could be gained from this setup. The Dresser coupling, which mechanically joined together pipe ends, was a more economical solution and largely replaced the gangs, but the couplings were not fully leakproof, even at relatively low pressures, much less as pressures approached one thousand pounds per square inch. Ultimately, and starting in the 1910s, electric arc welding pipeline segments together became the new industry standard, virtually eliminating leaks from the joints where one segment of pipe met another.

But the pipeline itself was still of fairly shoddy construction, and prone to failure at high pressures, even if the welds were sound. This is because the pipe was rolled from flat plates—first of iron, but after 1900 almost always from steel—and the seam was weaker than the rest of the line.

This led to some crude rules of thumb, where a joint factor was used to estimate how strong the seam was, relative to the rest of the pipe. Before 1900 the process used most frequently—furnace butt welding—yielded a joint factor of 0.6, meaning that the seam would split at just 60 percent of the pressure that the pipeline could otherwise bear. After 1900 a new process—lap welding—gave a higher joint factor of 0.8, and could also facilitate the construction of larger diameter pipe. But the seam strength was inconsistent, and the individual segments were still only twenty feet in length.

The technology that gave rise to long-distance, high-pressure pipelines was commercialized in the 1920s, and was called electric resistance welding. In this process, steel plate is fed into a set of rollers that form it into a tube. As the edges near each other, an electric current is applied to bring either edge to welding temperature, after which they are forced together, completing the weld. Importantly, the new electric resistance welding mills could produce forty-foot lengths of pipe, which cut in half the number of segments the field welders needed to join, and created a seam that was actually stronger than the parent material—the joint factor was consistently greater than 1.0.[41]

Like most incremental technological advancement, the development of pipeline technology in the early twentieth century was driven by market need, rather than by a eureka moment late at night in the lab. In fact, the eureka moment that led to electric resistance welding came all the way back in 1886, when Elihu Thomson invented the process, a full forty years before it was used in pipeline manufacture.[42]

If only to show how small the world really was during this era, Elihu Thomson's company, the Thomson-Houston Company, was bought in 1883 by Charles Coffin. Less than ten years later, J. P. Morgan bought Thomson-Houston and combined it with Edison General Electric, creating GE, forcing Thomas Edison out of the business, and appointing Coffin as Samuel Insull's new boss. It was Coffin who threw Insull his lavish dinner party at Delmonico's before he left New York for Chicago.

But back to the 1920s, and to gas pipelines that could now transport high volumes for hundreds, even thousands, of miles. Technological innovation and the uneven distribution of gas reserves in the United States—too much in the Southwest and too little in Appalachia—led to the construction of the "grand dames" of the long-distance pipelines. These systems were built in the late 1920s and early 1930s, and continue to operate today, although the original steel has been replaced.

The grand dames were built to transport gas long distances from one of two places—the Panhandle/Hugoton area or the giant Monroe field in Louisiana—to places as far flung as Birmingham and Atlanta, St. Louis and

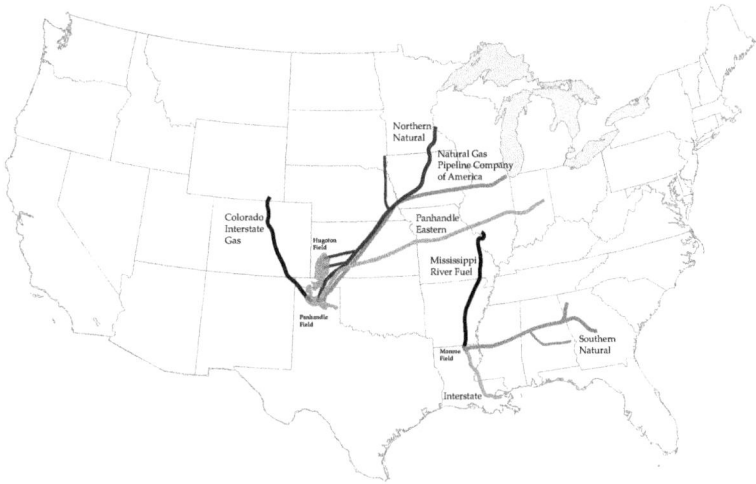

**FIGURE 3.2:** The grand dames: large interstate pipelines built between 1928 and 1931

New Orleans, Omaha and Minneapolis, Denver and Colorado Springs, and, of course, Chicago and Detroit.

Doherty's Cities Services, whose territory was geographically by far the closest to the Panhandle and Hugoton fields, had already established a position in the field by the late 1920s and built a 250-mile, 20-inch pipeline to Kansas City that began flowing gas in late 1928. This line was one of the first, and least grand, of the grand dames, and is now part of the Southern Star pipeline system.[43]

A line that sourced gas from the same properties as would NGPL, owned by Standard and a group of local Texas producers, was also built that year, but headed in the opposite direction: west, to Denver. Colorado Interstate Gas, which retains that name today, comprised four hundred miles of twenty-two-inch pipe when it began service in June 1928.

Mississippi River Fuel was another grand dame that delivered her first gas in 1928, through a twenty-two-inch line beginning in the Monroe field and ending five hundred miles later in St. Louis.

In 1929 another four-hundred-plus mile system called Southern Natural Gas (Sonat) brought yet more gas from Monroe to Birmingham and Atlanta, which marked the first natural gas delivered to the Southeast.

The grandest of the grand dames, though—the ones that targeted the large, industrialized markets of the Midwest, where winters were brutally cold—were not built until the 1930s.

North American Light & Power, the Illinois– and Missouri–based electric and manufactured gas utility half-owned by Insull, teamed up with Texas–based Lone Star Gas Company to build a large, long-distance pipeline from the Panhandle to the Twin Cities (incidentally, Lone Star was the fifth-largest natural gas company at this time, after Standard). North American and Lone Star had only just acquired this project from the Moody-Seagraves interests, which still owned the producing properties, but had intended to run the line only as far as Omaha. Northern Natural Gas, a twenty-four- and twenty-six-inch system that was the first pipeline to break the thousand-mile mark, began service in December 1931, less than two months after Insull, Doherty, and Payne's NGPL came online.

The period from 1928 to 1931 was the first pipeline boom in the United States, and was driven by huge imbalances in regional supply and demand and new technology that enabled economical long-distance transport. It was equally driven by a broader economic and financial boom—the Roaring Twenties gave holding companies an unlimited supply of cheap capital, and industry had expanded rapidly after World War I.

Tougher times lay ahead. The Great Depression would have led to massive changes in the corporate landscape and business practices no matter the preceding circumstances, and in any industry. But the utility industry, given how power was concentrated among just a few men, its clear and tangible affectation with the public interest, and the sordid nature of the highly publicized events that unfolded in the case of Mo-Kan versus the Power Trust, ensured that it would feel the effects of New Deal–era regulation more strongly than others.

## THE TRUST UNDOES ITSELF

By the time Gossler sat Maguire down and warned him that the Power Trust would not abide a raid into their territory, Mo-Kan had already acquired 150,000 acres in the Hugoton with estimated reserves of 1.5 trillion cubic feet of gas, more than enough to back a sizable line.[44] (While there was not a generally accepted practice at the time, in years hence it became standard to back sales contracts with twenty years of reserves, in which case Mo-Kan's 1.5 trillion cubic feet would amount to 205 million cubic feet per day, which would have represented just under 4 percent of the entire US market in 1930.)

Maguire must have been surprised by Gossler's threatening words, because he had used his substantial connections in the utility industry to try to avoid just such a confrontation. Initially, Mo-Kan's plans were for a pipe to the Twin Cities, but Maguire had worked closely with North American Light & Power's president and chairman, Clement Studebaker Jr. (heir to the automobile fortune), who told him that North American and Lone Star were building Northern Natural to serve that market, and that he should choose a different route. There was a consolation prize, though: in exchange for Mo-Kan rerouting its pipe away from the Twin Cities, North

American—again, half-owned by Insull—would buy gas from Mo-Kan in Missouri and Illinois.[45]

Maguire, Parish, and the rest of the Mo-Kan organization were happy to take the advice, and the sales contracts, and quickly redrew the route to the east, through Missouri and central Illinois toward Indianapolis, with an ultimate destination of Detroit. They christened the project the Panhandle Eastern Pipeline. As promised, they quickly and easily signed sales contracts with North American subsidiaries in Missouri and Illinois, but found it more difficult to extend service into Indiana, where local manufactured gas utility Citizens Gas Company was resisting a municipal takeover and anyway was uninterested in converting from manufactured gas, as it had just invested large sums in new, not-yet-amortized facilities. Mo-Kan's initial overtures to Detroit customers also went nowhere. The lack of markets in Indiana and Detroit were certainly disappointing to Mo-Kan, but all was not lost, as the Missouri and Illinois contracts were enough to go forward on. Indeed, the day after receiving them, Mo-Kan placed the order for pipe and contracted for Panhandle Eastern's construction. The system could always be extended at a later date.

Then came Maguire's fateful meeting with Gossler, which revealed that the Power Trust was not pleased with Mo-Kan after all. Several days later, Louis Fischer, one of Studebaker's lieutenants and president of Northern Natural, visited both Maguire and Parish and confirmed that Henry Doherty and Christy Payne wanted Panhandle Eastern dead, and were upset at Insull and Studebaker for letting North American sign the sales contracts. Fischer offered surrender terms on behalf of the Power Trust: if Mo-Kan sold its gas reserves to Standard and its existing pipelines to Cities Services, it would be awarded a 35 percent interest in Northern Natural and 10 percent of NGPL. When Parish, brash as he was, refused, Fischer took back the carrot and offered the stick: accept the terms of surrender, or the Power Trust will conduct a bear raid and bankrupt Mo-Kan.[46]

A bear raid meant coordinated short selling of a stock, accompanied by spreading negative and sometimes unfounded rumors. It is now illegal, but the Securities and Exchange Commission (SEC) was not formed until 1934, and these events took place in 1930. For a healthy company, bear raids can be a major nuisance, but not fatal. However, the Power Trust knew that Parish had been artificially propping up Mo-Kan's stock price by repurchasing all stock offered for sale and publishing financials that had only a tenuous link with reality. Only a month before the meetings with Gossler and Fischer, in May 1930, Parish issued a pro forma balance sheet indicating that Mo-Kan had $68 million in assets and $58 million in cash. The real numbers were closer to $4 million and $10 million, respectively.[47] Mo-Kan's financial position was far from healthy.

Parish thought he could rescue the situation by meeting with Christy Payne. He and Maguire boarded a night train from Chicago that was scheduled to arrive in New York early in the morning, before the market opened. But the train was late. Parish called Payne repeatedly, but got no response until after the market had closed for the day. By that time, the damage was done: Mo-Kan's stock had dropped from $36 to $15 dollars per share, before Parish's own, automatic purchases brought it back to $21. The way that Parish had organized his stock buying and selling enterprise meant that Mo-Kan now owed to the brokerage firm that had executed these automatic purchases $20 million that it did not have. It was, for all intents and purposes, bankrupt.

When they finally made it in to see Payne, his greeting was direct and brutal: "Well, Parish, how did you like it?"[48] Parish and Maguire left summarily. When they arrived at their hotel, though, Parish found several messages from Philip Gossler, who wanted to meet them both that very evening.

The perspicacious reader will have noticed that concerns controlled by Insull, Doherty, and Payne were all involved in the grand dame pipeline projects (actually, each had stakes in two of them). Even Lone Star was part of Northern Natural, which represented a substantial expansion from its Texas roots. Columbia, though, was conspicuously absent.

Perhaps Gossler felt betrayed by the other members of the Power Trust, or perhaps he merely saw the opportunity in Mo-Kan's weakness. Perhaps both. Whatever the case, Gossler offered to partner with Mo-Kan, and that evening in June 1930, the three men—Gossler, Parish, and Maguire—agreed that Columbia would take a 50 percent interest in Panhandle Eastern.[49]

That Gossler, one of their own, would conduct such a deal without first discussing it with the others was infuriating to the other members of the Power Trust, who thought they had succeeded in dealing a death blow to Panhandle Eastern. In a fit of rage, Doherty wrote to the president of the bank that issued bonds to pay for Panhandle's construction. His words were, in hindsight, astonishingly incriminating: "The Parish natural gas line" represented "an assault on other utilities," and Columbia and Cities Services had a "previous understanding," but Gossler had struck the deal "on grounds of fear that Parish would raid their territory." Doherty went on to accuse the bank of knowingly funding a competitor and insinuated that if it continued to do business with Mo-Kan, it might lose business with Cities Services. "I think your young men have made a bad mistake," Doherty wrote.[50]

As if this correspondence, which evinced corruption, extortion, and clear-cut monopolistic behavior, and which was later released as part of a massive federal investigation into utility practices, was not enough to convince the public and the Congress that the utility sector was badly in need of reform and regulation, Gossler's true plan in taking a 50 percent stake in Panhandle Eastern added an element of downright foulness to the mix. Columbia was not a white knight rescuing Mo-Kan—it was a wolf in sheep's clothing. Gossler's goal was not to create a partnership of equals, but to do all in his power to slowly bleed Mo-Kan of cash until it defaulted on its bonds, allowing him to take full ownership over the project. All of the construction bonds whose issuance drove Doherty so apoplectic? They were secured by a first lien on the pipeline, and Columbia owned every single one of them.[51]

The icing on this cake, which ended up yielding no useful information, but added to the public outrage once it was uncovered, was the fact Doherty had actually planted a spy in Mo-Kan's office. One of his associates chose a down-on-her-luck secretary to report back whatever internal information to which she was privy and also issue regular updates on the construction status, for despite everything else, Panhandle Eastern was being built.[52]

The next few years were not good for Mo-Kan, nor for almost anyone else in industry. While many would say that Black Tuesday—October 29, 1929—marked the start of the Great Depression, it was not until late 1930 that US banks began to fail, and most economic historians agree that early 1933 represented the trough of economic activity. The extreme downturn in industrial output meant many of the new pipelines built to move gas from fields in the Southwest to markets in the Midwest operated at less than full capacity. Panhandle Eastern, which began delivering gas in August 1931, was among them. The situation for Mo-Kan would have been difficult enough in good times, especially as its new "partner," Columbia, refused to purchase the twenty million cubic feet per day from Mo-Kan that it had agreed to, and appointed board members who stymied further expansions of the Panhandle system as part of its strategy to bleed Mo-Kan dry.[53] It was impossible in the throes of the Depression.

On March 6, 1932, Parish petitioned for receivership of Mo-Kan. The court obliged, deeming Mo-Kan in default on March 18, and appointed two receivers to direct operations. Less than a week later, Parish was indicted for mail fraud. Using a forged Canadian passport, he fled to Germany.[54]

Parish was in good company as a fugitive in Europe. Samuel Insull's holding companies, debt-laden as they were, were even more vulnerable to the downturn brought on by the Great Depression than was Mo-Kan. They suspended dividends in 1931, and were placed in receivership in April 1932. Insull, the great man, captain of captains, tendered his resignations—sixty-two of them in all, from all of his various operating companies—on June 6.[55] On October 4, Samuel and his brother Martin were indicted for embezzlement, but Samuel and his wife and son had already flown to France days before.[56]

Because France and the United States had an extradition treaty, Samuel then fled to Athens. (Martin was a British citizen and was left relatively undisturbed in Canada.) But even this was not enough for the government. Insull was too important a scapegoat to be left to senesce peacefully on the Aegean. President Franklin D. Roosevelt had decried Insull by name on the campaign trail, telling the people that the "Insull monstrosity" justified government intervention into private business.[57] The United States applied pressure to Greece to sign an extradition treaty. It did so in late 1933, but the treaty stipulated that extradition could only take place if the crime were illegal under Greek law, which Insull's embezzlement charge was not. Hell-bent on putting Insull on trial, the US government used its own extortion tactics: it revoked the money-transfer privileges of the Greek-American Merchants Association, an agency that sent remittances from workers of Greek descent in the United States to their families back in the Old Country. The United States assured the association that it would have its privileges restored as soon as the Greek government sent Insull back.

The effort was a success. While Greece did not officially extradite Insull, it forced him to leave the country. Insull boarded a ship bound for Turkey, where US agents tracked him down and were finally able to extradite him.[58]

When Insull made it back to Chicago, in October 1934, he stood trial. Parish had returned from Germany a few months earlier on his own volition, to visit his gravely ill brother, and had his day in court in April 1935. (Actually, Parish had several days in court. Angry at the accusations Maguire had leveled at him while he was a fugitive abroad, Parish and an acquaintance met Maguire in his hotel room and beat him and a former Mo-Kan treasurer senseless in 1934.)

In the end, both Insull and Parish were found not guilty. Parish effectively shifted the blame to the Power Trust, and Insull convinced the jury that, while his judgment was erroneous, he had never acted dishonestly. Parish went on to manage a firm that converted appliances in Detroit from manufactured gas to natural gas when Panhandle Eastern finally reached the city in 1936, though was never again officially affiliated with either Panhandle Eastern or Mo-Kan.[59] Later, he moved to Maryland and became a dairy farmer.[60] Insull spent the remainder of his days in Paris, where he died in the subway in 1938, bankrupt, but well-dressed, and with his good name restored.

Both men were right: it was the system that failed; or, rather, the lack of a system. The absence of regulation, in this case, did not lead to a competitive market situation, but a situation marked by a lack of competition. Insull was honest, but it cannot be denied that he acted to stifle competition, and that the debt-addled holding company model that he helped to pioneer fell to pieces during the Depression.

In the years that followed, new laws were written and government agencies were formed to break up the utility trusts. One consequence was that Columbia was forced sign a consent decree acknowledging its violation of the Sherman Antitrust Act of 1890, and, in essence, promise to allow Panhandle Eastern to expand and become profitable.[61] Panhandle Eastern began deliveries to Detroit in July 1936, and the profits allowed Mo-Kan to emerge from receivership in 1937. After several more years of litigation, Columbia was forced to divest its entire stake in Panhandle Eastern, and William Maguire became CEO and chairman of the board, a position he retained until his death in 1965 at the age of seventy-nine.

Philip Gossler stepped down as president of Columbia in 1936, the year that natural gas deliveries to Detroit via Panhandle Eastern began, but remained its chairman. Because of the geographical continuity of its operations, Columbia remained an integrated gas system whose layout has not changed much since the 1930s, though ownership has changed several times and Columbia, like all utility holding companies, was forced to separate its gas and electricity operations.

By the mid-1930s, Henry Doherty was so incapacitated by severe arthritis that he operated his empire from a specially equipped bed.[62] He died in 1939, at which point Cities Services was in the midst of divesting itself of all of its utility holdings, a process that continued for almost twenty more years. It retained its upstream positions, though, and became an integrated oil company during World War II. It was bought by Occidental Petroleum in 1982, and not long after its refinery operations were spun off as CITGO and sold to a company called Southland Industries, which owned all of the more than seven thousand 7-Eleven stores in the United States and thought purchasing a refinery would be a good complement to this business. It was not, and eventually CITGO was sold to state-owned Petroleos de Venezuela (PDVSA).[63]

Christy Payne may have fared the best among the senior members of the Power Trust. He retired from Standard in 1935, before the new regulations came into effect, and retired to a mansion that he had painstakingly custom-built in Sarasota, Florida. He lived there for twenty-seven years, before passing away in 1962 at the age of eighty-eight.[64] After Payne's departure, Hope was spun off from Standard in 1943 and renamed Consolidated Natural Gas, headquartered in Pittsburgh. Consolidated, in turn, was bought by Virginia–based Dominion Resources in 2000. The system is now known as Dominion Transmission, or DTI. East Ohio Gas was included in the purchase, and functions as a local distribution company and an intrastate gas transmission system, known as Dominion East Ohio, or DEO.

# CHAPTER 4

## REGULATION AND REJUVENATION

The Great Depression brought down a great many businesses, but few were as vilified as public utility holding companies. "Racketeers with their hands on the throats of the American people," was how they were described by supporters of federal oversight.[1] In fact, negative public perception had plagued utilities since the beginnings of the Progressive Era in the 1890s, when the federal government began to play a serious role in private business for the first time, regulating railroad rates and breaking up those trusts that it deemed wielded too much control over a certain industry, most famously Standard Oil in 1911. Anti-utility sentiment had already reached a fever pitch by the late 1920s, even before Black Tuesday and the bankruptcies that followed it. Recall that in 1927 Henry Doherty felt compelled to purchase the *Kansas City Star* after it ran an editorial lamenting the fact that Kansas City residents had given Cities Services a monopoly and allowed it to charge for its services whatever price it deemed appropriate.

And it was not just the popular media who cried foul. Gifford Pinchot, twice governor of Pennsylvania, wrote a thoroughly researched treatise in 1928 called *The Power Monopoly: Its Make-Up and its Menace*, in which he noted that those companies controlled by Samuel Insull alone accounted for 11 percent of total US electric output.[2]

No one was the subject of more personal enmity than Insull, whom newly elected President Roosevelt had personally lambasted on the campaign trail. Indeed, the mood of the public and elevation of Roosevelt to the presidency virtually guaranteed federal action to reform the utility business.

In actuality, the bankruptcy of Insull's holding companies was more attributable to their high levels of debt than to any inherent weaknesses in the corporate structure—after all, most of the other holding companies survived, by the skin of their teeth. And Governor Pinchot, though outspoken in his critique of holding companies, did recognize that utilities were most appropriately managed as natural monopolies: "Two telephone companies, two gas companies, two electric companies in one town usually amount to an economic extravagance if not a financial impossibility."[3] Pinchot, and others in government, could abide privately owned monopoly utilities. What they could not abide was the "financial inflation" that the holding companies commonly employed. Indeed, the primary aim of New Deal utility legislation was to eliminate, or at least greatly reduce, this financial inflation.

## A NEW LEGISLATIVE LANDSCAPE

At its simplest, what was known as financial inflation at public utility holding companies involved overpaying for acquired assets, like power plants. Since regulated companies earn a preset rate of return on their assets, which is referred to as their rate base, their profit margins are fixed. This cost-plus structure incentivizes the utility to spend in order to increase its rate base, and hence revenues, and hence absolute profits. A holding company acquiring a new utility would therefore purposely overpay to increase its rate base and profits. This would be impossible in an open market where several bidders competed for the asset, but a monopoly meant that utilities could assign value to an asset quite arbitrarily. A landmark Federal Trade Commission study, published in ninety-six volumes between 1928 and 1936, accused holding companies first and foremost of "loading the fixed capital account of public utilities with arbitrary or imaginary amounts in order to establish a base for excessive rates."[4]

This economic concept is known as "gold plating,"[5] and it is not unique to the utility industry. While the economic incentive for utilities to gold plate and grow their rate base has not changed to this day, New Deal reforms implemented standardized bookkeeping (the Uniform System of Accounts) and simpler operating structures, meaning that costs were less subject to flagrant inflation.

In the 1930s, three acts of Congress were passed that would change the gas business. The first two were reactionary, and sought to vanquish once and for all the sprawling utility holding companies and their opaque and highly nonstandard financial reporting. The Public Utility Holding Company Act of 1935 (PUHCA) mandated a maximum of three layers of ownership at any utility, and also that operations be confined to an integrated and geographically contiguous service territory. The Securities and Exchange Act (SEA) of 1934 was a landmark law that mandated financial transparency and standardized accounting for publicly traded companies across all industries, utilities included. Notably, it outlawed the practice of "stock pegging" that Frank Parish had used with Mo-Kan, and would have made impossible the sort of inflated numbers that both Mo-Kan and

Middle West had reported in their pro forma financials. The SEA also created a new enforcement agency, the Securities and Exchange Commission (SEC), which was charged not only with instituting and enforcing the new accounting measures across all industries in the United States but also with forcing public utility holding companies to break up and spin off their operating companies.

PUHCA was extremely controversial, as it effectively forced shareholder-owned companies to abandon a large portion—indeed, sometimes the very core—of their business. It was one of the New Deal–era regulations that brightly illuminated the split between reformers and conservatives in government, with the latter convinced PUHCA was ruinous for the economy and the country. "Government dictatorship over private business," one congressman said. "The most drastic and extreme measure of regulation ever offered in an American legislative body," said another.[6] The historian M. Elizabeth Sanders writes: "Not only the company directors, but the institutions and individuals holding stock in utility companies protested vigorously against [PUHCA]. This was perhaps the only New Deal measure that anticipated a genuine redistribution of wealth and corporate power, and political debate reflected that redistributive content."[7] President Roosevelt was "convinced that defeating the power trust required the destruction of the holding company structure," because he saw no way to regulate the holding companies themselves, and he went so far as to pledge publicly, in his 1935 State of the Union address, "the abolition of the evil holding companies."[8] So unequivocal was Roosevelt's view that he forced members of his famous Brain Trust, the close group of advisors who essentially created the New Deal, to abandon their preferred version of PUHCA in favor of a more radical bill, which include a "death sentence" clause that would have summarily forced the divestiture of operating companies of holding companies that were deemed too big (mandatory reorganization or dissolution after January 1, 1940). Only after several brutal months of back-and-forth did the conservatives manage to rework the death sentence, allowing the SEC to mandate divestitures on a case-by-case basis, based on whether the agency determined it "in the national interest," and without a deadline. Holding companies could live on, so long as they could demonstrate economies of scale.

This slightly amended version of PUHCA, which gave the SEC broad discretion over which holding companies could go on and which should perish, became law in August 1935, but it was far from clear to anyone that the New Deal regulatory apparatus would survive judicial scrutiny. The Supreme Court had struck down as unconstitutional another major New Deal regulation, the National Industrial Recovery Act (NIRA), just three months prior, and the holding companies held out hope that PUHCA would also be overturned by the high court. In fact, placing the power of life or death in the hands of the SEC may have been meant to set PUHCA up for such a failure, and the agency balked at being given such discretion.

James Landis, a member of the Brain Trust, was the chairman of the SEC at the time, and he devised a strategy that some historians credit with ensuring that PUHCA did not meet the same fate as the NIRA.[9] His approach combined grandiosity with restraint, pleasantry with duplicity. He did not begin by threatening to break up the companies, but simply by requiring them to register with the SEC, as the law provided. Here was the restraint. However, Landis knew that the companies were fiercely opposed to complying with any part of the law until a court ruled on the constitutionality of all of it, including the modified death sentence clause that allowed the SEC to force divestitures.

For his test case, Landis chose EBASCO, one of the nation's largest holding companies. Here was the grandiosity. Of this decision, Landis later said: "We came to the conclusion that, let's pick a big one, a top one, take a big one and topple that one, and then the little ones would fall into line."[10] He attended meetings and negotiated with EBASCO ahead of the registration deadline, but when the company missed the deadline, which Landis knew it would, he immediately filed a lawsuit that he and his team had drafted earlier. The Supreme Court was thus forced to rule on a very limited case in which it was clear that the SEC had not acted capriciously, invoking only the bare minimum of its authority, and where the defendant, EBASCO, was unwilling to comply even with a section of the law that required relatively little expense and effort. The court ruled in favor of the SEC, and PUHCA was not struck down.

The SEC's measured approach meant that holding company breakups did not start in earnest until 1940, and the divestiture process dragged on for twenty-some years.[11] Cities Service did not divest the last of its utility operating companies until 1958.[12] James Landis would continue to be a driving force in US regulatory matters for three more decades, and his advice would guide the regulatory agenda of the Kennedy administration in the 1960s.

The PUHCA-mandated breakup of the holding companies was a long slog and involved substantial litigation but, in the end, it achieved its goals: Between 1935 and 1954, 417 independent operating companies were spun out of the holding companies, of which 158 were gas local distribution companies (LDCs). The great breakup also decreased holding company control over interstate pipeline mileage to 18 percent from 80 percent—there was no longer a Power Trust, or a Gas Trust, or anything near the same degree of concentration in the industry, and there has not been since.[13] Whereas before, gas transmission companies that did not own utilities, like

Panhandle Eastern, were the exception, after PUHCA they became the rule.

After PUHCA and the SEA, the third act of Congress that affected the natural gas industry was substantially less controversial, as it did not involve the seizure or forced sale of property. It was the Natural Gas Act of 1938 (NGA), and it subjected interstate transport of natural gas to federal price and entry controls.

## THE NATURAL GAS ACT OF 1938

The Natural Gas Act was written to close a regulatory gap. State public utility commissions were empowered to regulate rates and grant exclusive franchises to utilities. These powers were substantially strengthened after PUHCA and the SEA ensured that the large holding companies, which before had operated across state lines, were not burdening their in-state operating companies with holding company-level expenses or using financial inflation to charge higher rates. But, especially after the flurry of new long-distance pipeline build between 1928 and 1931 (the grand dames), an increasing share of the gas bought by the state-regulated local distribution companies (LDCs) came from out of state. And the Supreme Court had consistently ruled that state bodies had no jurisdiction over interstate commerce. Hence the regulatory gap: state public utility commissions could make sure LDCs were not overcharging their consumers, but the commissions were clueless and powerless over whether LDCs were getting a fair price for the gas they bought from the interstate pipeline companies.

PUHCA actually made this problem worse, as the integrated holding companies that built and operated most of the long-distance pipelines—Insull's NGPL, North American's Northern Natural Gas, Standard's Hope Natural Gas—no longer had any direct ties to the distribution business.

The NGA was an important, but originally quite limited, act. In the decades to come it would morph into something entirely different from what its creators had intended, and then, in the more recent past, return to its original intent. It is not hyperbolic to say that the history of the natural gas industry since the passage of the NGA can be told in terms of how the courts and the administrative body whose task it became to enforce it interpreted the act's language.

The NGA set out to regulate interstate pipeline companies as monopolies affected with the public interest, not unlike how state utility commissions regulated LDCs. It set up a cost-plus price-control mechanism, which ensured pipelines would receive a fair rate of return on their investment, and created a barrier to entry by requiring any new interstate pipeline to apply for a certificate when entering a market already served by an incumbent. When the market was already being served, the onus was on the new applicant to prove that the market required new capacity that the incumbent could not provide via extensions to its system at lower cost.

Responsibility for issuing certificates and reviewing and approving interstate pipeline transport rates fell to the Federal Power Commission (FPC), which was created to oversee hydropower. The FPC and its successor organization, the Federal Energy Regulatory Commission (FERC), consist of five commissions with six-year terms, appointed by the president.

Being awarded an FPC certificate became extremely important after PUHCA, since financing could no longer be raised so easily at the holding company level. Now, pipeline companies needed to go to banks for loans. A certificate more or less guaranteed that the only way there would be competition—a raid, in the parlance of the Power Trust—was if the end market was growing so quickly that a single pipeline system could not keep pace.[14] This situation would not be such a bad thing: a rising tide lifts all boats, after all. In sum, with an FPC certificate in hand and a group of fairly competent company offices, bank loans for new pipelines were at the ready; without a certificate, there was no financing, either from within (holding companies no longer existed) or without (banks would not lend).

Surely another reason for the lack of controversy surrounding the passage of the NGA was that the prevailing nationwide supply/demand balance was such that no new pipelines were in the offing. In 1938 the United States was still recovering from the Great Depression, and the pipeline boom of 1928–1931 had created excess transportation capacity. While some existing systems were expanded in the late 1930s, very few large, new pipelines were built until the United States became involved in World War II in 1941.

The NGA was very straightforward and hewed closely to plain vanilla utility regulation. There were a few heated arguments that had to be settled in the draft-writing stages, but the NGA was strongly supported by the importing states, which were much larger and politically powerful than the producing states in the still sparsely populated Southwest.

Geography and population are extremely important in examining how the NGA was passed and later interpreted. In 1940 the gas-consuming Midwest states, who obviously favored price control, were home to thirty-two million people. Appalachia was home to another twenty-six million. It had substantial amounts of both consumption and production, so might have been fairly agnostic to regulation, but the region was also home to a very large coal industry, which was having its markets quickly usurped by natural gas, oil, and even hydropower, in addition to dealing with the broader negative effects of the Depression. The Appalachian coal industry employed almost half a million men—ten times as many as the natural gas industry[15]—and coal shipments accounted for about 20 percent of total railroad revenues.

Appalachian consumers, coal companies, and railroad companies formed an unlikely, but very politically powerful, alliance in favor of regulation. Appalachian consumers favored it for the same reason as their midwestern brethren: regulation meant price control. The coal and rail industries favored regulation for a different reason: entry control. Coal and rail concerns would be allowed to intervene in the certification process and argue that pipelines should not be allowed to expand to territories already served by coal.

So, since gas producers in the Appalachian states did not hold nearly as much political or economic influence as the alliance of consumers, coal, and rail interests, politicians representing fifty-eight million people in 1940—the combined population of the Midwest and Appalachia—can be said to have been broadly in favor of regulation.

The producer states were home to only thirteen million people, which meant they probably would not have managed to stop the NGA no matter how vehement their opposition. But actually, they were not against it. In fact, producer states voiced tepid support for the act, as it was thought to be positive for the growth of the industry in general. Gas was still very much a by-product of oil, and producers were wasting untold volumes of it, flaring it at the wellhead for lack of a market, so any regulation making it easier for pipelines to be built would be a welcome boon for sales. In 1938 the value of all the gas sold in Texas was $133 million, versus $559 million for oil sales; in Oklahoma, the corresponding figures were $27 and $224 million, respectively; in Louisiana, where the Monroe field boasted more connectivity to downstream markets than in most other regions, gas sales still totaled only $48 million versus $103 million in oil sales.[16] Oil was still the prize.

The Northeast states not yet served by natural gas had a combined population of thirty-one million, but no immediate interest in regulation, and the Southeast and West, each with around fifteen million people, were also fairly unimportant voting blocs at this time. In sum, the NGA passed without much acrimony.

## PRODUCERS AND THE NATURAL GAS ACT

The main reason that the NGA garnered only mild support from the producing states, rather than their ringing endorsement, is that a key provision in the act's original wording, common carrier status of pipelines, had been deleted from the final version.

At this point I should apologize for withholding some relevant information about the genesis of the NGA. It did not emerge after PUHCA, as separate legislation, but rather began its life alongside PUHCA, written into the same bill, which was submitted by the Texas Democrat Sam

Rayburn. The original Rayburn bill had three titles. Title I went on to become PUHCA. Title II became the Federal Power Act, which regulated interstate electricity rates and market entry, and, like PUHCA, became law in 1935. Title III was an early version of the NGA, but was staunchly opposed by pipeline companies, mostly because it would have treated them as common carriers.[17]

Common carriage is a legal concept that traces its roots to medieval England. Its two key provisions have not changed since the seventeenth century, when the lord chief justice of the king's bench —a fine title, to be sure—wrote that business "affected with a public interest" must serve any and all customers on a nondiscriminatory basis up to its capacity, and must not charge excessive rates for the service rendered.[18] In the United States, common carriage already had a long history by the 1930s. In passing the Interstate Commerce Act of 1887, Congress created the first federal regulatory agency, the Interstate Commerce Commission (ICC), and empowered it to regulate railroads as common carriers, requiring that their rates be "reasonable and just." The ICC's power was originally quite limited, however, as it did not have the power to set specific rates—it was still up to the railroads to determine what "reasonable and just" actually meant, numerically. This changed with the passage of the Hepburn Act of 1906, which was one of President Theodore Roosevelt's crowning legislative achievements. The Hepburn Act allowed the ICC to set maximum rates and extended common carrier regulatory status to several other industries, including oil, but not gas, pipelines.

Gas pipeline companies objected to being treated as common carriers for reasons both logistical and commercial. The logistical issues related to the extreme seasonality of gas demand and the relative difficulty and expense of storing gas, and the commercial issues mostly followed from those. At the time that the bill was being debated, in 1935, interstate pipelines bought almost all of their gas from wholly owned, or "affiliated," producers. Independent producers, not owned by a gas pipeline company, were focused overwhelmingly on producing oil and so were thought to be undependable, prompting one executive from NGPL, one of the largest integrated producer-pipelines, to remark in a Congressional hearing on the bill: "It would not only be impractical but virtually impossible to furnish dependable service to the 1.2 million domestic consumers if the pipeline company were dependent on the whims and fancies of independent producers to drill additional wells as required."[19]

A hallmark of common carrier status is nondiscrimination among shippers, which would have compelled pipelines not only to take receipt of gas from independent producers but to offer them the same rates as their affiliates. Logistically this threatened to put the reliability of the whole system at risk, as independents ramped gas production up or down for reasons wholly unrelated to, and likely out of step with, seasonal gas demand. Affiliated producers would have had to assume a balancing role.

The broader charge was that natural gas companies viewed the business as an integrated whole and thus saw their production assets as complementary to their transportation and storage assets—they had built the pipes to give life to their wells, on which they had risked their capital. They also argued that Congress would be acting unconstitutionally in regulating them as common carriers, since they had never held themselves out to the public as such. The legal precedent is that advertising an enterprise as a common carrier can force it to accept the terms of common carriage, but that no obligation can be forced on enterprises that have never held themselves out as common carriers. An attorney from the Federal Trade Commission stated simply: "The Supreme Court has said, of course, very definitely, that you cannot make a person a common carrier by declaring him to be one."[20]

Common carriage would certainly have been more feasible if independent producers assumed the same balancing responsibilities as affiliated gas producers. But even if they had, swings in gas demand are not just seasonal, but occur in real time as temperatures change. A representative from Hope Natural Gas testified that, as temperatures in Cleveland dropped from sixty degrees to ten, gas consumption rose from 50 million one day to 230 million cubic feet the next—a near quintupling, day over day![21] The only way to guarantee reliable wintertime gas service was thus to operate large storage systems, which are much more costly for gas than they are for almost any other commodity. The essential role of storage in the gas market, more than anything else, was why common carriage was thrown out as unworkable, and not included in the Natural Gas Act of 1938.

Gas storage is so expensive because the volumes being transported are so large as to make aboveground tanks impractical and uneconomical. For a very rough but illustrative example, let us say, just in Appalachia in 1935, that production was about 1 billion cubic feet per day, and that consumption over the 151 days of winter (November 1–March 31) averaged 2 billion cubic feet per day. The market would thus require 151 billion cubic feet of gas to have been stored before winter began. At normal conditions (atmospheric pressure and normal ambient temperatures), this would necessitate a structure as tall as the Empire State Building (1,250 feet) and slightly larger than two miles wide and two miles long. And, of course, the structure would have to be airtight. The largest building in the world by volume— Boeing's aircraft manufacturing facility in Everett, Washington—is 1/300 that size. And again, those numbers were just for Appalachia, in 1935: the total amount of gas stored before each winter nowadays is closer to 4 trillion cubic feet, or 8,500 of Boeing's Everett facilities.

This is not to say that aboveground gas storage wasn't tried—it was, in the early days of the industry. But storing large volumes is much more economical and practical by using the very same underground reservoirs from which the gas was produced in the first place: depleted fields.

Gas is injected into these depleted fields during the spring, summer, and fall, and withdrawn in the winter using the same sort of wells that produce oil and gas. The main difference is the large amount of on-site compression needed to inject the gas. Building an underground gas storage field is expensive and takes much longer than building aboveground tanks.

Regulating gas pipelines as common carriers could have worked, but would have required fundamental changes to business models without clear short-term benefits. Someone would have to pay for storage, but who? Producers did not want to, and why should they have? LDCs were the ones with the mandate to serve, and the corresponding ability to pass on the costs. But it hardly suited their business model to pay for assets far from their service territories, often across state lines. How would they share the cost with other utilities, since one large facility might be used to serve a dozen or more LDCs? Storage fields, after all, were usually located in the same places where gas had once been produced—it was the exception, not the rule, for LDCs to find suitable geology to build storage facilities within their own service territory. Distributors were not clamoring for pipelines to be regulated as common carriers, either.

Changing pipelines' regulatory status after the fact would have dealt another blow to the same companies that PUHCA was already forcing out of the utility business. Despite the scandals that had plagued utility captains during the age of the Power Trust, the captains and their lieutenants still held influence among legislators at all levels of government. They eventually accepted the fact that they would need to stay out of the local distribution business, but production and transportation had not been so shaken by scandal and fraud. Ultimately the public outrage that had spurred LDC reform was absent in production and transportation, and the small, independent producers did not form a large enough economic or political bloc to force Congress to shove common carrier down pipeline companies' throats.

The logic of the integrated pipeline-producer companies was clear: this is my pipeline, that serves my wells. If you want your own, go and build it yourself. But one cannot turn a deaf ear to the plight of the independents, who did not have the financial resources to even consider such an undertaking.

The situation was gravest in Texas, Oklahoma, and Kansas. Pipelines either flatly refused to do business with the independents or else offered such high transport rates as to be the equivalent of a refusal. This was bad enough, but there was often insult after the injury of not being served.

**FIGURE 4.1:** Pumpjacks, or nodding donkeys, applying artificial lift in the field

The rule of capture meant that pipeline-affiliated producers were drilling on their land but also draining gas out of surrounding acreage, including acreage owned by the independents. This does make the pipeline companies seem mendacious, as they denied transport service to the independents while stealing the gas from beneath their very feet—"drinking their milk-shake!"—to quote Daniel Plainview, the hard-nosed oilman whom Daniel Day Lewis portrayed in the 2007 film *There Will Be Blood*.

Drainage is an important concept, both geologically and economically. In most petroleum reservoirs, oil and/or gas are held under intense pressure, such that they will flow up to the surface on their own after the first wells are brought on. This can have spectacular results when the initial pressure is enormous—such as the blowout at Spindletop—or, at lower pressure, can lead mundane wells that flow a few barrels a day, such as the wells at Corsicana. Eventually, after the pressure equalizes, the oil and gas will no longer flow, and "artificial lift" must be used to coax out any more. The most common artificial lift device is the enigmatic and ever-present pump-jack, or nodding donkey.

In a conventional reservoir, like the ones being drilled in the 1930s, there is high enough permeability for the oil and gas to migrate through the reservoir to the wellbore, just as water migrates through the pores of a sponge. The well thus drains the extent of the reservoir, even if the reservoir extends to adjacent acreage. Since acreage rights are held over a defined area that may or may not correspond with the extent of the reservoir, there is opportunity for nearby producers to "steal" oil and gas. Eventually field rules would be developed that mitigated this rampant milkshake-drinking, but it was a cause for major consternation in the 1930s. (Incidentally, the defining characteristic of the unconventional reservoirs more common in US oil and gas production today (the blanket term has become *shale*) is that there is not high enough permeability for the oil and gas to migrate to the wellbore, so the permeability must be manmade, via hydraulic fracturing. There is no milkshake-drinking, because the milkshake is frozen solid.)

In sum, it is unsurprising that producing states were not passionate supporters of the NGA, having been stripped of the common carrier provision that Sam Rayburn included in title III of his original bill. But they nonetheless recognized that it would provide stability, predictability, and security to potential new pipeline ventures and their would-be financial backers. All else equal, this would allow more to be built, which would allow more acreage to be developed, which was a good thing for producers everywhere.

## THE ORIGINAL INTENT OF THE NATURAL GAS ACT

Because the language of the bill that would become the NGA had been amended not to regulate pipelines as common carriers, the act's intended impact on producers was limited and diffuse: the stability, predictability, and security it brought would support the industry's overall growth, and this rising tide would lift all producers' boats. In fact, there was even more targeted language in the act that specifically exempted production and gathering from its purview, with section 1(b) saying that the act "shall not apply to . . . the production or gathering of natural gas."[22]

In the years that followed, however, and despite what seems like a clearly worded production and gathering exemption, the NGA was interpreted as giving the FPC power to regulate the price of natural gas received by producers. At first, price regulation was limited to producers owned by interstate pipeline companies, but by 1954 it was extended to all producers, nationwide. The results were as sclerotic as they were predictable, leading first to shortages as the price was set too low, then to surplus as the government overshot and set prices too high. But these issues did not materialize for quite some time. The incredible bounty of already-discovered Southwest reserves meant that a shortage did not begin to develop until the late 1960s, and it would be another ten years until they became so severe as to compel the US government to fundamentally alter its policy. To use the terminology of addiction, rock bottom came in the winter of 1977. Temperatures dropped well below normal, and gas was in such short supply that deliveries needed to be curtailed not just to industrial consumers but also to schools. In Columbus, Ohio, public schools were closed for three weeks in February, and students met briefly only once per week in buildings that used coal, oil, or electricity for heat.[23] In 1978 the US government dramatically increased the price producers received for new gas supplies.

A more in-depth analysis of the changing meaning of the NGA follows in chapters hence. And this book criticizes many times both the FPC (later FERC) and, to an even greater degree, the Supreme Court over their overly broad interpretation of the NGA, specifically whether and how the act applied to producers. Lest the discussion digress into a series of semantic arguments, however, where the reader would clearly be wiser to defer to the judgment of a group of Supreme Court justices rather than a rookie author, we shall examine the act's original intent is examined through the congressional record. As it so happens, the senators discussing the bill that would become the NGA spoke very explicitly about how the act would affect producers.

Senator Warren Austin represented Vermont, which neither produced nor consumed natural gas, and was an anti–New Deal Republican. He was therefore against federal intrusion into private business in general, especially when states had the ability to regulate. During debate on the Senate floor in August 1937, he asked one of the bill's sponsors "whether the bill undertakes to gain control over the natural resource of gas—that is, the natural gas of any state—to enable the federal government to control it? If it does, of course I should object." One of the bill's sponsors, Senator Robert LaFollette, who represented Wisconsin and was generally supportive of New Deal legislation despite being a Republican, responded: "All [the bill] attempts to do is to give the Federal Power Commission the right to regulate interstate transportation and sale and resale of natural gas which moves in interstate commerce."[24]

Austin's line of questioning did not stop, however. He directed a more pointed version of his inquiry to another of the bill's sponsors, Senator Burton Wheeler, a pro–New Deal Democrat who was elected on the back of support from labor unions. He represented Montana, which was a producer of gas, but not on nearly as grand a scale as states farther south.

> Mr. Austin (R-VT). Does the bill undertake to regulate the production of natural gas, or does it undertake to regulate the producers of natural gas?
>
> Mr. Wheeler (D-MT). It does not attempt to regulate the producers of natural gas or the distributors of natural gas; only those who sell it wholesale in interstate commerce.
>
> Mr. Austin. Is the bill limited in its scope to the regulation of transportation?
>
> Mr. Wheeler. Yes; it is limited to transportation in interstate commerce, and it affects only those who sell gas wholesale.

Later during the same debate, Wheeler is asked a question by Senator Tom Connally, who is best known as the lead author of the Connally Hot Oil Act of 1935. This act allowed the president to fine and even imprison individuals who imported from abroad or moved interstate crude oil in excess of either state or federal quotas. As Texas Democrats, Connally and the bill's lead author, Sam Rayburn, wore similar (ten-gallon) hats: their Democratic status meant they supported regulation as a means to solve problems, and their Texan-ness meant they supported regulation that would benefit, or at least not hurt, producers.

> MR. CONNALLY (D-TX). I have not had a letter for this bill, nor, so far as I recall, a letter against it. I have no particular interest in the matter, except that the bill does affect a large industry in my state. We want to sell our gas, of course, but we do not want to sell it at a price that is not just and fair.
>
> MR. WHEELER. Let me say that Representative Rayburn called me up only yesterday and stated to me that he was very anxious that this bill be taken up . . . I do not think he would have called me up and asked me to have the bill passed, if possible, if it was going to hurt the state of Texas, or anybody down there.

Beyond these quotes, producers were not discussed very much during debate. The focus was on transportation, which was what the bill sought explicitly to regulate. Wheeler said to Austin during the debate:

> The purpose of the bill is to help the state [public utility] commissions and the people of the country find out what is the cost of transporting natural gas to the larger cities, such as Detroit, Chicago, Cleveland, and New York City. It does not do any good to give the Illinois commission the power to regulate the price of gas in the city of Chicago if they can be held up—and that is all it is—by pipeline companies which ship the gas to Chicago and be told, "It is going to cost the city of Chicago so much at wholesale for this gas." What good does it do the people of Chicago if they cannot reach the wholesale price and do something to regulate the pipeline companies which transport the gas into the city of Chicago?

The focus on transportation made perfect economic sense: in the 1930s, producers were only receiving around a nickel per thousand cubic feet at the wellhead, or less than 10 percent of the total cost to consumers.[25] And the nickel was an average that included higher cost producers in Appalachia, who were getting closer to 15 cents. In Texas the average price at the wellhead fluctuated between 2.1 and 2.4 cents from 1931 to 1939; in Oklahoma it was lower still.[26]

Moreover, and probably more importantly, there had never been the same amount of concentration in production as there had in either

**TABLE 4.1: GAS PRICES FOR VARIOUS GROUPS AND GEOGRAPHIES IN 1937, CENTS PER THOUSAND CUBIC FEET**

|  | Producer | Industrial | Commercial | Domestic | Average end-user |
|---|---|---|---|---|---|
| United States | 5.5 | 10.0 | 48.1 | 73.3 | 22.0 |
| Southwest | 2.5 | — | 36.5 | 66.1 | — |
| Appalachia | 16.8 | — | 52.0 | 56.6 | — |

Source: US Department of the Interior, US Geological Survey, *Minerals Yearbook*, 1938.

transportation or distribution. Whereas the big four public utility holding companies in 1930—Cities Services, Columbia, EBASCO, and Standard—accounted for more than 60 percent of all the gas both transported and sold to consumers across the country, they controlled only 16 percent of production.[27]

Lastly the Uniform System of Accounts passed as part of the NGA specified a line-item for "gas acquired for resale," and pipelines were not permitted to earn a margin on these resales. This suggests that the NGA saw pipelines primary role as "purchasing agents" for their utility clients,[28] passing the cost of purchased gas on to utilities without any markup. If they themselves happened to own producing properties, so be it—their financially separate pipeline businesses were prohibited from making a profit on any gas purchased, no matter from whom.

It is clear that the senators debating the bill did not believe that it sought to regulate producers, even if the wording did not categorically exclude such an interpretation. Then again, the NGA was written to close a regulatory gap and give state public utility commissions the ability to ensure that those same sprawling holding companies that PUHCA had just forced out of the local distribution business did not continue to gouge consumers via too-high interstate transportation rates. While the NGA closed that gap, it created another, in the field, where the prices received by producers were unregulated. Despite the fact that those prices were a small fraction of the total paid by consumers, and that there was substantial competition among many producers, some lawmakers saw the gap as an opportunity for monopoly and exorbitant profits.

# CHAPTER 5

## WARTIME PIPES
## AND THE POSTWAR BOOM

The onset of World War II brought the Great Depression to an end and breathed life back into the gas industry. It also brought to a halt the conflict between the US government and private business that President Roosevelt's New Deal regulations had engendered. After the attack on Pearl Harbor in December 1941, government and business were on the same team, and became increasingly indistinguishable.

The goal at home was to maximize wartime production at the expense of all else. This meant churning out tanks, ships, aircraft, rifles, and ammunition, all of which necessitated a great deal of energy. The grand dame pipelines built just before the Depression, which had operated well below capacity through those lean years, filled quickly. Gas shortages began to spring up, necessitating new supplies, but building new pipelines would require huge allotments of precious steel. It took serious cajoling from energy men to get these allotments, and only two major systems were built during the war.

After the war ended and raw materials could be had freely, however, North America's pipeline network expanded rapidly. By 1954, when pipelines finally made it to New England, no major region in the United States was without natural gas service (save for balmy southern Florida), and the manufactured gas industry was effectively dead. In fact, a map of US gas pipelines from 1954 looks almost identical to a map from 2018. While consumption has more than quadrupled over the six decades that have followed, most of the extra demand was served by expanding existing systems, rather than building entirely new ones.

The decade after the war also saw the rise of oil money in national politics, and the beginnings of the modern conservative movement. A small group of independent Texas oilmen had accumulated acreage in the 1920s, then increased their holdings drastically during the Depression, when the majors stopped investing. Soaring demand for oil and gas during and after the war created a new class of über-wealthy oilmen who became known as "the Big Rich." Natural gas, which before World War II was mostly a useless byproduct, was now "Texas Oil's new profit engine,"[1] turning fortunes of hundreds of millions into billions, and Texas, long considered a far-off colony, into a legitimate political force.

## OPERATION DRUMBEAT

One of the Allies' key advantages in World War II was the primacy of the US oil industry. As the war began, the United States accounted for 60 percent of total global oil production, with the Soviet Union and Venezuela making up most of the rest. Germany had virtually no oil, and was making due mostly with a substitute synthesized from something they did have a lot of: coal.

In the United States, the majority of crude oil was refined into gasoline, diesel, kerosene, and heating oil on the Texas Gulf Coast, whereupon it was barged to the East Coast and also abroad to fuel Allied efforts in Britain. Indeed, before the official start of hostilities between Nazi Germany and the United States, virtually all oil was transported via barge. Of the 1.4 million barrels per day shipped from the Gulf to consumers on the East Coast, only 50,000 barrels per day arrived via pipeline, and another 5,000 via rail. When the United States loaned Britain eighty of its tankers to replace the losses it had suffered from German U-boat campaigns in the spring of 1941, it resulted immediately in supply shortages at home, despite a heroic effort that increased rail transport to an impressive 140,000 barrels per day within a matter of months.[2]

Things went from bad to worse later that year. Germany declared war on the United States on December 11, 1941, four days after the Japanese attacked Pearl Harbor. Over the next two weeks, a fleet of five German U-boats left port in occupied France for the US East Coast. Their departure signaled the beginning of Operation Drumbeat, whose purpose it was to paralyze maritime shipping on the East Coast, which culminated in the almost total loss of crude oil and refined products shipments from the Gulf.

In charge of Operation Drumbeat was Admiral Karl Donitz, a shrewd and highly competent man who in 1943 was promoted to commander in chief of the navy, and who actually served as the last head of state of the Third Reich. After the war, while in Nuremberg on trial for war crimes, Donitz scored an impressive 138 on an Allied–administered IQ test, putting him in the top 1 percent of the world on this measure of intelligence.[3]

Impressive as Donitz was, the United States was so ill-prepared for unrestricted U-boat warfare that his acumen was largely unnecessary to the task. After the first wave of U-boats returned to their base in France in February 1942 to refuel and rearm, having run out of torpedoes, Donitz wrote that each commander "had such an abundance of opportunities for attack that he could not by any means utilize them all: there were times when there were up to ten ships in sight, sailing with all lights burning on peacetime courses."[4] The United States lacked the resources to detect the U-boats, much less to mount any effective counterattack. It was unwilling even to order nighttime blackouts in coastal cities, which meant that U-boat spotters could easily see the silhouettes of tankers passing in front of the brightly lit cities through their periscopes.[5]

The first tanker to go down was the *Norness*, on January 14, 1942, sixty miles off of Montauk Point in Long Island, New York. The German-built ship was leaving New York with a load of fuel oil, bound for Liverpool.[6] Over the next five months, Nazi submarines sank 171 ships between Florida and New York, and another 62 in the Gulf of Mexico.[7] Insurance companies began refusing to write policies on the ships; their crews began refusing to sail. By the end of the U-boat campaign, Gulf Coast shipments of oil had dropped from 1.4 million barrels a day to a paltry 100,000.[8] The US response was threefold: continue to increase rail shipments, build more tankers and give them armed escorts, and build a pipeline.

Enter a bewildering array of government corporations, public agencies, special wartime bodies, public-private partnerships, and industry committees that characterized the wartime economic landscape. By the end of the war, the US government owned nearly one-quarter of the nation's industrial capacity, including more than 50 percent of its aluminum smelters, 15 percent of its steelmaking, and almost 90 percent of its aircraft and ship-making facilities.[9] While the World War II era is generally remembered the most successful example of public-private enterprise in modern history, this simplistic description papers over the tangle of decision-making bodies that held overlapping authority over different aspects of the wartime economy.

Direct government investment was made chiefly through loans from the Reconstruction Finance Corporation (RFC), which was established in 1932 to act as a lender of last resort to failing banks during the Depression, but during the war sprouted eight subsidiary corporations that invested in strategic industries and commodities. These wartime subsidiaries doled out more than $20 billion, or more than $300 billion in 2020 dollars. The Defense Plant Corporation (DPC) lent the most of the eight subsidiaries—$8 billion to almost 2,500 separate projects, over half of which went, directly or indirectly, to the aviation industry. It also financed tugboats, barges, railroad tank cars, and a large pipeline to transport crude oil and refined products from the Southwest to the Northeast.

Getting this pipeline built, however, was not as easy as simply asking Uncle Sam for the $160 million that it ended up costing.[10] It required the blessing of several agencies, plus substantial support from industry. Luckily, the pipeline project from the outset had the support of the Petroleum Administration for War (PAW), an agency that President Roosevelt created to "formulate plans and programs to assure for the prosecution of war the conservation and most effective development and utilization of petroleum in the United States." Unluckily, the PAW could not overrule the decisions of the powerful War Production Board (WPB), which had ultimate authority over which projects were allocated supplies of strategic raw materials, such as steel and rubber.

The administrator of the PAW was Harold Ickes, who simultaneously served as secretary of the interior. Ickes had been with Roosevelt since he assumed the presidency, and had earned the nickname "Honest Harold" for his management of yet another New Deal agency, the Public Works Administration (PWA). At the PWA, Ickes oversaw the building of high-dollar infrastructure projects, such as dams, tunnels, railroads, and airports. This post presented Ickes countless opportunities to dole out favors to businesses and politicians alike, and even to enrich himself. He never wavered, though, and became famous for the close eye he kept on budgets, his lack of preference among contractors, and his dogged efforts fending off congressmen and senators who wanted pet projects for their states.[11] Thus, when Ickes began advocating for a cross-country petroleum pipeline in summer of 1941, after the United States had begun to lend Britain barges, but before Operation Drumbeat had begun, his voice carried great weight within the government.

Seven private oil companies joined the PAW in advocating for the pipeline, and were represented by a transportation committee led by Henry Doherty's successor at Cities Services, W. Alton Jones. But despite the impressive cadre of backers, the WPB shot the project down, giving it lower priority than artillery, aircraft, and warship manufacture and refusing to allocate the required raw materials. Ickes tried again only a month later with a new, larger group of eleven private companies, arguing that crude oil and refined products shortages were imminent where they already did not exist. But again, to no avail.[12]

It was not until the attack on Pearl Harbor, on December 7, 1941, that the WPB began to warm to the idea. Before then, the WPB may have judged that whatever shortages existed or came to pass in the Northeast could be cured simply by building new barges. Until that point, after all, the war had remained a series of sorties fought on foreign soil. The chief advantage of a pipeline, from a national security point of view, was to ensure that the flow of oil was safe from attack by sea. Indeed, while advising President Roosevelt on the need for a pipeline, W. Alton Jones offered, laconically: "No one ever sank a pipeline." But before Pearl Harbor, a naval war on America's shores was a fear, not a reality.

Thirty-eight days separated the attack on Pearl Harbor from the sinking of the *Norness* off Montauk Point and the beginning of unrestricted U-boat warfare in the Atlantic. But even after Admiral Donitz's crews began inflicting heavy damage in early 1942, the WPB refused to budge, continuing to focus on armament. Realizing the severity of the situation, Jones called a three-day, industry-wide meeting in March 1942 in Tulsa, Oklahoma, to devise a national pipeline strategy, which became known as the Tulsa Plan.[13]

The Tulsa Plan called for analyzing all existing pipeline systems to determine how to move petroleum to Northeast refineries most effectively, including by integrating the operations of separate, privately owned systems; digging up and relaying existing pipe; and salvaging old valves and meters—basically squeezing every extra barrel per day out of existing steel. It also called for new steel, specifically a twenty-four-inch line from Texas to the Northeast capable of transporting three hundred thousand barrels a day of crude, and another, parallel line to transport a smaller quantity of refined products. The PAW approved the Tulsa Plan, Ickes appealed directly to Congress, and Jones continued to advocate to his group of powerful acquaintances. Finally, in June 1942, the WPB allocated the required materials, including an initial outlay of 137,000 tons of steel, to build the pipeline from Texas to the East Coast.[14]

Although owned by the Defense Plant Corporation (DPC; an RFC subsidiary), the pipeline was built and operated by War Emergency Pipelines, Inc. (WEP), a nonprofit organization owned by the eleven private oil companies that had backed the project. Jones became WEP's president. The project would become known as the Inch Lines, and comprised two separate pipelines, just as the Tulsa Plan had recommended. The Big Inch, twenty-four inches in diameter, would move three hundred thousand barrels per day of crude from oilfields in East Texas. The Little Inch, twenty inches in diameter, would move one hundred thousand barrels per day of gasoline, diesel, kerosene, or heating oil in batches from refineries on the Gulf Coast.

Construction on the Big Inch began first, and in two phases. Phase one stretched 530 miles, starting in Longview, Texas, and ending in Norris City, Illinois, where a large railroad tank-car loading facility was being built to move the oil on to the East Coast. Phase two continued the line to refineries Philadelphia and Linden, New Jersey, almost another 800 miles. The Little Inch paralleled the Big Inch, except that it started farther south in Beaumont, Texas, where there were several Gulf Coast refineries. Pipe laying began in August 1942.

Technology had progressed quite a bit since the days of the daguerreotyped, antediluvian tong gangs, and even since the era of the grand dames.

Not only was the pipe larger and more uniform in quality, and the welds stronger even than the pipe itself (1920s-era advances), rotary-wheel ditching machines now filled in for men with pickaxes and shovels, producing ditches 4.5 feet deep and 3 feet wide, and side-boom tractors lowered the welded pipe gently into the ditch. While it was still frenzied and backbreaking work, the machinery allowed crews to lay more than two miles per day in good conditions. In fact, pipe laying progressed so quickly that the first phase was completed in January 1943, less than six months after work began, and before the pumping stations were finished, meaning the line laid idle for another month before commencing deliveries.

Phase 2 was sorely needed because there were not enough railroad cars or storage tanks at Norris City to allow the Big Inch to flow at its full three-hundred-thousand-barrels per day capacity. Work had begun on the Norris City to Philadelphia/Linden leg in November 1942 and finished in July 1943. Oil finally reached Philadelphia refineries in August (though some problems meant Linden did not receive Big Inch oil until October). The Little Inch commenced service in March 1944, and marked the completion of the megaproject.

Major J. R. Parten, a Texas-born oilman who was director for the transportation division of the PAW and helped oversee construction of the Inch Lines, called the projects "the most amazing government-industry cooperation ever achieved." At its peak, the project employed over fifteen thousand people, and the Inch Lines became the largest single consumer of electricity in the United States after its 243,500 horsepower of compressors began moving oil to the Northeast.[15]

## APPALACHIA BECOMES A GAS IMPORTER

While the Inch Lines were the largest and certainly most publicized of the wartime pipes, they represented only around one-quarter of the total pipeline miles that the WPB authorized. The second largest project was a natural gas pipeline that ran 1,265 miles from Texas to West Virginia that was, somewhat oddly, called the Tennessee Gas and Transmission Company.

Recall that the Depression had left the gas pipelines built in the late 1920s and early 1930s from the Southwest and Gulf Coast underutilized for lack of demand. But most of these pipelines terminated in the Midwest, outside the US industrial heartland in Appalachia, which was still very much a gas island, unconnected to the great fields in Texas, Oklahoma, and Louisiana. The Appalachian supply shortage that had cropped up in the boom years of the 1920s had dissipated only on the back of plummeting demand, not because new supplies had been found. Thus, the resurgence of Appalachian industrial activity during the war immediately reignited the natural gas shortage there.

**FIGURE 5.1:** Map of US gas pipelines, 1940

More than six hundred factories in Appalachia were designated as critical to the war effort, producing steel, aluminum, chemicals, and other strategic commodities. The Aluminum Company of America (ALCOA), which produced the lion's share of the metal needed for aircraft construction, wrote to the WPB in November 1941: "You understand we are engaged in the production of defense work 100% and it is very essential that we are furnished gas to carry on our production."[16] Still, natural gas was deemed less crucial to the war effort than oil. The immediate solutions, taken throughout 1942, were to curtail deliveries to nonessential customers and to initiate a massive Appalachian drilling program, which helped keep these factories supplied with gas through that winter. But these measures could not change the fact that local supplies were simply diminishing more rapidly than producers could keep pace—the new drilling barely moved the needle: total Appalachian production rose by less than 3 percent in 1942 and a bit less than 4 percent in 1943.[17] A new pipeline connecting the Southwest's massive reserves to Appalachia was sorely needed.

The WPB was slowly coming around to the need for new gas supplies to Appalachia, but even if it had been quicker to act, the Natural Gas Act, as it was originally worded, made it impossible actually to build pipelines to new markets. This was because the NGA required that any new interstate pipeline obtain a certificate from the FPC under section 7(c), but the act's original wording said the FPC could only grant a certificate when the market was already being served by natural gas. The NGA had been designed to regulate like-for-like (e.g., gas-on-gas) competition, and the 1938 wording in section 7(c) amounted to a de facto ban on entering new markets.

Pipelines seeking new markets, not yet served by natural gas, essentially found themselves in a regulatory no-man's-land.

Ironically, it was the coal and railroad industries—bitter competitors to natural gas—that lobbied to change this wording. Coal production had dropped by 10 percent over the course of the Depression, which contributed to financial woes at the railroads, who counted coal companies as their largest clients. Much of this decline was related to the economy-wide slowdown in industry, but it was also true that coal's share of total US energy consumption had dropped from 80 percent before the World War I to just 52 percent by 1933, the result of competition from oil, natural gas, and hydroelectric power. While somewhat deflated in stature, the coal industry still commanded substantial political power: it employed half a million people, at a time when the total US labor force only numbered about fifty-five million.[18] Meanwhile, the natural gas industry employed only about fifty thousand people.[19]

Big Coal was irked at the FPC for not considering its plight when the agency reviewed new gas pipeline certificate applications. During its first few years (1938–1941), it was the FPC's opinion that neither the coal nor rail industries had grounds to intervene against a prospective natural gas pipeline that would enter a market already served by natural gas, since coal and coal-derived manufactured gas were already being displaced. The FPC thought that, in these situations, the coal and rail industries were not "vitally affected." The powerful coal and rail lobbying groups obviously did not agree, and after repeated complaints to Congress, the FPC suggested a compromise solution. The FPC requested that Congress empower it to grant certificates to natural gas pipelines that proposed to serve new markets, where natural gas had not yet been introduced. In exchange, the FPC would consider both coal and railroads "vitally affected" by any proposed new pipeline, regardless of whether the pipeline would serve new or existing markets. Coal and railroad lobbyists would thus be allowed to participate in certification hearings, where they could do their best to halt new pipelines from being certified. The FPC's suggestion was embraced by all sides: coal companies and the railroads were happy because they could oppose the further growth of the natural gas industry, while natural gas pipeline companies were happy because they could at least try to enter new markets.[20]

The proximate threat that the coal industry perceived as this was happening, in 1941, was a large new pipeline that had been proposed to run from Texas all the way to New York City, the largest manufactured gas market in the country.[21] While the city relied almost wholly on manufactured gas, it did have a pittance of natural gas supply from wells in upstate New York. The FPC thus considered New York City an existing market for natural gas, which meant that, under the old rules, a new gas pipeline company could apply for a certificate and neither coal nor rail companies could intervene. As it turned out, the pipeline that so scared the coal and rail interests, backed by Hope and called Reserve Gas Pipe Line Company, would be delayed for several years, then resurrected as the Transcontinental Pipeline. But the threat was enough: coal and rail companies considered the right to lobby against gas pipelines that targeted major markets more important than preserving their monopolies over the relatively smaller markets not yet served by gas.

This strategy proved short-sighted, as has been almost all coal industry lobbying, in hindsight, but anyway was something of a foregone conclusion, as wartime needs necessitated a massive expansion in gas service. In February 1942 Congress amended section 7(c) of the NGA to allow the FPC to grant certificates to all interstate pipeline companies. The number of applications shot up almost overnight.

The amendment breathed new life into the industry as a whole, and also into Tennessee Gas and Transmission Company, a troubled company with no pipeline assets and what had been an exceedingly slim chance of corporate survival. Tennessee Gas was led by Curtis Dall, whose marriage to Franklin Roosevelt's daughter had ended in acrimony, but left him with some powerful political connections. Dall had tried in 1940–1941 to bring gas from Louisiana to Tennessee, but had failed to get authorization either from the FPC or from the state. For one thing, Tennessee Gas had neither the contracts ensuring a twenty-year supply of gas, nor the financing to build the pipeline, both of which the FPC required. For another, the pipe Dall proposed would both cross state lines and serve markets not yet served by natural gas, which the original wording of section 7(c) put in regulatory limbo.

Just two days after Congress amended section 7(c), in February 1942, Dall tried again, but now it was the WPB that balked. The pipeline only indirectly addressed the gas shortages in Appalachia, and was not dedicated exclusively to serving wartime industries. Dall's argument was that Tennessee Gas would serve factories that were currently utilizing electricity for needs that could be replaced by gas, such as heating, and that the pipeline would thereby "free up" electricity for war-related industries where it was essential, such as in aluminum manufacture. While the logic is sound, it was obviously too convoluted for the WPB, which anyway still had not given the green light even to the Inch Lines, which were considered more essential to the war effort than a gas line to Appalachia, and which had the strong support of powerful PAW administrator Harold Ickes.[22]

With Appalachia in the throes of a severe and worsening gas shortage, both the FPC and the WPB increased their activity in the gas space over the following months. The two agencies began working together to bring in new supplies, and went after the low-hanging fruit first, compelling the Panhandle Eastern Pipeline to be connected to the Ohio Fuel Gas Company. Panhandle's line terminated in a cornfield just eighteen miles from the Ohio Fuel system. This had been done purposefully to avoid direct competition in markets served by Columbia Gas, which owned 50 percent in Panhandle and had tried to bankrupt its "partner," Mo-Kan, in 1930–1931. (In fact, Columbia still owned a stake in Panhandle in 1942, although functional control now belonged to William Maguire, and Columbia finally officially divested its stake the following year.) The WPB authorized the necessary steel and other materials and the FPC fast-tracked the certificate, which it issued in October 1942. This allowed Panhandle to flow fifty million cubic feet per day into Appalachia, which was certainly welcome, but represented just 5 percent of combined natural gas production in West Virginia, Pennsylvania, and Ohio in 1943 of just over one billion cubic feet per day.[23]

During the winter of 1942–1943, studies were commissioned and testimony taken that made clear just how dire the situation of gas supply in Appalachia was. The chairmen of both Hope and Columbia, which together served about 60 percent of the region's needs, testified that neither company would be able to meet normal gas demand after the war. The FPC wrote in July 1943, "It is crystal clear that additional natural gas is needed in the Appalachian region." Finally, on August 28, 1943, the WPB preauthorized all the materials needed for the construction of a large new pipeline from the Southwest to Appalachia, a little more than a year after giving the go-ahead to the Inch Lines.[24]

The WPB did not back any specific proposal, leaving the project open for bidding. But it stressed urgency, stipulating that the winning bidder must have placed all its purchase orders by October 1, giving potential builders just thirty-four days to bid, win, raise the capital, and order materials for the entire project. The extremely tight timeline would ensure that the winning project could begin deliveries by the onset of the winter of 1944–1945, which was projected to bring large shortages without a new source of supply.[25]

It is useful to consider the timing in the context of the war itself. Some historians argue that an Allied victory was a foregone conclusion as soon as the United States and the Soviet Union entered the war, in December and June 1941, respectively. Winston Churchill remarked upon learning of Pearl Harbor: "So we had won after all." But this is a conditional argument, predicated on the nearly limitless industrial capacity the United States and Soviet Union could bring to bear. If Allied victory was clear after Pearl Harbor, it was only because it was also clear that the United States would massively increase its industrial output, which would necessitate a corresponding increase in energy production. Events tied more direct-ly with the turning of the tides were the Battle of Kursk in the summer of 1943, D-Day in June 1944, and the Battle of the Bulge in December 1944–January 1945. The point—as it relates to the US gas industry—is that when the WPB preauthorized the materials for a new Southwest to Appalachia gas pipeline in August 1943 and demanded that the winning bidder commence construction almost immediately, the United States was in the middle of an unprecedented industrial expansion, and the Nazis were far from a sedate enemy. An Allied victory was not certain.

There were two serious contenders for the Southwest to Appalachia pipe. The first was Tennessee Gas, as Dall had amended his application and route in November 1942 in order to better align with the WPB's pri-orities and exclusively serve wartime industries. The FPC had reviewed the project favorably in the summer of 1943, but Tennessee Gas had still not demonstrated that it had access to the necessary financing or twenty years of supply. This left the stage open for Hope Natural Gas, which submitted its own plan to build a line from the Hugoton field, where it owned a large acreage position, to its existing system of trunk lines in Cornwall, West Virginia. It owned the supply—and much more than twenty years of it—and, as a Standard Oil subsidiary, financing would not be an issue. In fact, Hope would be involved one way or the other, because after Dall's reroute, Tennessee Gas would also terminate at Hope's system in Cornwall.[26]

It was up to the FPC to decide which project would get built. The smart money was on Hope, and the FPC scheduled hearings on the Hope project to begin on September 21. But perhaps its advantages were not as clear-cut as appeared. The RFC was willing to offer loans for a large portion of construction costs, so the winning bidder would not be required to finance the project totally out of pocket. And finding someone to sell you gas in the Southwest was not a particularly difficult thing to accomplish in 1943, as the glut was still very much alive. Accordingly, Curtis Dall did not have much trouble seeking out and signing a gas supply agreement with the Chi-cago Corporation, a Chicago-based conglomerate that owned a large acre-age position in south Texas, near Corpus Christi.

The Chicago Corporation was producing both oil and gas from its wells, but then reinjecting the gas back into the reservoir, to keep the pres-sure high and allow oil to flow for longer and at higher rates. It had been interested in marketing its gas for some time, and had partnered with Hope in the failed Reserve Gas Pipe Line Company project to bring Texas gas to New York City, whose very potential threatened the coal and rail industries so much that they lobbied to have section 7(c) of the NGA reworded.[27]

After signing supply agreements with the Chicago Corporation, the only difference between Tennessee Gas and Hope's applications, from the

FPC's point of view, was financing. Both projects had ample gas supply, and both fed the same system, and, indeed, at the exact same point of interconnection. But an unexpected turn of events made the issue of securing financing more of a challenge for Curtis Dall than had been securing gas supply. Tennessee Gas's chief financial backer, who was meant to use his banking connections to raise capital and arrange loans, died suddenly on August 29—the day after the WPB authorized steel for whichever project that the FPC chose.

The Chicago Corporation viewed the pipeline as strategically important to its business, so it can be said that events played out expectedly after the unexpected death in the Tennessee Gas circle. The Chicago Corporation bought Tennessee Gas (for just $500,000, as it was mostly a shell company, with few tangible assets) and demanded, as part of the deal, the resignation of all of its officers, including Curtis Dall. On the morning of September 20, the day before the FPC was set to begin hearings on the Hope project, the Tennessee Gas board accepted the Chicago Corporation's takeover offer. Later that same day, Tennessee Gas's new, Chicago Corporation–appointed officers reported the news to the FPC, which immediately granted an oral certificate. Hope's project was dead, and Tennessee Gas began placing purchase orders.[28]

Construction began in December 1943, but after six months had passed, only 75 miles had been laid, as flooding and material shortages slowed the project almost to a standstill. It took a Herculean effort between May and October 1944 to string the remaining 1,200 miles of pipe—crews needed to complete an unprecedented 6.5 miles per day, more than triple the pace that the Inch Lines crews had sustained. Yet somehow the project began service on October 31, 1944, exactly as the WPB had requested, and in time for the winter of 1944–1945.[29] Throughput soon ramped up to the project's full capacity of just over 200 million cubic feet per day, which boosted total supplies in Appalachia by almost 20 percent. Between this and the additional 50 million cubic feet per day from Panhandle Eastern's interconnection with Ohio Fuel, Appalachia now had enough gas supply to avoid wintertime curtailments.

The Chicago Corporation had appointed one of its vice presidents to head Tennessee Gas, Gardiner Symonds, who would become a mover and a shaker in the gas world over the next two decades. He was singularly obsessed with bringing gas from the Southwest to the Northeast, and his aggressive management style meant he took a no-holds-barred approach to expanding Tennessee Gas's footprint throughout the 1940s and 1950s.

The Chicago Corporation actually divested its stake in Tennessee Gas in September 1945, having owned the functioning pipe for just under a year, in order to prevent its production assets from falling under FPC

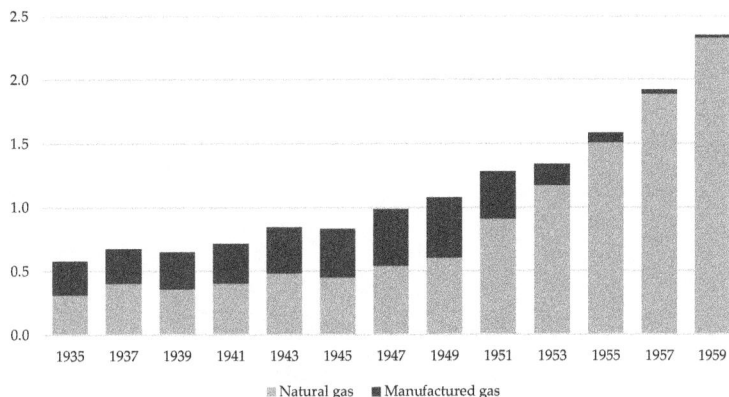

**FIGURE 5.2:** Natural gas versus manufactured gas sales in the Northeast, billion cubic feet of gas equivalent per day

regulation (the FPC had, by then, begun regulating gas prices received by pipeline-affiliated producers). Symonds remained chief executive and chairman, however, and quickly won remit from both the FPC and the WPB to add compressors to expand capacity by another sixty million cubic feet per day. But the shortage in Appalachia had been vanquished, and the war was over—Nazi Germany unconditionally surrendered on May 8, 1945, and the formal Japanese surrender was signed on September 2 of that same year. Symonds would have to wait for the postwar economic boom to continue with his grand expansion plans, but continue he would. The Tennessee Gas system today remains one of the largest and most important pipeline systems in North America, picking up and delivering gas to Appalachia, New York, New England, and Canada on the northern end of its system, and to the Gulf Coast, Texas, and Mexico in the south. It is unquestionably the jewel in the crown of the largest gas pipeline company in the United States—Kinder Morgan.

## THE END OF MANUFACTURED GAS

Immediately after World War II, the split between manufactured and natural gas in the Northeast—Pennsylvania, and all the states north and east of it—was about even, since the large cities of Philadelphia, New York City, and Boston had no natural gas of which to speak. Ten years later, the manufactured gas industry had been virtually destroyed, thanks mostly to three men and their pipelines.

Any gas industry professional will understand you when you talk about "the three T's," and, in fact, two of them have already been discussed. The three T's are the three arteries that were built to move gas from Texas

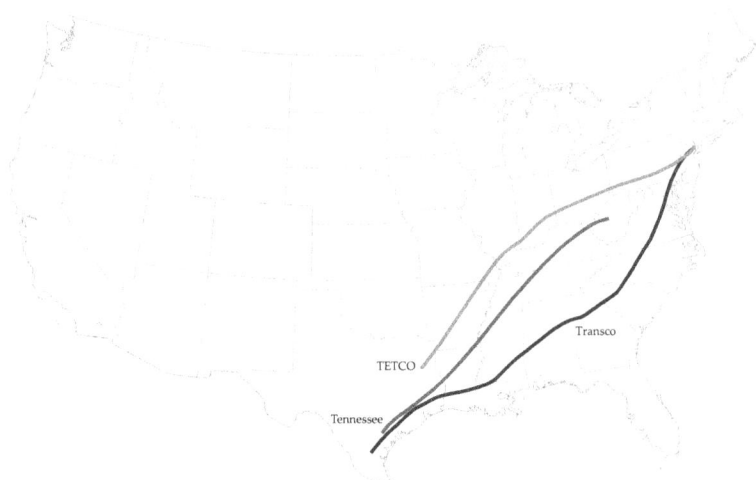

**FIGURE 5.3**: The three Ts: Tennessee, Texas Eastern, and Transcontinental

and Louisiana to northeastern markets. The first was Gardiner Symonds's Tennessee Gas Pipeline, which began deliveries in October 1944. The second were the Inch Lines, which were sold after the war by the US government to a company called Texas Eastern Transmission Corporation, or TETCO, and converted from wartime oil service to peacetime gas service in 1947. The last was a new pipeline that started its life as Reserve Gas Pipe Line Company, but was built as the Trans-Continental Pipeline Company, or Transco, on which construction began in May 1949 and service commenced in January 1951. Tennessee Gas, TETCO, and Transco have remained, since being built, three of the largest and most important systems in the United States.

The story of how these lines were built and expanded to serve the major Northeast markets reads like grade-school playground drama, where Tennessee Gas's Gardiner Symonds was the loudmouthed bully who wanted the swing set all for himself; TETCO's Reginald Hargrove was the negotiator, willing to ally with the other students to share a seat; and Transco's Claude Williams was content picking daisies, leaving the others to fight it out. In this narrative, the FPC was the teacher—at once all powerful, yet unable to control its students.

It was perhaps destined to be this way, not only because Symonds was a characteristic bully but because he felt, not unreasonably, that it was Tennessee Gas that should have taken control of the Inch Lines after they were converted from wartime oil service.

After the war ended, the long and chaotic process of privatizing government-owned assets began. The Surplus Property Administration

(SPA) was created and tasked with maximizing sales prices, while also maintaining the strategic advantage that the assets conferred. For the Inch Lines, this meant soliciting the highest bid from industry, but initially stipulating that the pipelines remain engaged in oil and refined products transportation—not converted to gas.

In September 1945 the SPA announced that the government would cease operating the Inch Lines, and intended either to sell or lease them to private industry. It was not until July 1946 that an auction was finally held, and although the stated goal of the SPA was for the Inch Lines to remain in oil service, the auction rules did not prevent natural gas companies from applying.[30] They also did not specify a format for bidding. Because of this, once the bids were in, the SPA had a hard time determining who had won. The highest bids were from companies intending to convert the Inch Lines to natural gas service, but the SPA did not have a numerical means of expressing its preference for keeping the lines in oil service. Some bids included only one, upfront price, while others also specified future payments based on throughput. One bidder even requested a long-term lease of the lines.

The SPA was left comparing apples, oranges, and apple sauce, so it planned another auction, for February 1947. This time, the SPA would consider only one factor, price, and, more specifically, a flat price. The only oil-related provision was that the winning bidder was required to maintain the oil-pumping equipment in case of a national emergency. Otherwise, the highest bidder would be free to use the lines as it saw fit.

The company that would become Texas Eastern Transmission Corporation, or TETCO, was cobbled together by E. Holley Poe, an insider who had served as the head of the natural gas division at the PAW and also as a member of the Committee on Postwar Disposal of Pipe Lines. Before the war, he was secretary of the industry's then-only lobbying group: the American Gas Association. Poe knew the players as well as anyone, but had never been an operations man. He thus sought an alliance with Norris McGowen, president of United Gas, which had earlier been a subsidiary of GE-owned EBASCO, and was, at the time, the largest gas transporter in the nation. (It was also one of its largest producers, having a large position in the Monroe field in Louisiana.) McGowen supported the idea of converting the Inch Lines to gas service in concept, but was fearful of monopoly, as he was still in the process of divesting utilities to comply with PUHCA. McGowen was content with his profitable empire in the Gulf South region. While neither he nor United joined forces with Poe directly, McGowen gave his second-in-command, Reginald Hargrove, a leave of absence to help Poe organize a bid for the Inch Lines.[31]

Poe and Hargrove then joined forces with some well-connected Houston businessmen, George and Herman Brown, who owned the large engineering firm Brown & Root, and a Houston lawyer to complete the board of the proto-company.

Poe's group had submitted a bid during the first auction, for $80 million upfront, plus another $20 million to be paid over time when annual deliveries surpassed 120 billion cubic feet. When the SPA tore up these first bids (there were sixteen of them, altogether) and organized the second, final auction, TETCO knew it had to swing for the fences. One of its chief competitors in the bid for the Inch Lines, Trans-Continental Pipeline Company, had offered an upfront price of $85 million, and was sure to pencil up this amount in the second round.[32] Even more concerning, however, was that in December 1946 Gardiner Symonds's Tennessee Gas had convinced the SPA to allow it to lease the Inch Lines on a short-term basis and begin supplying Appalachia with gas, just two months before the final auction. While this never gave Tennessee Gas any long-term rights or ownership, it did prove to all parties concerned that it had the technical and organizational ability to run a pipeline. It also meant that, while Tennessee Gas did not bid in the first round, it was sure to in the second.[33]

The Inch Lines were originally built at a total cost of $145.8 million. The day before the bids were due, on February 7, TETCO's organizers gathered in a hotel room in Washington to deliberate and agree on a price. Encouraged by their banker, they agreed on $143 million, at 2 o'clock in the morning on February 8. In fact, the final number was a bit higher—George Brown feared that another bidder might also choose an even $143 million, so the group agreed on $143,123,456. But Poe wanted to make sure his lucky number, seven, was among the digits, so the number TETCO submitted to the SPA was $143,127,000. It was enough to beat out Trans-Continental, the second-highest bidder, and Tennessee, the third.[34]

While Poe, Reginald Hargrove, and the rest of the TETCO founders were rejoicing in their victory, Gardiner Symonds was furious. He now had competition for northeastern markets. On May 1, 1947, when TETCO's personnel officially took over from Tennessee Gas's, a TETCO executive remarked that "it was like coming into an enemy camp."[35]

More competition was on the way. With Tennessee Gas serving Appalachia and TETCO setting up to serve Philadelphia, the next big market was obvious: New York City.

Hope and the Chicago Corporation had tried to build the Reserve Gas pipeline to New York City, but the effort faltered for two reasons. First, the original proposal had come during the war, and the WPB, which was responsible for allocating steel and other necessary materials, gave higher priority to supplying factories in Appalachia engaged in wartime production than to extending natural gas service to a new region, where it would mostly be used for heating homes. Second, New York City was served by manufactured gas, which meant both that coal and rail companies would fight hard to keep natural gas out of the city, and that the city's large and politically powerful utilities had vast sums of unamortized capital tied up in manufactured gas plants. Peace had solved the first problem, but the second remained.

Claude Williams founded Trans-Continental Gas Pipeline Company with the original aim of buying and operating the Inch Lines, but even before he was beaten out by TETCO, he had begun to plan a new line from south Texas to New York City. After TETCO won the bid for the Inch Lines, Transco was left with no choice but to press ahead with the new pipeline, and the race for New York was on.

Transco began its application process with the FPC in 1946, whereupon notices to intervene began pouring in.[36] Coal and rail companies behaved exactly as expected, opposing the pipeline at every turn. It was, after all, an earlier version of Transco that had caused them to lobby for section 7(c) to be reworked to allow them to intervene. But the most strident opposition came not from coal or the railroads but from another interstate gas pipeline, TETCO, which argued that Transco would duplicate services that it intended to offer. Recall that the Little Inch pipeline had two termini— one in Philadelphia, the other in Linden, New Jersey, just across the river from Staten Island. TETCO wanted to serve New York City in addition to Philadelphia, and Reginald Hargrove quickly announced his intentions to build a new line parallel to TETCO's existing system to transport an addition 400–450 million cubic feet per day for New York and New England markets.[37]

The FPC denied TETCO's motion to intervene, which set the tone for how the agency treated one of its two main functions—entry control—for decades to come. In denying the motion, the FPC noted that it was clear that TETCO sought a monopoly in the mid-Atlantic and that the FPC saw no reason why such a monopoly would be in the public interest. The denial established the natural gas pipeline industry as an oligopoly, rather than a monopoly, where large markets such as New York City could and should be served by more than one pipeline.[38]

The last major group of intervenors, New York City's manufactured gas utilities, whom Transco was being built to serve, had shot down Reserve Gas Pipe Line's advances years earlier. But demand had been growing at a rapid clip, and the cost of operating their manufactured gas plants had increased to the point where many were suffering losses. Rather than oppose the pipeline, the utilities supported it, and signed sales contracts with Transco for such a large amount that lead engineer Raymond Fish had to resize the system to handle throughput of 505 million cubic feet per day, versus the initial design capacity of 325 million.[39]

Throughput, billion cubic feet per day                    % of total US gas demand

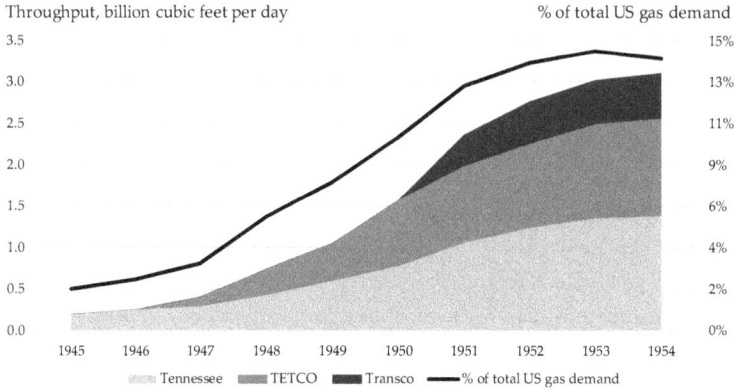

**FIGURE 5.4:** Three T's throughput and percent share of total US gas demand

Transco began construction in May 1948 and delivered its first molecules to New York City in January 1951.[40] For all intents and purposes, Transco had won the race for New York, although, technically, it was not the first interstate pipeline to deliver natural gas to the city. TETCO had extended its system from Linden and began deliveries of relatively small volumes of natural gas to Staten Island utilities in 1949.[41] But Transco was the sole supplier to Manhattan, Brooklyn, Queens, the Bronx, Westchester County, and Long Island, which were much larger markets. Commemorating the arrival of natural gas from the Southwest, the *New York Times* wrote: "While this is not the first natural gas to get here—Staten Island has had it since 1949—it is the first to arrive in any volume."[42]

New York City's conversion from manufactured gas to natural gas was rapid, and all but complete by 1955. It was a devastating blow to the manufactured gas industry, which now had only one large market left, in New England. Of course, that is precisely where the Reginald Hargrove and Gardiner Symonds intended to take their systems next.

While Transco was designed as a New York City pipeline, its route also enabled it to serve markets in the Southeast (indeed, Southern Natural Gas was another intervenor that opposed Transco's construction). It also passed by Philadelphia, and made some sales there, much to the chagrin of TETCO and Reginald Hargrove. But as the next great competition began, for markets in New England, Claude Williams decided to sit back and pick daisies. He had captured the prize of New York City, and also held what was essentially a monopoly from Charlotte, North Carolina to Baltimore. Indeed, despite its route terminating in New York instead of continuing into New England, Transco today is the single largest pipeline system in North America, measured by annual deliveries to customers.

By the mid-1950s, the three T's were bringing a bit more than three billion cubic feet of gas from Texas and Louisiana to the Northeast each day, which represented about 15 percent of total national demand.[43] This increased the price of gas that producers received at the wellhead, with the Texas average price rising from around three cents per thousand cubic feet before the three T's began service to above seven cents in the mid-1950s. That meant that these three pipelines alone were paying some $80 million per year to Southwest producers like the Chicago Corporation and the famed oilman H. L. Hunt, who supplied roughly one-quarter of TETCO's gas from his Texas fields.[44]

## THE BATTLE FOR NEW ENGLAND

Symonds was sore at Hargrove for "stealing" the Inch Lines from Tennessee Gas, and Hargrove was sore at Williams for "stealing" New York City from TETCO. Both were determined to expand their systems into New England. But the two men used different strategies to get there.

When TETCO first proposed extending its system into New England, it encountered resistance and failed to sign sales contracts with local utilities, who wanted partial ownership of the project. Hargrove decided to acquiesce, and in September 1949 formed the Algonquin Gas Transmission Company. Algonquin was majority owned by the utilities—TETCO owned only 28 percent—but sourced all its gas from TETCO, and so would enable the company to build a new, wholly owned line from the Gulf to where the two systems met, in New Jersey.[45]

Symonds, forsaking New Englanders' strong cultural desire for local participation and ownership, proposed extending Tennessee Gas from its erstwhile terminus in Buffalo to the New York-Massachusetts border, then further into New England via a wholly owned subsidiary, Northeastern Natural Gas. Deaf to the criticism that followed, Symonds said: "We had no intention of giving away part of [Northeastern] for the privilege of doing business in New England."[46]

The FPC began separate hearings on the projects in early 1950, but by the summer had determined that both would be needed to serve the region's long-term needs. As it had when it denied TETCO's motion to intervene in Transco's application, the FPC favored oligopoly over monopoly for large new markets. In its opinion, issued in October 1950, the FPC railed against both companies for competing with each other instead of working together to serve New England in a complementary manner, and it denied both applications. The FPC wanted both pipelines to be built to serve the region, but not to overlap so much as to be needlessly duplicative.[47]

Algonquin sought compromise and suggested to Northeastern that they jointly redesign their systems and split up the market. But Symonds would have none of it, and, without replying to Algonquin, filed a new application with the FPC to serve the entirety of the New England market. Moreover, Symonds accused Algonquin as being "under the domination of persons whose primary interest is the protection of the coal industry," seeking to turn one of Algonquin's key advantages—the fact that local manufactured gas utilities owned a majority of the project—into a potentially fatal flaw.[48]

This was a ruse, however, and one that the FPC saw through. The agency did grant Northeastern a certificate, but only to serve slightly more than half of the New England market. It also demanded a response within thirty days, and in November 1950 Northeastern accepted the FPC's decision. Algonquin got its certificate a few months later, and although it was authorized to serve a slightly smaller portion of the region as a whole, it won a monopoly over the city of Boston.[49]

It would seem that the battle was over, but Symonds refused to accept the compromise. He filed for another certificate to serve those cities the FPC had carved out for Algonquin, to no avail. He then filed to increase the size of the pipeline to twenty-four inches from twenty inches, claiming that there was a lack of twenty-inch pipeline on the market. (In reality that was because Symonds had called his pipeline suppliers and told them to stop rolling twenty-inch pipe and start working on twenty-four-inch sections.) He figured this would make it easier to expand service to the whole region, after he had won. The FPC reluctantly authorized the larger gauge.[50]

Symonds then upped his ante and took his case to a federal appeals court, claiming that the FPC had denied Northeastern due process. This procedural move forced the FPC to reconsider its rulings and invalidated the certificate it had granted to Algonquin until it held new hearings, which only began in November 1952, almost two years after the original certificate was issued.

And while Symonds was doing everything in his power to delay Algonquin, he was also tirelessly pressing ahead with Northeastern, which began delivering gas to some communities in September 1951, even as construction work continued farther downstream. Full service began in early 1952.

Symonds's delaying tactics began to bear fruit. Local politicians clamored for gas, here and now, without strong preference for who and how. But they also worked against him, as he could not expand his system with the situation thus in limbo. Northeastern and Algonquin finally settled their differences in the summer of 1953, agreeing to split the market almost exactly as the FPC had initially suggested.[51] When Algonquin finally began deliveries to Boston in 1954, the last stronghold of manufactured gas was overrun.

## WEST TEXAS GAS FINDS A HOME IN CALIFORNIA

Outside of briefly mentioning Colorado Interstate as one of the earliest and least grand of the grand dames, we have paid no attention to the West. In fact—large, albeit mostly local—markets existed there well before World War II. While California is better known today as a leader in producing renewable energy, it was the largest oil producer in the United States in the early 1900s, and remained either number one or number two until 1930, trading places with Oklahoma. Gas accompanied oil in many cases, and California was second only to Texas in terms of customers served and overall volumes sold in the mid-1930s, before the three T's were built. (Appalachia, taken as a region, consumed more than either Texas or California.)

Drilling was initially concentrated in Southern California, near the coast. Oil production began in the Los Angeles Basin, which underlays the city of the same name, all the way back in 1890, and continues to this day. Many Los Angelenos would be surprised to learn that urban oilfields remain in production, even in the affluent city of Beverly Hills, where the rigs are hidden in windowless buildings with artfully decorated façades.

While the Los Angeles Basin and coastal fields were good producers, it was the Kern River field in the San Joaquin Basin, about 120 miles north of Los Angeles, that catapulted California into the petroleum big league. The Kern River discovery well was drilled in 1899, and though it did not blow out prolifically like Spindletop would two years later, it nevertheless started a boom in rural Kern County. Los Angeles began receiving natural gas from a neighboring field, Midway-Sunset, in 1913, and within a few years, Southern California Gas Company was serving more than two hundred thousand customers. It and several smaller gas utilities were consolidated in the 1920s by a holding company called Pacific Lighting Corporation, which is now once again called Southern California Gas Company.[52]

Meanwhile, California's other major population center, San Francisco, remained fueled by manufactured gas. Midway-Sunset was 150 miles from Los Angeles, but closer to 270 miles from San Francisco, which, at least in 1913, made running a pipeline uneconomical. Two things happened in the late 1920s to change this. First, pipeline technology advances—namely, electric resistance welding—allowed gas to be transported at much higher pressures, and at lower cost. Second, by the late 1920s, drilling in the San Joaquin Basin had expanded northward, closer to San Francisco. A large oil field with substantial reserves of associated gas was discovered about 60 miles north of Midway-Sunset, called Kettleman North Dome. The field was so named for its location in the Kettleman Hills, which are a misspelt paean to the pioneer, failed gold prospector, successful mercantilist, and legendary cattle-driver David Kettelman.

Kettelman was one of the thousands of young men who flocked to California during the Gold Rush of the late 1840s and 1850s. Originally from Prussia, but who had come by way of New York, he arrived in San Francisco in 1849 and sought fortune in the gold-laden hills. Like so many others, Kettelman was destined to be disappointed. But he soon found success in selling supplies to other miners. Using the wealth generated from this business, he founded a cattle-trading operation, which proved perfectly timed, as the steady influx of immigrants was stressing the ability of California's farms to provide them with meat. Seeing the opportunity, Kettelman rode his horse to Missouri and bought a tremendous herd of cattle, which he and a group of hired hands drove back to California. The prior winter had been extremely wet, making some grasslands unfit for grazing, so Kettelman and his drovers found a high valley where the grass was dry.[53] This area was later christened Kettleman Hills, and sits about 220 miles south of San Francisco.

Dry grass was not the resource that would make the area famous, however. About ten million years before Kettelman arrived with his cattle, the entire San Joaquin Valley was an inland sea, squeezed between California's coastal ranges and the volcanic Sierra Nevada. Tectonic events deepened the waters, and plankton, diatoms, and other detritus accumulated rapidly on the sea floor, creating layers of organic sediment that were, in some areas, thousands of feet thick.

Then, as now, the San Andreas Fault was the meeting point of the Pacific Plate and the North American Plate, and then, as now, the area was seismically active. Slips and strikes along the fault created a fold in the crust, which covered the sediment with a large domelike structure, known as an anticline. As millennia passed, the dome and the sediment beneath it sank deeper into the earth, where growing pressure and heat began the slow process of cooking the sediment, cracking its complex molecules into simpler structures: oil and gas. In October 1928, after nineteen months of effort aboveground and ten million years of pressure cooking, the drill bit of a rig belonging to the Milham Company reached a depth of seven thousand feet, piercing the dome that held oil and gas beneath the Kettleman Hills and unleashing four thousand barrels of oil and ninety million cubic feet of gas per day.

Pacific Gas & Electric (PG&E), which had begun consolidating gas and electric utilities in northern California in 1905 and whose forebearer had been serving San Francisco with manufactured gas since 1854, wasted little time running trunk lines to the Kettleman North Dome field. In fact, Milham Company had discovered a gas field nearby the year earlier, and although it was much smaller, PG&E had already planned to connect it to the city's mains via a long-distance trunk line. The company's engineers redesigned the system, increasing its size notably, and finished construction in the summer of 1929. Owing to what was then an extremely high system capacity, San Francisco's full-scale conversion from manufactured to natural gas was almost immediate.[54]

Indeed, PG&E's trunk lines boasted capacity of 250 million cubic feet per day. For context, NGPL and Panhandle Eastern both had initial capacity of only 175 million cubic feet per day. Not only that, but PG&E's system was built before NGPL and Panhandle, which would seem to make its omission from the list of grand dames an egregious error.

But California was an energy island: both PG&E and Pacific Lighting's systems were located wholly within the state's borders. And because the utilities confined themselves to their geographically contiguous, intrastate territories, and only bought from in-state producers, neither PUHCA nor the NGA affected the gas business in California, which remained highly local.

Until, that is, the population grew so quickly—and gas demand with it—that the state needed more. California was home to fewer than six million people in 1930, but over nine million by the end of World War II, and would continue to grow rapidly for decades thereafter. Just as important as its rapid population growth was the intense clip at which the state industrialized. California became the center of the aerospace industry during World War II and, indeed, through the end of the Cold War. And there was no better industry to be tied to—aircraft manufacturing grew from the forty-first largest industry in the United States before the war to the largest by the war's close. The United States, having produced just over two thousand aircraft in 1939, against more than ten thousand in the Soviet Union and eight thousand in both Germany and the United Kingdom, ramped production up to almost one hundred thousand aircraft by 1944, almost as much as those three countries combined. Building planes was energy intensive, requiring large amounts of rolled, bent and milled aircraft aluminum, copper for wiring, rubber for gaskets, and other raw materials that the WPB made sure arrived to the factory gates.

Because of industrial demand growth, and despite its large in-state resources, California found itself short on gas by 1944, and Pacific Lighting's main subsidiary, Southern California Gas, began circulating its desire to buy from out of state.

A thousand miles away from Los Angeles is a town—if it can be called that—named Jal, in New Mexico. One might think that this name has Spanish provenance, but the true origin is with a group of cowboys who brought to the area a herd of Texas longhorn cattle branded with the initials of their original owner: John A. Lynch. The cowboys considered rebranding the longhorns, but decided it would be too much work. A store was built near the JAL ranch in 1910, and the town of Jal (pronounced like "pal") was born.

Jal was home to the wells that supplied a medium-sized regional pipeline, Paul Kayser's El Paso Natural Gas. This system began as an idea in early 1928 when Kayser and associates went to New York to see a stockbroker about raising capital for a royalty trust. This is an upstream business, whose model is to accumulate leases and own mineral rights, but contract the actual work of drilling and operating out to production companies. The stockbroker told Kayser that it would be a tough sell, but that if he was able to put together a natural gas company instead, "We will finance it for you just like that."[55] While still in the broker's office, Kayser and his associate found a map of Texas and could identify only one major city that lacked natural gas service: El Paso.

Jal is in the southeast of New Mexico, only a few miles from the Texas border. It overlies a petroleum province known as the Permian Basin. The Permian Basin is a rugged and unforgiving tract of sunbaked desert. The little vegetation that has adapted to survive there are cacti and scrub, all of it more brown than green, none of it rising above the height of an average man. In summer, the heat is oppressive. The winter is freezing. The winds are relentless, and on the rare occasions that it rains, it does so in torrents, so that dry creeks—or arroyos—overflow, flooding the cracked caliche ranch roads. There is an austere beauty to it, especially during the hours just after the sun rises and before it sets, when the desert takes on a palette of orange and purple that would have left even Monet spellbound. Mostly, though, it is a wasteland.

In an important way, the Permian Basin's virtual uninhabitability works to its advantage. If it was a beautiful place, there would certainly be more fuss about the tens of thousands of oil wells strewn across its vast plains and rocky hillsides.

Oil is the lifeblood of the Permian Basin—its raison d'etre. There is no farming to speak of, neither of agriculture nor of livestock. The endless cotton fields several hundred miles to the north, around Lubbock, give way to mesquite, and pumpjacks take the place of cattle. The pumpjacks, also called nodding donkeys, are the Permian Basin's livelihood. They tilt back and forth ceaselessly, with a regularity and utter ignorance that make them amiable, even lovable, just like the cattle ubiquitous across the rest of the state. If amiability is a difficult trait to assign to an oil well, given that most people only hear of them when they have blown out or exploded, it helps to know that the wells that nodding donkeys adorn are far past their prime. Each nod downward applies pressure to the reservoir, coaxing a bit more oil out of the ground. Initially many of these wells were monsters, spurting out oil without any need for artificial lift, but pressures inevitably decline with time. A pumpjack is like Botox for an oil well.

If these tired, old wells only produce a few barrels of oil per day, their lackluster performance is offset by sheer numbers. In the heart of the Permian Basin, if you are standing on high ground, over a thousand are visible to the naked eye. From a less elevated vantage point, you may not perceive the donkeys directly, but you can see power lines overhead, which supply the electricity that keep them atilt. This is a decidedly more eerie view, of a grid of electric lines and transformers large enough to blanket a large neighborhood, only without any houses.

Although the first truly commercial well in the Permian Basin did not start flowing oil until 1923, within just a few years the boom had achieved truly epic proportions, with tent towns popping up seemingly overnight. By 1928, when Paul Kayser had the idea to pipe gas to El Paso, some two hundred miles from the Permian Basin's western boundary, several thousand wildcatters, engineers, roughnecks, and saloon owners had migrated to this windswept, sun-scorched patch of earth.

These thousands were not near enough to provide a market for the Permian Basin's oil and gas production, though. Even today, West Texas and southeast New Mexico are sparsely populated. Millions of barrels of oil and billions of cubic feet of gas were being wasted, so Kayser had little difficulty securing the supply in Jal. Gas reached El Paso in June 1929, after just six months of construction, and just in time for the Great Depression.[56]

Despite the timing, El Paso actually managed to grow during the Depression, albeit from a small starting stature. In 1929 the pipeline had capacity of just 10 million cubic feet per day. Sales to copper smelters in Arizona and power plants in Mexico allowed Kayser to build another three hundred miles of pipe two years later, and another two hundred miles when he signed a contract with the city of Phoenix two years after that.[57] By 1941 the system had reached a respectable size of 130 million cubic feet per day,[58] but the company was still horse trading: even small new business was hard-won, and most producers in the Permian Basin were content flaring their gas, not clamoring at El Paso's doorstep.

This all changed after World War II. El Paso would quickly grow to become the country's largest pipeline system, and Paul Kayser a preeminent and nationally respected businessman.

In 1945 Kayser met with the president of Southern California Gas (SoCal), Frank Wade. At the time El Paso was selling just under 140 million cubic feet per day across its whole system, so it would have been quite a substantial achievement had Kayser gotten Wade to sign for the 50 million that Kayser proposed. Instead, Wade responded that he needed 75 million immediately, another 50–125 million shortly thereafter, and 300 million within a few years. Kayser was floored, and almost immediately filed an application with the FPC to construct a new, twenty-six-inch line that would run parallel to the existing system before extending to the California border at Ehrenberg, Arizona. As ever, the California utilities wanted to keep control of everything within the state, so SoCal built their system to pick up gas once it had crossed over the border.[59]

The stars were aligning for Kayser on the west end of his pipe: California represented a virtually insatiable market for gas. They were also, and for unrelated reasons, aligning on the east end, a few miles from Jal, in the Texas Permian Basin.

The Federal Power Commission had, by the mid-1940s, extended its powers substantially. And the war had shown that natural gas was a precious commodity, one that should not be wasted. But the war was also a time of practicality before all else—oil was key to the effort and was prioritized over gas. The massive increase in wartime oil production led to a correspondingly massive increase in the amount of gas flared at the wellhead, but neither the FPC nor the PAW dared do anything in the name of conservation of gas that would, as a consequence, limit production of oil. Thus, by 1944, 70 percent of gas produced in association with oil was being flared, nationwide,[60] which equated to a bit less than one-quarter of the total volumes of natural gas that producers were bringing up from the ground. Peacetime brought the rowdy gallop of development to a more controlled canter, giving regulators the opportunity to introduce conservation measures. In 1946 the FPC began hearings on the subject of gas waste.

The idea of Washington bureaucrats inserting themselves into matters historically overseen by the states does not sit well in many places or in many industries, most of all in oil and gas in Texas.

The oil and gas business in Texas is regulated by the Texas Railroad Commission (RRC), so named because its original purpose was to regulate intrastate railroads, where the Interstate Commerce Commission had no jurisdiction. In 1917 it began to regulate common carrier oil pipelines, and by the 1920s regulating the oil and gas sector had become its primary function. From its humble—and legally contested—beginnings, the RRC grew to become one of the most powerful agencies in the world, as it essentially dictated the price of oil globally from around 1931 until the 1960s. (The definitive end of the RRC's price-setting abilities came in 1972, when the newly formed Organization of Petroleum Exporting Countries, or OPEC, took over that role.)

The RRC acquired this incredible ability after it began establishing and enforcing production quotas in the wake of the East Texas oil boom that began in 1930, which quickly vaulted Texas above both California and

Oklahoma in terms of oil production but also led to a dizzying crash in prices to just 10 cents a barrel, from over a dollar the year before. The RRC gives an unattributed quote in its archives: "There was much unrest, with talk of threats of blowing up wells and pipelines." Quotas, or proration orders, were meted out en masse, and after many lawsuits and court rulings, it became clear that the RRC had the authority to establish and enforce quotas both to limit physical waste and to limit production above what it determined to be "market demand." The RRC's ability was even recognized at the federal level after the Connally Hot Oil Act passed in 1935, declaring all oil produced or transported in violation of RRC quotas or statutes as contraband and enabling the president to impose hefty fines on all violators.

The RRC was less active in the gas industry than in oil. It had, in 1935, been empowered to forbid gas production that caused waste, and it used this power to stop flaring in the Panhandle field. But this authority was limited to gas flared from gas wells (nonassociated gas) rather than gas flared from oil wells (associated gas). Wells produce different quantities of the two products across a spectrum defined by the gas-oil ratio, measured in cubic feet per barrel. On one end of the spectrum is "dry gas," where a well flows more than one hundred thousand cubic feet of gas for each barrel of oil. On the other is "dead oil," where there is almost no gas dissolved in the oil. Most wells are somewhere in between (although there are far more dry gas wells than dead oil wells), and defining what constitutes a gas well versus an oil well is an arbitrary choice of gas-oil ratio.

In the late 1930s, the RRC issued an order trying to limit the flaring of associated gas by threatening to prorate oil production from wells that flared more than two thousand cubic feet of gas for every barrel of oil they produced. In theory, this would have done at least some good, but in practice, the flared gas was not being measured by producers, and, even if it was, the RRC lacked the staff to go well-to-well to check the measurements (in 1940 there were ninety-five thousand producing wells in Texas).[61]

The RRC did allocate some resources to this effort, including a young, Texan engineer named William Murray. In 1939 the twenty-four-year-old Murray began testing gas-oil ratios in the field. He was appalled both by the level of waste, and by how helpless he was to do anything about it. In 1941 he left the RRC to work for the PAW, in Washington, where he made contacts at the FPC and told them about the enormous volumes of gas being flared in his home state. The FPC demurred from taking any action, however, saying the RRC was the world's preeminent oil regulatory body, and that it would know how best to regulate hydrocarbon production in Texas. But this was not the end of Murray's battle for conservation. Back in Texas after resigning from the PAW, he attended, as private citizen, a special RRC hearing on gas flaring in December 1944. Upon hearing the official statistic that less than 1 percent of gas produced in Texas was being flared, Murray "stood up and stated that he knew from personal experience that the received figures were gross underestimations; that from ten to twenty-five times the official estimates was being lost."[62]

To investigate the problem more thoroughly, Murray was placed in charge of a committee of engineers, which in November 1945 released a report showing that around 1.5 billion cubic feet per day, or 57 percent of total Texas gas production, was being flared or vented into the atmosphere.[63] This won him both friends and enemies. Among the former group was the erstwhile RRC commissioner, Beauford Jester, who was campaigning to be governor of Texas. Jester won the governorship, and appointed Murray to serve out his unexpired term and become the youngest-ever RRC commissioner.

In January 1947, Murray was sworn in. Two months later, the RRC issued the Seeligson order, targeting the flaring of associated gas from oil wells in the Seeligson field and forcing producers to cease all production until the gas was either sold to an end user—or, more likely, reinjected into the reservoir—where it would keep pressures high, lead to more oil production, and ultimately another round of gas recycling. After the Texas Supreme Court upheld the order, the RRC issued sixteen more just like it, took a hard tack against producers who threatened noncompliance, and was supported by an extremely deferential judiciary, who ruled time and again that the RRC was a body of experts dedicated to preserving the public interest.[64]

Of course, Murray had plenty of enemies, mostly producers who thought the extra costs of recycling the gas would be ruinous or that no markets existed for the gas to be sold. But things were changing rapidly in the late 1940s. Gardiner Symonds's Tennessee Gas Pipeline sourced gas from South Texas and was expanding relentlessly. In 1947 TETCO started receiving gas from East Texas from United Gas and H. L. Hunt. The grand dames were moving it out of the Panhandle field in record amounts every passing year. And now, producers in the other great petroleum province, the Permian Basin of West Texas, had their eyes set on California, and, by extension, on Paul Kayser and El Paso. The more forward-thinking among the producers saw the writing on the wall well before the Seeligson order, and recognized that natural gas represented an important and growing stream of profits. Those who did not reacted very quickly when the RRC threatened to shut down production.

Very rapidly, Kayser found himself bombarded by producers who desperately wanted to sell him gas, asking if they could pay him to run El Paso trunk lines into the heart of the Permian Basin, and even extend it north, to the Panhandle field.

El Paso's system reached the California border, where it met up with Southern California's, in November 1947, and as it filled, the great lines of flares that had kept the barren, West Texas flatlands lit so brightly that you could read a book in a moonless night, went out.

# PART II

## A MANUFACTURED SHORTAGE
## OF NATURAL GAS, 1954–1992

Part I of this volume ended on a high note for the natural gas market: it has now completely displaced manufactured gas, and the grid of pipelines in the United States is without par anywhere else in the world, touching nearly every population center in the country. Moreover, the FPC has set a policy that encourages pipelines to compete for business in large markets, preferring oligopoly over monopoly.

But there have also been ominous portends from the FPC, which issued several landmark decisions affecting gas producers that seem to contravene the original intent of the NGA. The FPC has been prodded down this path by an increasingly activist Supreme Court and by Congress. It is tempting to ascribe the movement toward tighter regulation to a single political party, or even individual justices or lawmakers, but it is more accurate to interpret it as a consequence of the industry's great expansion in the postwar era. Whereas before the three Ts, Algonquin, and El Paso there were only a few states that imported natural gas from the Southwest, now there were many, which tilted the political balance. Every new pipeline that delivered gas from Texas, Oklahoma, Kansas, or Louisiana to the East Coast was greeted by producers, as it gave them a larger market. Equally, however, every new state served by an interstate pipeline meant new votes in favor of regulations to ensure low prices.

# CHAPTER 6

## REGULATION AND RUIN

The Natural Gas Act was written by Congress and enforced by the Federal Power Commission. Where FPC decisions were challenged, the courts became the final arbiters. In these disputes, the courts ruled on two issues: (1) whether Congress had granted the FPC the jurisdiction to regulate the matter at hand, and (2) whether the FPC's decision was appropriate and lawful. The courts had ultimate authority in a dispute to decide both who and how the FPC should regulate.

When laws are well written, court involvement is minimal, as there is little question of both jurisdictional and functional authority. Said another way, a clearly worded law both minimizes suits and leaves the court little room for interpretation. But the sections of the NGA that defined both who the FPC was meant to regulate and how it should do so were worded so ambiguously as to plunge the entire industry into chaos for more than four decades. Beginning in 1940 and for fourteen years after, the courts made a number of rulings—some contradicting others—that culminated in the Supreme Court's 1954 *Phillips Petroleum Co. v. Wisconsin* decision, which brought all gas producers under federal, cost-based price regulation. In the 1940s and 1950s, the chaos and contradictions on the regulatory front had little impact on the industry's growth. Gas supplies were cheap and abundant, and demand was rising rapidly in the postwar boom. New pipelines were built at a furious clip, allowing every major US city access to natural gas, as I detailed in the preceding chapters. But by the late 1960s, supplies were dwindling, and by the 1970s, shortages became so severe that large-scale curtailments were required, forcing the temporary closure of factories, businesses, and even schools, which could not heat their classrooms.

### IS A TOMATO A VEGETABLE? AND, IF SO, HOW BEST TO EAT IT?

The early days of FPC regulation were seen as successful and were mostly uncontroversial, but the confusion stemming from the NGA's ambiguous wording meant that the Supreme Court was able to assume the role of de facto policymaker by the mid-1940s, interpreting the intent of the law and

generally taking the broadest possible view of the FPC's jurisdictional authority. The FPC, for its part, was stuck in the middle: whatever it thought,
it had no authority to question the Supreme Court. The only permanent
solution would have been for Congress to amend the NGA, but this did not
occur. Part of the reason for this was a rapidly changing political environment, which also turned what otherwise might have been simply an impo-
tent regulator into an sycophantic regulator, which tried to reinterpret the
Supreme Court's interpretations in a manner that would please Congress.
To use a somewhat atavistic metaphor to boil the dynamic down to its essence, the Supreme Court was a mother who had just told her son, the FPC,
to finish his vegetables. The son finishes his broccoli, but not his tomatoes,
claiming that a tomato is not a vegetable. Both mother and son look to the
father, Congress, for a final ruling. But the father merely gets up and goes
into the other room to watch TV. The boy grudgingly eats his tomatoes,
all the while maintaining that they are fruit. At the heart of the confusion
over who to regulate (is a tomato a vegetable?) was section 1(b) of the NGA,
which states: "The provisions of this Act shall apply to the transportation
of natural gas in interstate commerce, to the sale in interstate commerce of
natural gas for resale for ultimate public consumption for domestic, commercial, industrial, or any other use, and to natural gas companies engaged
in such transportation or sale, but shall not apply to any other transportation or sale of natural gas or to the local distribution of natural gas or to
the facilities used for such distribution or to the production or gathering of
natural gas."[1] What cannot be argued is that the NGA was written to close
a regulatory gap, leaving states unable to ensure that their utilities were
paying just and reasonable prices for the gas they bought from interstate
pipeline companies. What can be argued, and was for almost forty years, is
whether the act allowed the FPC to regulate natural gas producers.

The exemption in section 1(b) for production and gathering was what
caused the consternation. Did the exemption mean that the FPC could not
regulate the physical market (i.e., how many wells were drilled and how
far apart) but could regulate sales by producers? Did it matter when the
gas changed hands—before or after crossing the state line? This first set of
vagaries made it difficult to answer the question "who to regulate?" (i.e., is
a tomato a vegetable?).

Supposing there was consensus on that—which there was not—the
other question was "how to regulate?" (i.e., how do I eat this vegetable?).
Sections 4(a) and 5(a) of the NGA said only that the FPC should set rates
that were "just and reasonable." Did this mean that, if it was to regulate
production and gathering, that the FPC should employ an original cost
standard? Or would a methodology based on replacement value be better
suited for an extractive industry, defined by depletion and hence prone to

cost inflation over the long term? Should producing properties be included in the same rate base as pipeline assets, even though natural gas must be found, rather than built?

These are not easy questions, and history showed that easy answers were not forthcoming. If anything, it showed that the United States tried every wrong answer before finally arriving at the right one.

I will describe this tumultuous, but seldom boring, period by detailing the key decisions from the FPC, circuit and supreme courts, and congressional and executive actions. For those of you who want the short story, or whose eyes glaze over when reading about policymaking (admittedly, you might have done better to purchase a different book), here it is:

The FPC began by hewing to Congress' original intent of not regulating producers. Then it switched course, but only regulated those producers who were owned by interstate pipeline companies (i.e., companies they would otherwise have regulated anyway). They did this using an original cost standard, which caused angst among producers, as it did not reflect inflationary market conditions or reward more efficient producers. The main argument, for a while, was about "how to regulate." Then the courts started interpreting the NGA more broadly, and encouraged the FPC to regulate all producers, including independents, changing the main argument to "who to regulate." The FPC initially demurred and urged Congress to make the production and gathering exemption more explicit. But it took Congress too long to pass amendatory legislation, and by the time that it had, it was met with a presidential veto because the timing was very poor in the court of public opinion. Various scandals prevented any legislative amendments, and in its 1954 *Phillips* decision, the Supreme Court officially ordered the FPC to regulate the price paid for gas to all producers, nationwide. The FPC quickly found it was not up to this Herculean task, however, and it took eleven years to set the first rate, and another nine years to finish the process, by which time the surplus of the 1940s and 1950s had turned into a severe and worsening shortage. In 1978, after a harsh winter that led public schools to close for days and even weeks for lack of gas to heat classrooms, President Jimmy Carter passed the Natural Gas Policy Act, which began the process of first increasing, and eventually deregulating, prices received by producers.

## WHO TO REGULATE: *COLUMBIAN FUEL* AND *PEOPLES* CASES

The first case of note involved Columbian Fuel Company, which produced gas in Kentucky and sold it at the termini of its gathering lines to a pipeline engaged in interstate commerce. It was not affiliated with that, or any other interstate pipeline. In a 1940 ruling, the FPC decided it did not have the authority to regulate sales by Columbian Fuel because the company

was not already subject to FPC jurisdiction—it was an independent producer, not affiliated with an interstate pipeline. In coming to its decision,
the FPC reviewed, among other things, the *Congressional Record*, where it
doubtless paid close attention to the comments of Senator Wheeler, who
said, "[The Act] does not attempt to regulate the producers of natural gas or
the distributors of natural gas; only those who sell it wholesale in interstate
commerce."[2] But the ambiguity of section 1(b) was clear even at this early
juncture: the FPC admitted that Columbian Fuel was selling gas "wholesale in interstate commerce," but that it was covered under the production
and gathering exemption.[3]

Importantly, strong dissent came from Commissioner John Scott, who
believed that section 1(b) only denied the FPC jurisdiction over the physical process of production and gathering. In Scott's reasoning, the NGA
prohibited the FPC from telling producers where or how they could drill,
but did not prohibit it from regulating the resulting transaction, or "sale,"
if the gas was eventually to be used in interstate commerce.[4] In later cases,
the Supreme Court would adopt Scott's position as its own.

While the FPC chose not to regulate Columbian Fuel, it did not categorically refuse to regulate independent producers, writing that if a producer, unaffiliated with a pipeline though it might be, held such a dominant
position in the field as to allow it to "maintain an unreasonable price despite
the appearance of competition," the commission "will decide whether it can
assume jurisdiction over arbitrary field prices or should report the facts to
Congress with recommendations for such broadening of the Act . . . to close
this gap in effective regulation of the natural gas industry."[5]

The aforementioned, plus the fact that *Columbian Fuel* came so early
in the FPC's history, when the congressional testimony was still fresh, in-
dicates that the FPC interpreted its mandate rather cautiously, and that it
may have doubted whether it had been given the power to regulate producers. In other words, in 1940 it seemed that the production and gathering exemption in section 1(b) of the NGA was being interpreted as an exemption
from any regulation, including of sales, Scott's dissent notwithstanding.

In 1942 the FPC tested the limits of its jurisdiction when it ruled that
Peoples Natural Gas, which sold gas at the termini of its gathering lines in
Pennsylvania to an affiliated pipeline that had sales contracts in New York,
was subject to rate regulation. As an affiliated producer, Peoples was owned
by an organization that unambiguously qualified as a "natural gas company"[6] subject to FPC jurisdiction, unlike Columbian Fuel. Peoples disagreed
with the ruling and invoked the *Columbian Fuel* decision, arguing that
it was exempt as a production and gathering company. Essentially, Peoples wanted the FPC to treat it as two, separate entities—an unregulated
production and gathering company whose business stopped at the termini

of its gathering lines, and a regulated interstate pipeline, whose business started at the receipt point of its trunk line. The DC Circuit Court threw out Peoples's "geospatial" argument, siding with the FPC and writing that Congress had intended to regulate all sales of gas that were about to move interstate, even if they had not already done so. "The particular point at which title and custody of the gas passes to the purchaser without arresting its movement to the intended destination does not affect the essential interstate nature of the business. . . . We cannot disregard the plain language of the statute because the Commission at one time interpreted it narrowly."[7]

This was an important validation for a young agency. In a review of this and other cases, a law student who went on to become an FPC commissioner and judge, Alfred Scanlan, writes that the decision gave the FPC so much more confidence in its broadened jurisdiction that it "overruled, or at least substantially undermined, the Columbian Fuel case,"[8] where the FPC had hewed to a much more conservative reading of its own power.

After *Peoples*, the convention at the FPC became known as the "otherwise-a-gas-company" standard, which stated that the FPC would regulate production and gathering assets but only at companies that would otherwise fall under NGA regulation because they were owned by or affiliated with interstate pipeline companies.

It would not stay this way for long. Indeed, a deeper reading of the circuit court's opinion shows that it took little note of the fact that Peoples was an affiliated producer.[9] Its ruling was based wholly on the fact that Peoples sold gas "in interstate commerce," a characteristic that it shared with many independent producers. Already the courts were beginning to adopt Scott's dissenting opinion in Columbian Fuel.

## HOW TO REGULATE: THE *HOPE* AND *CANADIAN RIVER* CASES

Recall that Hope Natural Gas was wholly owned by Standard Oil of New Jersey, which also owned Hope's largest customer, East Ohio Gas, a state-regulated local distribution company (LDC). Regulators in Pennsylvania and Ohio had grown suspicious of what they thought were excessive and unreasonable rates for the gas their LDCs were paying to Hope, so they petitioned the FPC to investigate Hope's rates almost immediately after the NGA was passed. After a lengthy investigation, in May 1942, the FPC found that Hope's rates were indeed unjust and unreasonable, and ordered the company to reduce them by almost half.[10]

Hope appealed the case, arguing that the FPC's methodology was deeply flawed, as it used an original costs standard, rather than the company's preferred method of using replacement costs to determine the rate base. Original costs meant that Hope would receive a return on the amount it had spent drilling and operating the wells, whereas replacement costs would be calculated as what it would cost to produce the same amount of gas in contemporary conditions. Replacement costs were determined by the market, and reflected the supply and demand for gas, along with the quality and abundance of the resource, the cost of the drilling rig, the labor of the men who operated it, laying gathering lines, and all of the various other costs associated with production.

The circuit court ruled in Hope's favor, but then the FPC appealed, and the case went to the Supreme Court in 1943–1944, marking the highest court's first, but far from last, foray into NGA interpretation. In *Federal Power Commission v. Hope Natural Gas Co.*, the court ruled that the FPC was justified in using actual costs as the basis for setting rates.

A case so similar that it should be considered practically as an extension of *Hope* was that of Canadian River Gas Company, Colorado Interstate Gas, and Cities Services (*Colorado Interstate v. FPC*). In this triumvirate, Canadian River was the producer, owning reserves in the Panhandle field in Texas; Colorado Interstate—also owned by Standard—was the pipeline company; Cities Services was the LDC for Denver and Pueblo, Colorado.

In addition to selling its gas to Colorado Interstate, Canadian River sold some to markets in Texas (intrastate). And, besides reselling gas to Cities Services, Colorado Interstate sold to a large industrial user upstream of Denver. Both of these activities explicitly did not fall under FPC jurisdiction.

In *Colorado Interstate*, the companies took issue with how the FPC had divvied up the costs of all the facilities and split them into regulated and unregulated buckets. This case made it to the Supreme Court in 1945, and the justices often repeated, sometimes elaborated, but never refuted the stances they had taken in *Hope*.

In both cases, Justice William O. Douglas wrote the majority opinion. Douglas was appointed by Franklin Roosevelt and regarded generally as a civil libertarian and not overly concerned with judicial restraint. He wrote in *Hope*, and repeated in *Colorado Interstate*, the following, which can be taken as the prevailing philosophy of the court with respect to FPC rate-making ("how to regulate") over the next several decades.

> It is not theory, but the impact of the rate order, which counts. If the total effect of the rate order cannot be said to be unjust and unreasonable, judicial inquiry under the Act is at an end. The fact that the method employed to reach that result may contain infirmities is not then important. Moreover, the Commission's order does not become suspect by reason of the fact that it is challenged. It is the product of expert judgment which carries a presumption of validity. And he who would upset the rate order under the Act carries the heavy burden of making a convincing showing that it is invalid because it is unjust and unreasonable in its consequences.[11]

In sum, the court did not care what standard the FPC employed—whether it used original costs, replacement value, some combination of the two, or something completely different. The question of "how to regulate" was entirely discretionary, and that discretion rested entirely with the FPC's body of experts.

In this approach, the ends justified the means, but what were the ends? Obviously to keep costs as low as possible for consumers. There was also recognition that producers needed to make a profit to survive and remain as going concerns—the court was not so short-sighted as to create precedent that would immediately bankrupt natural gas companies. But, as is clear from the following quote from the ruling, the overriding concern was to keep rates low: "Rates which enable a natural gas company to operate successfully, to maintain its financial integrity, to attract capital, and to compensate its investors for the risks assumed cannot be condemned as unjust and unreasonable under the Natural Gas Act, even though they might produce only a meager return on a rate base computed on the 'present fair value' [e.g., replacement value] method."[12]

Some context is useful here, lest you become too jaded by what appear to be activist and profoundly anti-business stances. The Supreme Court and the FPC were cognizant both of Hope's ownership and its financial history. Standard was hardly short on capital, or profits, and it was noted in the *Hope* decision that, since being formed, the company had netted Standard a return of more than 600 percent on its initial investment.[13] Also keep in mind that the breakup of the holding companies was still in its early stages—it had not begun in earnest until 1940, and the *Hope* and *Colorado Interstate* decisions were not issued until January 1944 and April 1945, respectively. The evils of monopoly were still very fresh in the minds of the courts, the FPC, Congress, and the public at large. (Standard completed its utility spin-offs in October 1943, such that, by the time the Supreme Court issued its decision, a new company called Consolidated Natural Gas owned both Hope and East Ohio Gas. But the Supreme Court was ruling on an FPC decision made in 1942. Cities Services still owned the LDCs in Denver and Pueblo at the time of the *Colorado Interstate* decision.)

The companies were not the only ones opposed to the Supreme Court's decisions in *Hope* and *Colorado Interstate*. West Virginia collected a production tax based on the value of the gas sold, and wrote the Supreme Court in an amicus curiae, or "friend of the court," brief that it would be unjust to deprive the state of these revenues, which elicited the decidedly unfriendly retort from Justice Douglas: "If the Commission is to be compelled to let the stockholders of natural gas companies have a feast so that the producing states may receive crumbs from that table, the present Act must be redesigned."[14]

*Hope* was a landmark case not just for the natural gas industry but for public utility policy. The question of whether and when to use replacement value, and how it should be calculated, had vexed regulators since the Supreme Court issued a ruling in *Smyth v. Ames* in 1898, where it said that regulated companies were entitled to a "fair value" for their property, and that fair value should be calculated by estimating the present cost of construction, among other things. The lack of specificity in *Smyth* led regulatory commissions to "wing it," to at least some extent, applying "judgment whose processes defy analysis or description" and selecting rates that bore "no derivative relation to any figures in evidence."[15] This, in turn, invited intervening parties in rate proceedings to submit reams of cost and performance data from other projects and industries, sometimes only tangentially related to the issue at hand, which turned what should have been a technocratic process into a lengthy and costly "politico-legal-mystic" ordeal.[16]

*Hope* overturned *Smyth*—no longer was the regulatory agency bound to consider replacement costs (although it could, if it chose to). The *Smyth* case bears mentioning not only because it helps to explain the significance of *Hope* but because it illustrates the primacy of profit over principle. One might assume that in *Smyth*, as in *Hope*, industry was advocating for using a replacement cost standard, to reflect prevailing market dynamics. Private business, after all, favors a free market and, when regulations are necessary, advocates for market-based regulations.

In *Smyth*, though, industry—in this case, railroad companies—argued for using original costs. It was the regulatory agencies, charged with protecting consumers, that argued for replacement value. This was so because, as the case was being heard in the 1890s, price levels had dropped consistently, the result of post–Civil War deflation and economic depression. Original cost was higher than replacement value at the time of *Smyth*, whereas the opposite was true as *Hope* was being considered, during the post–World War II economic boom.

The lesson here is that industry does not abide ideology. Private business lobbies for the sort of regulations that yield the highest profits.

On the surface, the key "how to regulate" issue in both the *Hope* and *Colorado Interstate* cases was which cost standard to use: companies favored a replacement value standard, while the FPC favored an original cost standard. However, while I have summarized these two cases in the context of "how to regulate," the question of "who to regulate" also came up. The transport operations of interstate pipeline companies, such as Hope, were clearly subject to FPC regulation, but the companies argued that the production and gathering exemption in section 1(b) of the NGA precluded the FPC from regulating their production assets, instead arguing that a fair field price be used.

But to little avail. The Supreme Court approved the inclusion of production assets in the FPC's rate base calculation in *Hope*. In *Colorado Interstate*, it elaborated, going so far as to quote Senator Wheeler's position that the NGA "does not attempt to regulate the producers of natural gas," but providing the following justification for why upstream operations were subject to regulation: "Congress of course might have provided that producing or gathering facilities be excluded from the rate base and that an allowance be made in operating expenses for the fair field price of the gas as a commodity. Some have thought that to be the wiser course. But we search the Act in vain for any such mandate."[17]

Indeed, the Supreme Court—and, by extension, the FPC—adopted the stance that the production and gathering exemption applied only to the physical business of producing and gathering the gas, not to any transactions. Commissioner Scott's dissenting opinion in the 1940 *Columbian Fuel* case had become official policy.

> That does not mean that the part of Section 1(b) which provides that the Act shall not apply 'to the production or gathering of natural gas' is given no meaning. Certainly that provision precludes the Commission from any control over the activity of producing or gathering natural gas. For example, it makes plain that the Commission has no control over the drilling and spacing of wells and the like. It may put other limitations on the Commission. We only decide that it does not preclude the Commission from reflecting the production and gathering facilities of a natural gas company in the rate base and determining the expenses incident thereto for the purposes of determining the reasonableness of rates subject to its jurisdiction.[18]

In sum, the majority of the court believed that all parts of an interstate pipeline company's business were subject to FPC rate regulation, and that the FPC could choose whatever standard it found most appropriate. It was, however, not clear from either case whether the NGA applied to independent producers. The presumption was that the otherwise-a-gas-company standard prevailed, and market behavior reflected that. Recall that the Chicago Corporation divested itself of its stake in the Tennessee Gas & Transmission Company in September 1945, and did so specifically to avoid regulation as a natural gas company.[19]

The Chicago Corporation was not the only company to do so, and the otherwise-a-gas-company standard enhanced the division between the business of producing—the "upstream"—and the business of transporting—the "midstream." With PUHCA having mandated the separation of distribution—the "downstream"—companies in the three different segments of the gas industry were becoming increasingly specialized, and found their fundamental interests diverging. Clearly, the NGA was written to protect consumers from unjust and unreasonable prices. As such, it was meant as a foil to potentially monopolistic behavior from companies in the upstream and midstream segments. But whereas pipelines could be built as needed, gas could not be produced with such certainty.

In his dissent in *Hope*, Justice Robert Jackson proved incredibly far-sighted about the potential negative long-term consequences of cost-based rate-setting for an exhaustible commodity like natural gas. Justice Jackson, also appointed by President Roosevelt, is remembered for his commitment to due process and disdain from regulatory overreach, as opposed to Justice Douglas. He wrote:

> The heart of this problem is the elusive, exhaustible, and irreplaceable nature of natural gas itself. Given sufficient money, we can produce any desired amount of railroad, bus, or steamship transportation, or communications facilities, or capacity for generation of electric energy, or [manufactured gas]. In the service of such utilities, one customer has little concern with the amount taken by another, one's waste will not deprive another, a volume of service and be created equal to demand, and today's demands will not exhaust or lessen capacity to serve tomorrow. But the wealth of Midas and the wit of man cannot produce or reproduce a natural gas field.[20]

In other words, Jackson was taking issue with the concept of including upstream assets in the rate base. In *Colorado Interstate*, he developed this position further, detailing how the FPC's own calculations perversely assigned several different values to gas from the same wells.

A simple illustration of Jackson's point is as follows. Say one company, in the search for gas, spends $20 million over several years drilling hundreds of dry holes, finally finding a productive field with reserves of 20 billion cubic feet. For the sake of avoiding any difficult math, forget about the return on investment and economic lifespan of the asset, and just say the gas is worth $1 per thousand cubic feet, using an original cost basis. Now say another producer gets lucky and finds the same amount of gas straight away, having spent only $5 million. On the same cost basis, the second company's gas would be worth only 25 cents per thousand cubic feet.

What justification is there for a fourfold difference in prices for the same commodity, with the same value to society? And what if it was not just good luck, but decades of experience in prospecting and perfecting drilling techniques that led the second company to produce gas so much more cheaply? Should that company not be rewarded for its hard-won mastery?

Furthermore, imagine a situation in years hence, where gas has become increasingly difficult and expensive to find. Now it is thought to take $50 million to find the same 20 billion cubic feet, which requires a price of

$2.50 per thousand cubic feet. But the prevailing price has been set at just $1. What incentive is there to go and find more?

You can see how cost-based price regulation is extremely troublesome for a commodity in which unit costs are unknown and unpredictable, and which entails serious finding and development risk. Jackson concludes: "Far-sighted gas-rate regulation will concern itself with the present and future, rather than with the past, as the rate-base formula does. It will take account of conditions and trends at the source of the supply being regulated. It will use price as a tool to bring goods to market—to obtain for the public service the needed amount of gas."[21]

## THINGS HEAT UP: *INTERSTATE* AND CRIES OF OVERREACH

The year is now 1947 and the Supreme Court has interpreted the NGA to mean that affiliated producers are subject to whatever rate-making standards the FPC deems reasonable. The court had not explicitly embraced this otherwise-a-gas-company standard, but neither had it challenged it. Another case had been working its way through the FPC and the courts for several years and was now on the Supreme Court's docket.

In many ways, the Interstate Natural Gas Company case was similar that of *Peoples*, and, had the Supreme Court considered the case in 1942, when the DC Circuit Court had issued its ruling in *Peoples*, it may not have aroused as much controversy. But political and market dynamics had both changed considerably. The timing of *Interstate Natural Gas Co., Inc. v. Federal Power Commission* caused a furor.

President Roosevelt had died in 1945, elevating Harry Truman to the presidency. Truman commanded relatively little popular support, and his growing unpopularity led to a "Republican wave" in the House and Senate elections of 1946. Democrats, generally in favor of regulation, lost majority control of both houses of Congress. Despite their newfound majority, though, Republicans could not immediately shift the FPC's course. Its commissioners serve five-year terms, and its chairman, Leland Olds, had only just been reconfirmed in 1944.

Texas oilmen hated Leland Olds. To them, he was a communist, intent on destroying freedom and liberty wherever he found them. But then, so was practically anyone who limited the amount of profits they could make. Many in Washington still regarded these rabblerousing, ultra-conservative Texans as an unimportant, if loud, group of nouveau riche ne'er-do-wells. But their dollars spoke more loudly than their words.

In the late 1930s and throughout the 1940s, a handful of oilmen began the slow-motion takeover of the Texas Republican Party. They were vehemently against Roosevelt and anything that smacked of socialism. Despite Texas still being a solidly Democratic state, they donated heavily to

Wendell Wilkie, a midwesterner who ran, unsuccessfully, against Roosevelt in the 1940 presidential election. Forced to meet with his largest financial contributors, Wilkie came away decidedly unimpressed, saying: "You know the Good Lord put all this oil in the ground, then someone comes along who hasn't been a success at anything else, and takes it out of the ground. The minute he does that he considers himself an expert on everything from politics to petticoats."[22]

The crass, uneducated oilmen were just as repugnant to fellow Texans. Sam Rayburn was the Democratic congressman from northeast Texas, which produced little oil. He had served as Speaker of the House before the 1946 Republican wave, and went on to hold the title again after Democrats retook Congress. To this day, he has held the speakership for longer than any other congressman. He was close personal friends with a number of Republicans, and was also referred to as the "bridge between the northern and southern members of the Democratic party." Rayburn was, in other words, an easy man with whom to get along, but he could hardly abide speaking with the new "Big Rich" Texas oilmen. He constantly refused their bribes, and, in some cases, voted against their favored policies. But when the oilmen supported a competing Democrat with anti-Roosevelt views in the primary race of 1944 who almost beat Rayburn, he was forced to take a more amenable stance. From that point forward, Rayburn "held his nose as he back the oilmen's initiatives," but shared his true feelings with close colleagues, saying of the oilmen that "all they do is hate."[23]

The Big Rich really were rich. By 1950, H. L. Hunt, who served as inspiration for the protagonist in the popular TV series *Dallas*, had surpassed Henry Ford to become the richest man in America. Several other Texas oilmen were just a few hundred million behind Hunt. They were a motley crew, dissimilar in many ways, but they were uniformly against regulation of private enterprise, and began donating heavily to politicians who shared this view.

Whereas before the Big Rich in Texas had concerned themselves only with oil regulation, which was primarily the purview of the Texas Railroad Commission, they began to take a keen interest in natural gas regulation as their profits soared during the postwar years. The natural gas business was growing up quickly, and was now nothing to sneer at. It was still an ugly cousin to oil, but had started pulling out of its awkward teenage years.

The maturation of the industry meant that more dollars were at stake, but this dynamic cut both ways. Producers had grown more vociferous, while at the same time consumers had grown more numerous. The Tennessee Gas Pipeline had been built, and the Inch Lines were being converted to gas from wartime oil service. Southern Natural and several other systems had also undergone major expansions, which brought natural gas service to new communities in new states. Every new consumer was a new vote in favor of regulation. FPC chairman Leland Olds was a dogged consumer advocate, which pit him squarely against the Big Rich.

Interstate Natural Gas Company produced its own gas and also bought from smaller producers in the Monroe field in Louisiana. It then sold to three interstate pipelines—United, Southern Natural, and Mississippi River Fuel—with which it shared some corporate ownership. Important-ly, Interstate's own pipeline system crossed into Mississippi before serving customers in Baton Rouge and New Orleans.

In 1940 the Louisiana Public Service Commission ordered an investi-gation into whether Interstate's rates were excessive. Interstate argued that it was not subject to Louisiana state regulation, since the gas was sold in interstate commerce.[24] (This is at least symbolically important, as many producer states tend to think that the northern consuming states insisted on full regulation, and decry the attempts of these "Yankee imperialists" to bleed the southern states of their natural resource wealth. As the *Interstate* episode shows, this characterization is not always accurate: in this case, a Louisiana producer asked Washington for protection from Baton Rouge.)

The company itself thus admitted both that it was an affiliated producer and that it made sales in interstate commerce. Interstate hoped to qualify for the production and gathering exemption, citing the *Columbian Fuel* decision and arguing that the FPC jurisdiction was limited to its transport operations.

The passage of time worked against Interstate. By the time the FPC issued a ruling, it was already 1943, and *Peoples* had established a prec-edent for regulating the upstream operations of affiliated producers. The FPC agreed with Interstate that it should fall under federal regulation but also considered its production and gathering assets as subject to regulation. Future FPC commissioner Scanlan wrote of the case: "Re-enforced by the clear-cut decision in the Peoples case, the Commission had crossed the Ru-bicon; the Columbian Fuel case was presumably buried by inference, no mention of it being found in the opinion of the Commission."[25]

Interstate appealed the decision, and in 1947 the case made it to the Supreme Court. By this time, the Supreme Court had already issued its *Hope* and *Colorado River* decisions, affirming that the FPC's jurisdiction included sales of gas by affiliated producers. Independent producers across the nation, meanwhile, were up in arms, accusing the FPC of having se-verely overstepped its authority. There was even panic among oil producers that they would come under cost-based regulation, as a large percentage of gas came from wells whose primary product was oil (by value, at least).[26]

Prodded along by their new financial backers in Texas, the Republican-led Congress introduced three bills in the House and one in the Senate in rapid succession to amend the NGA and explicitly exclude from the FPC's jurisdiction any sale made at or prior to the interconnection between the gathering lines and an interstate pipeline. The aim was to exempt production and gathering at both affiliated and independent producers.[27]

This marked the beginning of FPC schizophrenia (again in the sense of "split mind"). Its masters in Congress were growing angry with its regulation of production and gathering, but the courts had been, consistently, forcing its hand.

The FPC had been, for nearly two years, conducting an internal Natural Gas Investigation. It began releasing conclusions from the investigation between March and May 1947, where, in deference to the Republican, anti-regulation Congress, it suggested that *Columbian Fuel* was still very much valid, and that it would not regulate the price of gas received by independents, only affiliated producers.[28] In essence, the FPC was saying that it did not want to regulate producers, but that it was being forced to when they met the otherwise-a-gas-company standard.

To show it meant what it said, the FPC then quickly issued three rulings where it found that independent producers were not natural gas companies as defined by the NGA and therefore not subject to rate regulation.[29] One of these was Chicago Corporation, which had divested its stake in Tennessee Gas Pipeline throughout 1945. This showed very clearly that the FPC had heard, loud and clear, the complaints from Congress and industry, and that it was prepared at the very least to hew to the otherwise-a-gas-company standard and not regulate independents.

Why was this happening? Because the FPC is a bureaucracy, and as such derives its power from, and is, ultimately subservient to elected officials, who had become increasingly displeased with the agency after the *Interstate* decision. In the words of the Oklahoma Republican congressman Roscoe Rizley, in March 1947: "It occurs to me, however, that the FPC, like numerous other agencies and bureaus which have been for many years interpreting the laws enacted by the Congress in such a way as to give them the broadest possible power without restraint, merely wants to head off legislative action that would curb its power. . . . There is only one proper answer to all of this confusion, and that is to let the Congress write into law the standards and make sure the FPC carries out the Congressional mandate."[30]

Cognizant of the political firestorm it had caused, the FPC sought to calm the waters. It is no coincidence that all three of the FPC's rulings not to regulate producers were issued in a single week in May, a short time after Rizley's comments, even though the cases had been filed with the FPC years earlier, and years apart from one another. That same week in May, the FPC issued an official notice of proposed rulemaking whereby section 1(b) would heretofore be interpreted as exempting sales made by independent producers.[31] It was disclaiming some of the authority that the court had given it.

This is the natural way of bureaucracy, so should not be interpreted as a failing of the system. Indeed, under ambiguous circumstances, a bureaucracy should bend to the will of elected officials, until the circumstances are rendered unambiguous. And, as may be evident, the true fount of jurisdictional uncertainty under section 1(b) was not the FPC but the courts.

The judiciary, however, cared little about how its rulings were interpreted by the new, Republican Congress. On June 16, 1947, the Supreme Court issued its decision in the *Interstate* case and affirmed that the company's production and gathering assets were subject to FPC regulation.

The court employed a clear "regulatory gap" argument, which said that the purpose of the NGA was to regulate all those portions of the natural gas industry that the states could not, and that the purpose of the production and gathering exemption in section 1(b) was not "to free companies such as [Interstate] from effective public control" but rather "to preserve in the States powers of regulation in areas in which the States are constitutionally competent to act."[32] Once again, the Supreme Court's decision hewed to the dissent of FPC commissioner Scott in the *Columbian Fuel* case, in that it viewed the production and gathering exemption as applicable to the physical processes of production and gathering, rather than to any transaction that occurred, no matter how far upstream of the terminus of the gathering system. Therefore, states could decide things like production quotas, field rules, or conservation measures, and would have the ability to subject producers to construction or reporting standards, but could not set prices when the gas would eventually find its way to markets out of state.

The importance of the *Interstate* decision to future court and FPC rulings, particularly the landmark *Phillips* case, was twofold.

First, the court's wording in *Interstate* came to serve as a basis for dismantling the production and gathering exemption altogether, no matter whether the producer was independent or affiliated. The court wrote: "By the time the sales are consummated, nothing further in the gathering process remains to be done."[33] That is, sales do not qualify for the production and gathering exemption, as the point of sale is, by definition, after the point at which the production and gathering processes have ceased.

Second, in what can be seen as a parting shot to the FPC, the Supreme Court also wrote that it did not agree with or see the logic in the otherwise-a-gas-company standard, writing: "We express no opinion as to the validity of the jurisdictional tests employed by the Commission in [Columbian Fuel and Billings]."[34]

The court's decision in *Interstate* thus left open the possibility that, given a similar case involving an independent producer, it might well vote to classify it as a natural gas company.

## THE FIGHT GOES BACK TO CONGRESS

Whether the FPC, the courts, or both were acting wisely or foolishly was beside the point. It was clear that there was a disagreement about the production and gathering exemption in section 1(b), and that the only permanent solution would be to amend and clarify the law. The judicial branch, after all, exists to interpret the meaning of laws, not to create them. Many argue that the Supreme Court's decisions in both *Interstate* and later in *Phillips* were activist, and I do not wholly disagree. But it must be admitted that, unlike some of the more ethereal questions that the court is forced to ponder, the section 1(b) exemption could have easily been clarified with a very straightforward amendment; it was clear what the disagreement was about. The blame for the confusion must, in the end, be placed with the other two branches of government: legislative and executive.

The four bills that were reported to the House and Senate in 1947 were attempts to settle the disagreement. The one reported out for vote, just one month after the Supreme Court's *Interstate* decision, was the Rizley-Moore bill, sponsored by two Oklahoma Republicans: Congressman Roscoe Rizley and Senator Edward Moore. Perhaps unsurprisingly, given that both sponsors represented a producer state, this was the most producer-friendly bill of the four. Not only did it seek an independent producer exemption, it sought to exempt pipeline-affiliated producers from wellhead price regulation when it suited them, and allowed them to choose cost-based rates when it did not.[35] In other words, if the prevailing market price was lower than costs, affiliated producers could choose a cost-based standard—heads I win, tails you lose.

Presciently, the FPC had urged Congress first to amend the NGA to exempt independent producers, and submitted its own bill to Congress that would make official the otherwise-a-gas-company standard it had been using. It must have known that expanding the exemption to pipeline affiliates, as Rizley-Moore envisioned, would have a tougher time becoming law. To quote at length from the FPC's own Natural Gas Investigation: "In view of the present unsettled state of this matter, it is desirable, as the Commission has heretofore recommended, that the Congress should adopt appropriate amendatory legislation to make it clear that independent producers or gatherers of natural gas, and their sales thereof to interstate pipe lines, are not subject to the provisions of the Natural Gas Act. Such action will confirm what clearly appears to have been the original intent of Congress when it enacted the Natural Gas Act in 1938."[36]

Against the wishes of the FPC, the Rizley-Moore bill was reported and debated in the summer of 1947. As expected, it aroused emotions from both sides of the aisle. "Eugene Worley (D-TX): If this legislation is not passed, there is absolutely no limit whatsoever to how far the FPC will go in reaching out to get more and more powers which the Congress never intended them to have. Adolph Sabath (D-IL): The FPC is acting for the Government, for the protection of the American people."[37] Rizley-Moore passed the House in July 1947, but stalled in the Senate for several months, until congressmen went home for their holiday recess.

As Congress reconvened in January 1948, the FPC released its final conclusions from the three-years-long Natural Gas Investigation. Recall that, in initial conclusions released less than a year earlier, the agency had supported an independent producer exemption. Lawmakers must have expected to hear more of the same on returning from their holidays. Instead, they received two reports that were diametrically opposed from each other in substance, tone, and style. Each was written by two commissioners, and as the fifth commissioner's seat was open, neither represented a majority opinion. While the two commissioners who remained in favor of an independent producer exemption took an evidence-based, balanced tack, recommending a market price, but one that was "subject to necessary safeguards against abuse," the two commissioners who now favored universal wellhead price regulation, including Leland Olds, spoke in rhetorical terms, warning of the looming destruction of competition. Olds's report "made no pretense of scientific rigor or adjudicatory balance; it was a vigorous brief against big business, sounding the alarm that the foxes were about to raid the chicken coop."[38] It would come back to haunt him.

The views espoused by these latter two commissioners stirred enough emotion among consumer advocates that they effectively killed whatever chances Rizley-Moore had of passing the Senate, though in truth, the bill was probably too strongly anti-regulation to pass even without the fire and brimstone. It died in committee in April 1948.

It was not until the summer of 1949 that a more limited bill was reported. The Harris-Kerr bill sought only to exempt independent producers from FPC regulation, and presumably would have had little difficulty passing had it been considered instead of Rizley-Moore. But over the two years between the introduction of the Rizley-Moore and Harris-Kerr bills, Congress had become less opposed to producer regulation. For one, Democrats won back both the House and the Senate. But perhaps more importantly, new pipelines—TETCO chief among them—had introduced gas service to new constituencies, creating support for regulation where before there was neutrality.

The Harris-Kerr bill passed the House in August 1949, but prompted "a wave of protests" to the Senate from mayors of major cities in the consuming North and East.[39] The debates that followed in the Senate were intense, but the bill still managed to pass, narrowly, in March 1950, after an amendment was added requiring the FPC to report any unfair competition among producers to Congress and the president.

At long last, the legislative branch had rewritten and clarified the bedeviled production and gathering exemption, putting a smile on the face of independent producers across the country. The bill was sent to be signed by President Truman.

This time, it was the executive branch that stymied reform: Truman had been won over by those in favor of regulation. He vetoed the bill on April 15, 1950, saying, "There is a clear possibility that competition will not be effective, at least in some cases, in holding prices to reasonable levels. To remove the authority to regulate, as this bill would do, does not seem to be wise public policy."[40]

The confusion thus continued.

## THE *PHILLIPS* CASE

The case that would end the confusion was filed by the state of Wisconsin against Phillips Petroleum Company. On the surface, it seems very odd indeed that Wisconsin, which was still being served almost exclusively by manufactured gas at the time that it filed suit, came to be the prosecutor in the most important court case in the history of the natural gas industry. As it happened, Wisconsin found itself a bystander in a struggle between two powerful businesses, led by two ambitious and unyielding chief executives with two very different ideas about how the gas market should look. On the one side was William Maguire and Panhandle Eastern Pipeline (over which he now had full control, since Columbia had fully divested its stake). On the other was William Woolfolk and Michigan Consolidated Gas Company (MichCon), the utility serving greater Detroit.

The struggle was an odd one from the beginning, as MichCon was Panhandle's largest customer. Panhandle enjoyed its first year of profitability in 1936—the year it began sales to MichCon[41]—and its origin story was that of a pipe built to bring Southwest gas to Detroit, as the Insull–backed Natural Gas Pipeline Company of America (NGPL) had a lock on the Midwest's other great metropolis: Chicago. One would think this would have put Woolfolk in Maguire's good graces, but in fact the two men were increasingly bumping heads.

Because of the seasonality of Detroit's gas demand, MichCon was not purchasing its full contracted volumes, taking deliveries to the pipeline's capacity only during the winter, when demand peaked. This left Maguire with spare capacity, so, starting in 1940, he tried to sell gas directly to industrial consumers that MichCon was serving.[42]

The point of interconnection between a long-distance, high-pressure pipeline and a utility's local system—between transmission and distribution—is known as a city gate. The utility serves all of the customers "behind" its city gate and charges rates that are overseen by state public utility commissions. The transmission pipeline, meanwhile, is overseen by the FPC when it crosses state lines, or the state commission if it is entirely intrastate. Both the utility and the pipeline are allowed a return, meaning that customers behind the city gate are subject to "pancaked" rates: they must pay not only for the gas delivered by the pipeline to the city gate but then also pay their LDC for its costs, plus a return. For small consumers, such as businesses that do not use much gas, or individuals, the extra costs are well worth it: the utility takes care of maintaining infrastructure, ensuring system pressures remain within safety limits, and transacting with the pipeline on a daily basis. For large consumers, though, the cost savings that come from establishing a direct connection with a pipeline, and thus paying a lower rate, are often worth the extra efforts. It was these large consumers—such as Ford's giant manufacturing facility in Dearborn, Michigan—that Maguire was trying to sign.

MichCon viewed Panhandle's attempt to serve customers behind its city gate as a raid. Whereas the word *raid* conveyed both malevolence and corruption when it was used by the Power Trust in the 1920s, its use by MichCon in the early 1940s was both more just and justifiable. Utilities like MichCon, after all, were regulated monopolies, and the entirety of their costs was monitored by the state. The public utility holding companies of the 1920s, on the other hand, used their unregulated, interstate businesses to saddle their in-state utilities with inflated, out-of-state costs, beyond the reach of the utility commissions. Closing this regulatory gap was the entire reason for creating the NGA in the first place; according to the Commerce Committee, "The basic purpose of the present legislation is to occupy this field in which the Supreme Court has held the states may not act."[43]

But before we start feeling sorry for MichCon over the injustices that William Maguire sought to perpetuate, know that William Woolfolk's intentions were not so pure as simply defending the territorial integrity of his regulated utility. In fact, MichCon's president—a former "trouble shooter" for Samuel Insull[44]—had plans to run Panhandle out of Detroit completely and re-create a Power Trust–style, vertically integrated enterprise by building its own long-distance pipeline from the Southwest. In the fall of 1944 MichCon announced a new pipeline, the Michigan-Wisconsin line (Mich-Wisc, now known as ANR), that would run a path very similar to Panhandle's, except that it would fork just west and south of Chicago, with one leg

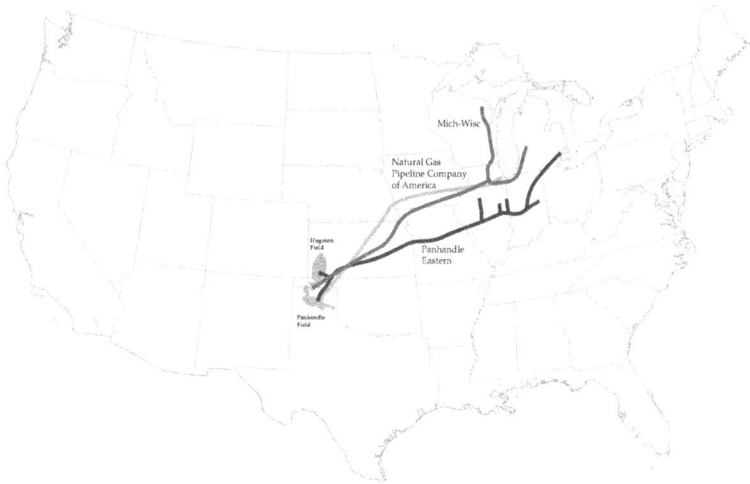

**FIGURE 6.1:** Panhandle Eastern, Mich-Wisc, and NGPL

running to Michigan, the other to Wisconsin. Mich-Wisc filed for an FPC certificate in February 1945.

Maguire intervened to try to stop construction of Mich-Wisc on the grounds that Panhandle was ready and able to serve all of greater Detroit's needs, and at lower cost than Mich-Wisc. This was a strong argument, as MichCon had not even been taking its present contract volumes, and section 7(c) of the NGA precluded the FPC from issuing a certificate to a new pipeline to a market that could be readily and more cheaply served by an incumbent. This was entry control at its most basic.

But several things worked against Maguire that caused the FPC to approve Mich-Wisc. First, Mich-Wisc would also serve Wisconsin, which was at that point a virgin market, still relying almost wholly on manufactured gas. Second, Detroit-area demand was growing rapidly, and it was quite clear that a great deal more capacity would be needed over the coming years. Third, MichCon argued that while it may not have been taking its full volumes all year round, it needed more gas than Panhandle could deliver in the winter. This last reason was technically true, but only because of Woolfolk's duplicity: MichCon had been delaying the completion of a storage field that it had bought all the way back in 1941 that would have allowed it to take more gas from Panhandle in the summer and deliver more behind its gates in the winter.[45] The FPC approved Mich-Wisc's application in November 1946.

With that background, we now arrive at how Wisconsin came to file a suit against Phillips Petroleum, an independent producer that owned neither interstate pipelines nor in-state utilities, leading to a Supreme Court decision that would, over the coming decades, bring the entire industry to its knees. As previous pages have hopefully illustrated, however, Wisconsin did not begin this process—Congress did—in passing a poorly worded piece of legislation and then bungling attempts to amend it.

Phillips was the largest independent gas producer in the United States and the sole supplier to the Mich-Wisc pipeline, to which it had dedicated over six hundred thousand acres in the Panhandle field.[46] After receiving its FPC certificate, steel shortages and poor weather caused several months of delays in construction, forcing Mich-Wisc to renegotiate the terms of its contract with Phillips.[47]

With the benefit of hindsight, what Phillips did next was exceedingly stupid. Its actions provided ammunition to those who favored regulating the price received by producers, arguing that the NGA was toothless if monopolistic producers were left free to charge whatever they thought they thought their captive customers would pay.

Phillips demanded a 70 percent price increase. Outraged, the city of Detroit and the state of Wisconsin quickly filed a motion to have the FPC classify Phillips as a natural gas company, thereby subject to rate regulation. The FPC took up the matter in October 1948.[48]

As it reviewed the *Phillips* case over 1949 and 1950, the FPC was being actively and repeatedly chastened by Congress, which criticized its regulatory overreach on many fronts, but especially on meddling in the affairs of independent producers. Behind the voices of congressmen like Roscoe Rizley and senators like Lyndon Johnson were clearly worded directives from the Texas Big Rich, who were about to win their most substantial political battle yet: getting rid of that no-good communist, FPC chairman Leland Olds.

Lyndon Johnson was, in his own words, "born into politics"—Texas politics, to be more precise. His father, his paternal grandfather, and his maternal grandfather all served in the Texas legislature, which is how he came to be close with Sam Rayburn. What catapulted Johnson onto the national stage, and eventually to the presidency, however, were his relationships with the Big Rich. In Washington, Johnson came to be recognized as a fundraiser without peer due to his status as the sole conduit to the vast reservoirs of Texas oil money. The act that solidified Johnson's standing among the oilmen was the smear campaign against Leland Olds that he led in September 1949.

Until that moment, Johnson was not fully trusted by the arch-conservative Texas oil crowd. He had, after all, ran his first congressional race on a New Deal platform, and his mentor, Sam Rayburn, was on only lukewarm terms with the Big Rich. Johnson went onto prove that he was neither as ideologically nor morally encumbered as Rayburn, and willing to pay back those who had funded his political rise, when he ambushed Olds in a routine reconfirmation hearing, painting him as a communist unfit to serve in any government office. The timing could not have been worse for Olds: the Soviet Union had, only a month prior, detonated their first atomic bomb, and Communist Party chairman Mao Zedong had only just established the People's Republic of China as a Communist state. The Red Scare was in full effect, and the fact that Olds did lean socialist on many issues may have been enough to doom his reconfirmation. Johnson left nothing to chance, though, and pulled out all the stops, quoting radical passages that Olds wrote all the way back in the 1920s, which he borrowed from the House Un-American Activities Committee and took completely out of context, and calling witnesses to testify that Olds was a flip-flopper on the question of whether independent gas producers should be subject to price regulation.[49] (This last part, at least, was relevant to the discussion at hand.)

Olds's liberal supporters were caught totally off guard and could do nothing to beat back Johnson's successful propaganda campaign. Olds lost the reconfirmation vote in the full Senate by a wide margin, and a new FPC commissioner was installed who took a much more pro-business approach to natural gas regulation (assuming that business was producing gas).

In sum, by 1950, the FPC had come around to the argument that independent producers, such as Phillips, should be exempt from rate regulation. That said, since Truman had vetoed the Harris-Kerr bill, the law did not guarantee that exemption. The commission would therefore need to make case-by-case decisions.

After almost three years of deliberation, during which time the Mich-Wisc line began to flow gas to customers in Michigan and Wisconsin,[50] the FPC issued its decision in August 1951: Phillips was not a natural gas company, and hence was not subject to rate regulation.

The Big Rich and producers across the United States breathed a collective sigh of relief. But the FPC—knowing that the decision would be excoriated by the city of Detroit, the state of Wisconsin, and others interested in producer regulation, and still unsatisfied by the unclear regulatory framework in which they were operating—"encouraged dissatisfied intervenors to seek redress through the judicial system," which they did.[51]

Wisconsin's attorney general, Vernon W. Thomson, wrote to one of Wisconsin's senators:

> It looks to me as though a gross deception has been practiced upon the people of Wisconsin by the Michigan-Wisconsin Pipe Line interests…Essentially, the fact as I see it is that Wisconsin is being deprived of its right to natural gas at a fair price, not only because of the tight pipe line monopoly here, or be-

cause of seeming lack of arm's length dealing for natural gas from the Phillips interests, but principally because [MichCon] are using the gas consumers of Wisconsin as mere tools or weapons in their fight to stand off competition [in Panhandle].[52]

Thomson's argument holds water. While it cannot be confirmed, it is likely that MichCon drew up Mich-Wisc's Wisconsin leg in order to serve a market not touched by Panhandle, thereby forcing the FPC to issue a certificate. Clearly, Detroit was the pipeline's main market. And its dealings with Phillips could hardly be called arm's-length transactions: an earlier incarnation of the FPC, less chastened by Congress, may well have ruled that Phillips's dominance in its vicinity of the field qualified it as a natural gas company, subject to rate regulation.

Wisconsin appealed the FPC's decision, and in January 1943 the case was heard by the DC Circuit Court. After several months of deliberation, it reversed the FPC's decision and ruled that Phillips was a natural gas company.

The DC Circuit Court used a regulatory gap argument, saying that the states had jurisdiction over the physical processes of production and gathering, but not over "the sale in interstate commerce of natural gas for resale." It cited the basic purpose of the bill that became the NGA: "To occupy this field in which the Supreme Court has held that the States may not act." It further argued that the Supreme Court's decisions in *Interstate* and *Colorado Interstate* invalidated the FPC's argument that Phillips qualified for the production and gathering exemption in section 1(b) because its sales were "closely related" to its gathering business. If the Supreme Court had decided that both Interstate and Colorado Interstate qualified as natural gas companies, then the production and gathering exemption meant nothing, or at least could not serve as the basis on which to regulate.[53]

This caused consternation for some obvious, and some not-so-obvious, reasons. Clearly, it was extremely negative for Phillips, which now found itself under FPC regulation, and of course appealed the circuit court's decision, along with the states of Texas, Oklahoma, and New Mexico. Other producers were frightful that they, too, would have the federal government regulate the prices they received. And while the FPC did not have profits at stake, it was also dissatisfied with the decision because it applied only to Phillips and provided no clarity as to how other producers should be treated. It was thus with applause that both producers and FPC officials greeted the news that the Supreme Court would hear the appeal. One FPC official said: "The appeals court's ruling was a horrible decision because it didn't give us any guide posts. Now we can hope for a ruling that'll give us some ground rules."[54]

*Phillips Petroleum Co. v. Wisconsin* was heard by the Supreme Court in April 1954 and decided in June 1954. It was hoped that this decision would settle once and for all the matter of who the FPC could regulate.

It did. The court affirmed the DC Circuit Court's decision and extended it, dealing a death blow to the FPC's otherwise-a-gas-company standard, and reinforcing the opinion first expressed by Commissioner Scott and then used in *Interstate* that the production and gathering exemption was, for all intents and purposes, meaningless. The Supreme Court in its *Phillips* decision made all producers of natural gas sold in interstate commerce subject to FPC rate regulation, no matter how small. After processing 1,600 pipeline certificates since the passage of the NGA, sixteen years earlier, the FPC received 6,000 applications from independent producers in the twelve months following *Phillips*.[55]

Justice Sherman Minton wrote the majority opinion. He was a close personal friend of President Truman,[56] who had vetoed the Harris-Kerr bill because he felt that the public interest required producers, independent or not, to be subject to regulation.[57] Perhaps the friendship influenced his opinion in *Phillips*, which seems somewhat at odds with his legacy as a Supreme Court justice who tried to ascertain and uphold original intent.[58]

Minton first tore apart the otherwise-a-gas-company standard: "Petitioners attempt to distinguish the Interstate case on the grounds that the Interstate Company transported the gas in its pipelines after completion of gathering and before sale, and that the Interstate Company was affiliated with an interstate pipeline company, and therefore subject to Commission jurisdiction in any event. This Court, however, refused to rely on such refinements, and instead based its decision in Interstate on the broader ground that sales in interstate commerce for resale by producers to interstate pipeline companies do not come within the 'production or gathering' exemption."[59] Minton continued, echoing President Truman's view:

> Regulation of the sales in interstate commerce for resale made by a so-called independent natural gas producer is not essentially different from regulation of such sales when made by an affiliate of an interstate pipeline company. In both cases, the rates charged may have a direct and substantial effect on the price paid by the ultimate consumers. Protection of consumers against exploitation at the hands of natural gas companies was the primary aim of the Natural Gas Act. Attempts to weaken this protection by amendatory legislation exempting independent natural gas producers from federal regulation have repeatedly failed, and we refuse to achieve the same result by a strained interpretation of the existing statutory language.

Two dissenting opinions were written. The first was from Justice William O. Douglas, who had concurred with the majority in both *Hope* and

*Colorado River*, supporting the FPC's decision to use actual costs as a basis for rate making. In those cases, Douglas acknowledged the FPC's expertise, and wrote that any claim that the commission had acted unjustly or unreasonably would carry a high burden of proof. These earlier views, combined with his dissent in *Phillips*, shows that Douglas preferred to let the FPC, as the expert agency, handle matters as it saw fit.

> The legislative history is not helpful. Congress was concerned with interstate pipelines, not with independent producers. . . . If one can judge by the reports of the Federal Trade Commission that preceded the Act, and the hearings and debates in Congress on the bills that evolved into the Act, little or no consideration was given to the need of regulating the sales by independent producers to the pipelines. The gap to be filled was that existing before the pipelines were brought under regulation. . . . That was the view of the Commission in a decision that followed on the heels of the Columbian Fuel Corp. That decision exempted from regulation an independent producer to whom Phillips is in all material respects comparable. It was a decision made by men intimately familiar with the background and history of the Act.
>
> That construction by the Commission . . . is entitled to great weight.

Douglas did not necessarily oppose the majority's decision in *Phillips* on the grounds that it went against Congress' original intent but rather because he felt that the court, and the federal government in general, lacked the expertise required to regulate producers: "There is much to be said in terms of policy for the position of Commissioner Scott, who dissented the first time the Commission ruled it had no jurisdiction over these sales. But the history and language of the Act are against it. If that ground is to be taken, the battle should be won in Congress, not here. Regulation of the business of producing and gathering natural gas involves considerations of which we know little, and with which we are not competent to deal."

The second dissent, written by Justice Thomas Clark, was perhaps more poetic than Douglas's, and focused more on original intent. Clark argued that Phillips was a "pure-play" producer, which the NGA had not meant to regulate, and that the majority's decision had stripped the production and gathering exemption completely of its meaning. The following excerpts are taken from Clark's dissent, but are not continuous: some paragraphs have been removed for brevity's sake.

> The natural gas industry, like ancient Gaul, is divided into three parts. These parts are production and gathering, interstate transmission by pipeline, and distribution to consumers by local distribution companies. A business unit may perform more than one of these functions—typically, production and gathering in addition to interstate transmission. But Phillips' natural gas operations

are confined exclusively to the first part—production and gathering. It has no interstate transmission or high pressure trunk lines, and does not sell to distribution companies—and it does not, of course, distribute to the ultimate consumer.

The Federal Trade Commission did not find abusive pricing by independent producers and gatherers; if anything, the independents at the producing end of the pipelines were likewise the victims of monopolistic practices by the pipelines.

By today's decision, the Court restricts the phrase 'production and gathering' to 'the physical activities, facilities, and properties' used in production and gathering. Such a gloss strips the words of their substance. If the Congress so intended, then it left for state regulation only a mass of empty pipe, vacant processing plants, and thousands of hollow wells with scarecrow derricks, monuments to this new extension of federal power.

So ended the battle over producer regulation. There was no more regulatory gap: every molecule of gas that crossed state lines was now subject to FPC rate regulation, from the wellhead to the city gate.

# CHAPTER 7

## STOPPING A FREIGHT TRAIN

The political reaction to the Supreme Court's 1954 *Phillips* decision was predictably partisan. Lawmakers from the consuming states were gleeful, while representatives from the producing states were livid. The decision's immediate impact on the gas market, however, was muted, if not entirely invisible. The market was in the midst of its fastest-ever expansion, growing by 6 percent per year throughout the 1950s, such that total consumption in 1960, at thirty-three billion cubic feet per day, was double what it had been a decade earlier. Growth did not end until 1972, by which point consumption had almost doubled again.[1]

This is not to say that *Phillips* was benign. It was a landmark decision that redistributed hundreds of billions of dollars. The first wealth transfer was from producers, who had their prices capped, to consumers. A second-order effect of *Phillips* was the development of two, parallel gas markets— the interstate market and the (federally) unregulated intrastate market. This led to a second redistribution of wealth, this time among consumers. Over time, shortages developed on the interstate market, which prompted energy-intense industries to leave their traditional homes in Appalachia, the Midwest, and the Northeast to relocate to Texas and Louisiana, where they could more easily source gas from intrastate pipelines. This second redistribution brought billions of dollars of investments back to producer states, and, ironically, may have meant that *Phillips* ended up inadvertently benefiting them over consumer states, over the very long term.

*Phillips* directly contributed to the energy bedlam of the 1970s and 1980s, which, in rapid succession, saw gas shortages so severe that they put a noticeable dent in overall US GDP growth, then a surplus so large that it crippled and in some cases bankrupted producers and pipelines. But there were two decades between the *Phillips* decision, in 1954, and the first major shortages, which began in the 1970s. Can a decision truly be called monumental if its effects are not felt for so long a period?

The answer is yes, but only because there were several mitigants, all of which forestalled a crisis. *Phillips*'s delayed impact can be explained by these mitigating factors, which are listed in their order of importance.

1. *Phillips* ordered the FPC to regulate wellhead prices in 1954, but the agency did not even begin until 1960, had not come to any final determinations until 1965, and did not finish its initial caseload until 1973.
2. The FPC's initial prices were not set unduly low—they were near market levels. Over the years, however, they did not rise enough to keep supply growing at pace with demand.
3. Oil production rose rapidly over the period, bringing substantial volumes of both associated and nonassociated gas to market. This was due to:
   (a) Tremendous growth in offshore production after the Tidelands Act of 1953 defined who held authority over submerged lands—the federal government, or the states;
   (b) The imposition of oil import quotas in 1959, which closed the US market to cheaper Middle Eastern oil and kept US prices higher than they otherwise would have been.
4. In the late 1960s, pipelines began to draw down their reserves at a faster than anticipated rate, effectively stealing future gas for present needs.
5. A direct sales loophole in the NGA allowed some—although very few—producers and consumers to contract directly and avoid FPC wellhead price regulation, even across state lines.

*Phillips* did not prevent the FPC from adjusting gas prices to match market conditions, but it made the procedure time-consuming, backward-looking, and subject to short-term political concerns. The FPC seemed always to be playing catch-up. When gas shortages did finally materialize in the fields, in the late 1960s, they were compounded by economy-wide inflation, which averaged 4.1 percent over 1967–1970, versus just 1.7 percent from 1955, the year after *Phillips*, through 1966. This made the FPC's job all the more difficult, as it was trying to hit a rapidly moving target.

The demand side also contributed to the crisis. Even as rates rose and pipelines bought more expensive supplies, these volumes were rolled in to the weighted average cost of gas, blunting the effect on consumers and thus delaying any meaningful demand-side response to the shortage. To make matters worse, an OPEC–imposed oil export ban in late 1973—the first global oil shock—sent oil prices skyrocketing, making gas much more economical by comparison and increasing demand further.

In sum, *Phillips* meant that the gas market would have no self-correction mechanism if and when it transitioned from supply-push—a buyer's market—to demand-pull—a seller's market. But neither was the ruling enough to cause such a transition in anything like the short term. In the wake of *Phillips*, US natural gas regulation tended to push the market toward shortage and made regulators poorly suited to deal with such shortages when they occurred, but this was not enough stop the freight train that was US oil and gas production over the following years.

## FEDERAL WELLHEAD PRICE REGULATION, OR A LACK THEREOF

Democratic President Truman, who had vetoed the Harris-Kerr bill—which would have amended the NGA and exempted independent producers from price regulation—was replaced by Republican Dwight Eisenhower in January 1953. Eisenhower, who espoused mostly free market principles, was in favor of an independent producer exemption from wellhead price regulation under the NGA, and appointed FPC commissioners who shared this view.

It was certainly the case that the Big Rich had Eisenhower's ear. While he wavered whether to run for the presidency after the war, it was Texas oilmen who persuaded him most ardently, seeing in Eisenhower a lion of laissez faire economics. In fact, Sid Richardson, one of the wealthiest Texas oilmen and a close friend of Eisenhower's, was the first civilian to learn that the former general would, definitively, stage a run.[2]

Presciently, Eisenhower's FPC warned advocates of regulation—mostly from the consuming states in the Midwest, Appalachia, and Northeast—that federal price controls would eventually cause the industries on which their communities relied to move elsewhere in search of gas.[3] (This, of course, happened when energy-intense industries relocated en masse to Texas and Louisiana to secure gas more readily available on the intrastate market.)

The Eisenhower administration obviously had no control over the *Phillips* decision, but could press for amendatory legislation, like the failed Harris-Kerr bill, that would render the decision moot. When such a bill passed the House and Senate in early 1956, it seemed a near certainty that Eisenhower would sign it into law. It came to light, however, that an oil company representative had offered a senator from South Dakota a $2,500 campaign contribution in exchange for his vote. Weary of the appearance of impropriety, Eisenhower vetoed the bill, even though he was "in accord with its basic objectives."[4]

Incredibly, producers shot themselves in the foot again the next year, when a Texas independent oilman planned an "appreciation dinner" for the Republican congressman who was drumming up support for yet another bill, this time one that would have allowed the FPC to use a market-based pricing standard. The dinner plans were circulated in the press before a vote was taken on the bill, and it quickly died due to the negative publicity.[5]

Despite the failure of both of these last-ditch attempts at amendatory legislation, the Eisenhower years saw the FPC do very little in the way of regulating producer prices, which suited both the president and his appointees at the agency. The political climate during the age of Eisenhower put

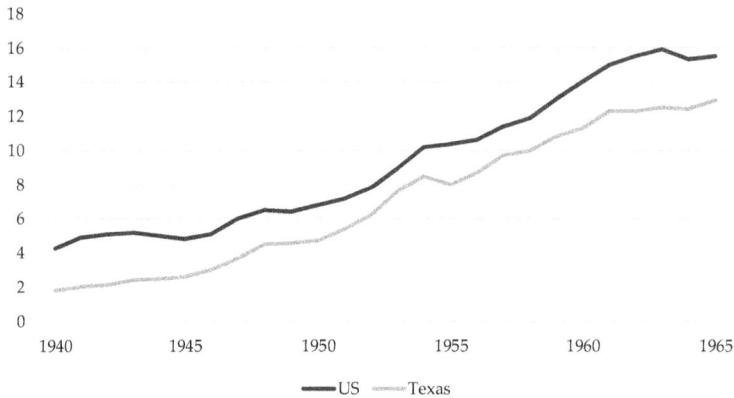

**FIGURE 7.1:** Wellhead natural gas prices, 1940–1965 (cents per thousand cubic feet)

the FPC under no pressure to move quickly to regulate wellhead prices. But even if it had, it faced a daunting logistical challenge. In the one year following the *Phillips* decision, it received over ten thousand rate filings by producers.[6] Every sale by every producer now needed approval.

For the six years following *Phillips*, the FPC appeared essentially para- lyzed with regards to wellhead price regulation. It seemed to many that the agency was shirking its obligation to regulate producer prices, waiting for legislation that would relieve it of this responsibility. The backlog of cases built and built, and by 1960 the agency itself forecast that it would not finish its present caseload until 2043.[7] During this time, field prices were rising, with producers receiving fourteen cents per thousand cubic feet in 1960, up from ten cents in 1954.

After eight years of the Eisenhower administration propagating a fairly hands-off approach to private enterprise, John F. Kennedy assumed office in January 1961, intent on regulating business more vigorously.

Reenter James Landis, whom you may recall as the SEC chairman that quietly but firmly took down several powerful public utility holding com- panies after PUHCA was passed in 1935. After the SEC, Landis was dean of Harvard Law School, and occupied various high-ranking public service and academic roles thereafter. He had also, over his long and prestigious career, grown close with the Kennedy family: it was John's father, Joseph Kennedy, who had preceded Landis as SEC chairman and who at one point was the largest single client of Landis's private law firm.[8]

On being elected president, Kennedy asked Landis to prepare a report on regulatory agencies, which he did in late 1960, while Kennedy was still president-elect. In his report, Landis levied criticism against the United States' entire regulatory complex, which was, in his words, plagued by long backlogs of unresolved cases, a lack of leadership, and inefficient and ineffectual procedures for enforcing current legislation. It had nobody with the expertise, courage, or creativity required to formulate new policy. In other words, the agencies had been left to fester for a decade or more, starting even before Eisenhower was elected, and had become less than useless. The Landis report would form the basis of the Kennedy administration's regulatory agenda.

His criticism was not at all limited to the FPC, but Landis reserved a special contempt for that agency, saying: "The Federal Power Commission without question represents the outstanding example in the federal government of the breakdown of the administrative process. The complexity of its problems is no answer to its more than patent failures. . . . [Its] defects stem from attitudes, plainly evident on the record, of the unwillingness of the Commission to assume its responsibilities under the Natural Gas Act and its attitude, substantially contemptuous, of refusing in substance to obey the mandates of the Supreme Court of the United States and other federal courts."[9]

After being sworn in, Kennedy hired Landis as a White House special assistant. In this new role Landis continued to lambaste the FPC and its eighty-year backlog of producer rate proceedings, going so far as to accuse the agency of deliberately overcomplicating its duties, presumably with the intent of spurring Congress to pass amendatory legislation exempting producer price regulation: "The incidence of this problem of rate backlogs could largely have been avoided. The FPC, however, trod water by removing federal controls over rates of natural gas producers. After this was defeated, I believe that the FPC deliberately sought to prove that rate regulation of natural gas production was an administrative impossibility."[10]

There is both truth and exaggeration in Landis's critique. It was indeed the policy of the Eisenhower FPC to issue certificates without rate conditions. In an opinion issued in 1958 regarding rates proposed by Hope Natural Gas, and reiterated several times thereafter, Eisenhower's FPC said that a sales contract was proof enough that there was a market for the gas, and that the commission would issue a certificate at the contract price unless it was presented with evidence that such a price would adversely affect the public convenience.[11] Thus, without disclaiming its power to review rates and order them set at lower levels, the Eisenhower FPC essentially said that it would not conduct full cost investigations for each application that it received—that if both parties had agreed to a contract price, then it could be said to be just and reasonable. This policy remained in effect until the Supreme Court issued a ruling in June 1959 in the so-called CATCO case, which required the commission to investigate all prices that were materially higher than other contracts in the area.

In 1957 Continental Oil Company, Atlantic Refining Company, Tidewater Oil Company, and Cities Service Production Company—collectively known as CATCO—began an application to sell Tennessee Gas Pipeline a whopping two trillion cubic feet of gas from a large block of acreage off the Louisiana coast, which would be the largest volume ever committed in one sale. The agreed upon rate was 22.4 cents, which was significantly higher than the company had ever before paid, and also higher than contract prices in southern Louisiana. The high rate, plus the enormous volume, guaranteed that extra attention would be paid to this case, and paid it was.

Four of Tennessee Gas's utility customers intervened, along with the New York Public Service Commission, contending that the FPC could not grant a certificate for application with such a high rate without a full cost investigation. The commission initially agreed with the intervenors, and offered to grant only a temporary certificate at the agreed upon price of 22.4 cents. It would then conduct a full cost investigation, which would presumably take a year or more, before issuing a permanent certificate, potentially with a lower price.

CATCO was unimpressed. The companies said they could not commit such large sums of capital without a permanent certificate and threatened not to sell to Tennessee at all and instead seek intrastate buyers within Louisiana. The FPC quickly changed its position and sided with CATCO, defending its about-face by arguing that such a large sum of gas was worth the higher price.[12] The case was eventually appealed to the Supreme Court, which ordered the FPC to conduct an investigation before issuing a permanent certificate, writing that "the fact that prices have leaped from one plateau to the higher levels of another, as is indicated here, does make price a consideration of prime importance."[13]

This established what became known as the "in-line" standard, which meant that the FPC would be required to examine prices thoroughly when they were not in line with existing prices in the same area.

These episodes evince the truth in Landis's critique of the Eisenhower FPC; namely, that the commission deliberately tried to avoid considering price when considering new applications until the Supreme Court compelled it to do so. However, another case begun during the Eisenhower era that shaped FPC rate-making policy was under way even before CATCO and Tennessee Gas signed their initial deal, which does not support Landis's view that the FPC was shirking any and all responsibility for regulating producer prices.

Since 1956 the FPC, Phillips, the Wisconsin Public Service Commission, and several eastern LDCs had been working to establish a cost-based rate in the "second Phillips case." In its landmark 1954 *Phillips* decision, the Supreme Court had ruled that Phillips was indeed a natural gas company, subject to rate regulation. But the court itself did not set a rate. That task remained the FPC's responsibility.

The second Phillips case dragged on for four years with little success. The competing groups each calculated their own "just and reasonable" price with expected results: consumers arrived at the lowest prices, producers at the highest, and the FPC in the middle.[14] Even after the years-long effort, no consensus emerged on how properly to account for the economic subsidization of gas when it was produced in association with oil; how and whether to include the cost of unsuccessful exploration wells; and, even if a cost base could be agreed on, what levels of return to apply to reflect the high risk that producers shouldered.

In September 1960, just a few months after *CATCO* and just a few months before Kennedy took office and Landis wrote his scathing report, the FPC terminated the second Phillips case without reaching a conclusion and immediately began developing a more practical regulatory framework. It had decided—a better word might be *conceded*—that producer-specific price regulation was simply unworkable, largely due to the failure of the second Phillips case. That, combined with the in-line standard that the Supreme Court had created in *CATCO*, led the FPC to begin formulating "maximum area rates," specific to a region, rather than a single producer, or even a single field. This was meant to be a way out of the regulatory morass, and showed that the FPC recognized that the Kennedy administration would not tolerate continued inaction, no matter the logistical challenges.

First, the FPC defined geographically contiguous production areas. Then it froze prices in each of these areas for all new gas dedicated to interstate pipelines after January 1, 1961, at 1958–1959 levels. It then began a detailed cost study in its first area rate proceeding, for the Permian Basin of West Texas and southeast New Mexico, which in 1960 produced almost 20 percent of the nation's crude oil and 10 percent of its gas.[15] The proceeding endured for five years before the FPC issued a decision in 1965, in which it established a two-tiered rate system.

Importantly, the tiers established in the *Permian* decision showed that the FPC recognized both the essential function of the gas price as an incentive to explore for new reserves and also that associated gas was subsidized by oil. The lower tier was an attempt to withhold windfall profits from producers whose gas would have made it to market anyway, either because it was a byproduct of oil or had already been discovered and developed. This pleased consumer advocates.

The higher price tier, of 16.5 cents per thousand cubic feet, was given to "new gas well gas." This repetitive-sounding name indicated two features of the gas in question. Because it was "gas well gas," it came from a well

Price

Higher

These fields are marginal, generally smaller, and require a high price to produce

These gas fields require different prices for operators to break even, depending on how productive each well is, and how much they cost to drill

These will produce at very low prices - basically, the cost of installing gathering systems

These fields will produce even at negative prices, because it is worth it to take a loss in order to produce oil

Lower

Less

More

Quantity

Price

The marginal field breaks even but makes no profit

Fields to the left of the clearing price will make profits

Fields to the right of the clearing price will not be drilled

Demand

P*
Clearing price

Producer profit

Q*
Clearing quantity

Quantity

Price

Lower tier reduces windfall profit to producers who would have produced anyway

Higher tier incentives the marginal resource to produce

Quantity

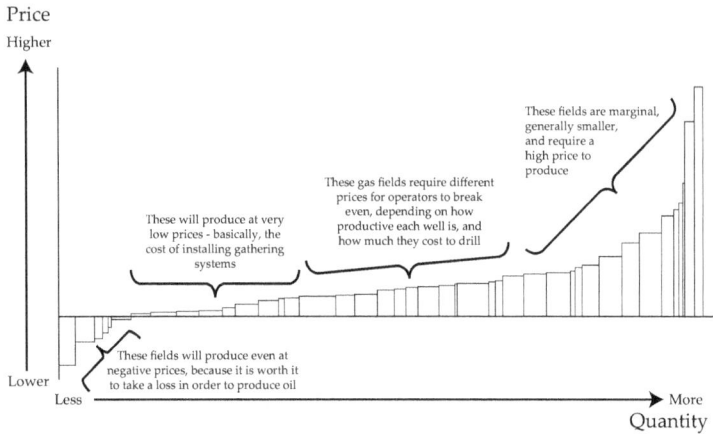

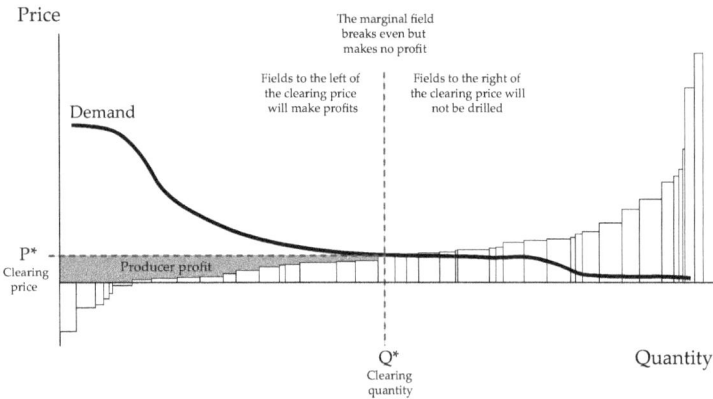

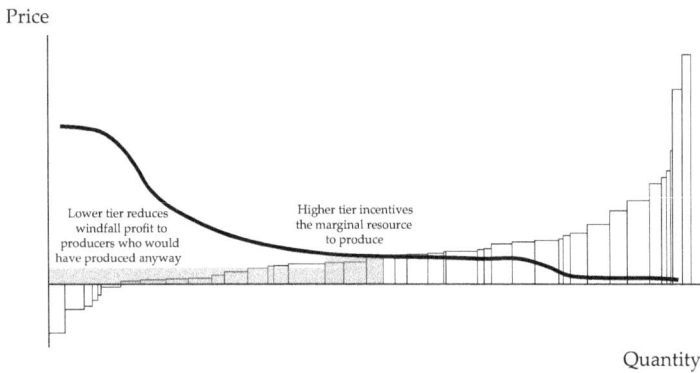

**FIGURE 7.2:** Supply and demand charts, and simplified explanation of FPC two-tier area rate policy

that produced relatively little oil. Because it was "new," it came from a well brought online after January 1, 1961. Meanwhile, both "old gas well gas" and associated gas of any vintage could receive a maximum of only 14.5 cents.[16]

Essentially, the FPC's two-tiered system tried to set the higher tier at the market clearing price and the lower tier at the weighted-average cost of associated and legacy production. In giving up on producer-specific rates, it instead tried to guess at the general shape of the "supply stack" in a given producing area.

This is an important economic concept. Economics 101 tells us that the supply and demand curves are two lines that slope in different directions. As prices move up from zero, supply increases and demand drops. The clearing, or equilibrium, price and quantity is where the two lines meet.

In reality, both the supply and demand functions are very oddly shaped and are not constant over time. Focusing on the supply side, each well has its own, unique cost of drilling and produces its own quantity of oil and gas. Assuming, for simplicity's sake, that wells within a given field have similar costs and produce similar amounts of oil and gas, if we calculated the price at which producers would "break even" at each field, and then plotted each field on a chart, we would have a supply stack that might look something more like figure 7.2. The height represents the breakeven price and the width represents how much gas the well will produce.

Then we overlay a demand line—less simple than a straight line, but not as complicated as the real world. The concept is now the same as in Economics 101: the intersection is the market clearing price and quantity, and the last well needed to satisfy market demand is referred to as the marginal well. The marginal well sets the price, and the producer who drills it breaks even but does not reap any profit. In a perfect economic world, all wells to the right of the intersection, whose breakeven prices are higher than the market clearing price, are not drilled, while all wells to the left of the intersection earn profits.

The FPC's higher tier was its guess at the clearing price, and thus represented its attempt to encourage new gas supplies to come online to meet market demand. The lower tier tried to reduce profits from all wells that cleared below this price, and thus represented its attempt to save consumers from paying producers more dollars than were needed to keep up with demand.

The Supreme Court noted the FPC's attempt to mimic the market clearing price when it upheld the *Permian* decision in 1968: "The FPC found that price could be an incentive for exploration and production of new gas well gas, while supplies of associated and dissolved gas and previously committed reserves of gas well gas were relatively unresponsive to price variations."[17] As important as methodology was the number itself: the higher tier in *Permian* was in line the going market rate, which signaled that the FPC was not seeking to set prices at confiscatory levels. The unregulated, intrastate price of gas in the Permian Basin was 17 cents in 1966 and 16.3 cents in 1967.[18] Indeed, the Coastal States Gas Corporation, which operated in intrastate commerce in Texas and so was not subject to FPC wellhead regulation, in 1962 signed twenty-year contracts with the cities of Austin, San Antonio, and Corpus Christi at a fixed price of 20 cents.[19] This was a delivered price, and the company turned large profits during the first years of these contracts, implying that its average purchase price was well below 15 cents. As demand rose and Coastal was forced to purchase new supplies, it found that it could do so in the Permian Basin for between 16 and 20 cents as late as 1971.[20]

The economist Paul MacAvoy takes a different view, suggesting that the FPC's rates in the *Permian* case and for other producing regions were set too low. The study of regional costs that it used to calculate the rates was conducted over the course of the price freeze that the agency instituted beginning January 1, 1961. During this freeze, MacAvoy argues, producers would only explore for and develop reserves that would be profitable at or below these frozen prices. This led to circularity: the price level before the freeze determined the area rate because the costs that the study examined were spent developing acreage that producers had decided would be profitable at the frozen prices.[21] While the logic is sound, it does not explain why intrastate prices were in line with the FPC rates. And no one would argue that the issue that led to shortages years later was not that the initial rates were too low by a few cents, but that they did not rise by the many cents—and, eventually, dollars—that the growth in demand required.

After the Supreme Court approved the *Permian* decision, the FPC began to formulate area rates for other major producing regions. By 1973 every producing region of the country had maximum area rates, or else fell under a national price cap,[22] and the FPC seemed finally to have landed upon a coherent, practical, and judicially approved framework for regulating producer prices.

## STILL SECOND PRIZE

It is worth elaborating on the relationship between oil and gas in the post–*Phillips* era, especially because such analysis is often absent from accounts of the gas industry. Indeed, while there is a great deal of literature on both the causes and effects of *Phillips*, many accounts seem to fast-forward from 1954 to the shortages of the 1970s without explaining why it took almost two decades for shortages to form. When explanations are given, attribution is thrust mostly on the FPC's long delay in actually setting prices, after which point the narrative moves on to explain the measures that the pipelines began taking in the late 1960s to increase their gas supplies, sometime by unorthodox means.

This narrative is not incorrect. Indeed, as I have detailed, the lag between *Phillips* and *Permian* and later rate cases is the most important reason that shortages did not crop up sooner. But nor is it complete. The gas market grew up in the shadow of the more profitable and strategically important oil market, and had not stepped out from under it by the time the court ruled in *Phillips* or the FPC set its rates for the Permian Basin. Some would say it still has not.

Oil generally comes out of a well along with gas. The proportions vary widely, but while only a minority of the new gas brought to market in the United States over the 1950s and early 1960s was truly associated with oil—that is, came from wells that produced mostly oil—a vast majority came from gas fields that had been found in the search for oil.

Investment into exploration and development was directed overwhelmingly toward oil during this era—very few producers, and even fewer independents—went looking for gas.[23] This meant that, while not all new gas was physically a byproduct of oil, as some of it came from wells that produced nothing but gas, almost all of it was economically a byproduct of oil, as the initial capital had been sunk in hopes of finding oil. The historian Richard Vietor summed it up well while detailing some of the problems that the FPC faced in the second Phillips case of the late 1950s: "Like other companies, Phillips purchased leases based on the hope of finding oil and the expectation that hydrocarbon resources were present. Exploratory wells were unidirectional, at least in the 1950s, making no distinction between oil and gas."[24] The economic concept of sunk costs is important to understanding the evolution of gas prices over the 1950s and 1960s. In a competitive market, producers will produce down to their variable, go-forward costs. That is, they will ignore monies already spent and produce so long as their revenues cover the immediate cost of doing so.

For an oil well that produces associated gas, where gas is truly a physical by-product of oil, the producer has three options: bring the gas to market, flare it, or reinject it into the reservoir (also sometimes called recycling). The decision to reinject is reservoir specific, and is usually made without regard to the prevailing gas price, so we can say that the producer's decision is binary: market the gas or flare it. In the case of associated gas from an oil well, neither the producer's upfront, sunk capital nor variable, go-forward expenses are directed toward gas. The gas price that gives the producer the economic incentive to market the gas instead of flaring it, then, is just enough to build and operate the aboveground gathering infrastructure required to move the gas from the wellhead to the pipeline.

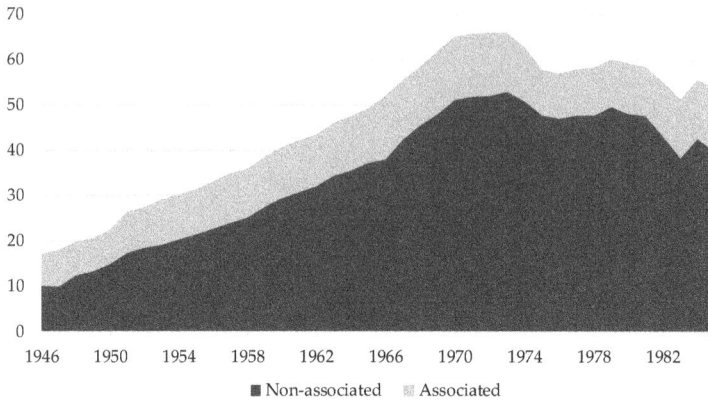

**FIGURE 7.3:** US gas production, 1946–1985 (billion cubic feet per day)

In reality, since producers in many fields are prohibited from flaring gas, and were even as early as the late 1940s,[25] in many cases they are willing to take an economic loss, effectively selling gas for a negative price so that they could turn a profit on the oil. A business born from a *Seinfeld* episode illustrates this type of cross-subsidization nicely. Suppose you find that the top of a muffin is more delicious than the bottom of a muffin, and that people are willing to pay more for just the top than for the whole thing. But you cannot bake only the muffin tops because the texture will not be the same—you need to bake the whole muffin, pop the top off, then dispose of the bottom. If it costs you twenty-five cents to make the whole muffin, and you can sell the muffin tops for one dollar apiece, you would be glad to pay five cents to get rid of the muffin bottoms, even though they are a useful product in their own right, because you'd still be making a profit of seventy cents.[26]

Assuming the decision is made to market the gas, the aboveground infrastructure required—gathering lines and processing facilities, as well as equipment on the drilling site itself—represents a small fraction of the costs of locating the reserves and drilling the wells. And, of course, there is no risk involved in getting associated gas to market. While many millions have been spent exploring for oil that was simply not there, when a producer spends money to buy pipelines and processing facilities, he is fairly certain that he will get what he paid for.

All this meant that associated gas could be gotten essentially for free. Producers did not demand any profit from gas sales, and the incremental costs of bringing the gas to market were very low. This situation characterized the market into the early 1950s, until the three T's and other long-distance pipelines connected the giant, but theretofore stranded, Southwest fields with large northern markets.

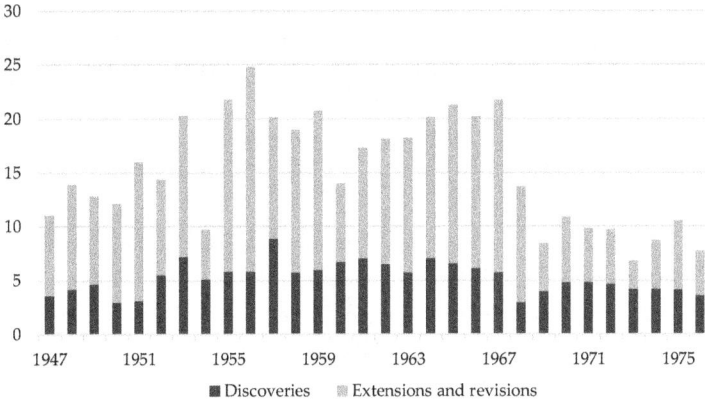

FIGURE 7.4: US natural gas reserves additions: new discoveries versus extensions and revisions at existing fields, 1947–1976 (trillion cubic feet)

The FPC's two-tiered rates recognized that the gas price did not serve as an incentive for producers to bring associated gas to market, which was why associated gas received a lower price.

This economic logic obviously does not hold for nonassociated gas wells. Since these produce little or no oil, gas needs to be priced high enough for producers to break even and recoup all their go-forward costs, not just the costs of gathering. And while production of associated gas did grow by more than four billion cubic feet per day between 1955 and 1970, production from nonassociated gas wells grew by around thirty billion over the same period. Thus, the direct subsidization of gas as a physical byproduct of oil cannot be said to have materially forestalled the gas shortages of the 1970s.

Nonassociated gas was still indirectly subsidized as an economic by-product of oil, however, because much of the initial capital had already been sunk in search of oil.

This is a more controversial argument. And certainly, placing associated gas and nonassociated gas into two economic buckets paints a black-and-white picture of a more subtly hued reality. The gas-oil ratio is continuous, not binary. But it is helpful to think of nonassociated gas in this era as an indirect, economic byproduct of oil because, even if the characterization was not wholly true in any one case, it was partially true in nearly every case.

Here we must separate sunk and variable costs once again, this time to draw the economic distinction between gas already found in the search for oil, and gas that had not yet been discovered. In the ten years following *Phillips*, more than two-thirds of reserves additions were via extensions and revisions at existing fields, versus less than one-third from newly discovered fields. This means that while producers of new gas from old fields required a price high enough to recoup the entire cost of drilling and production, they did not require a price high enough to take on the extraordinary risk of wildcatting to find truly new fields.

This situation characterized the 1950s and the early 1960s, when producer prices rose substantially above where they had been before the three T's but were still substantially below the equivalent price of oil. Over this period, gas functioned mostly as an economic by-product of oil, inasmuch as oil de-risked the search for new reserves. The risk was substantial, as around 40 percent of wells drilled during this period were unsuccessful—dry holes, or dusters, in industry parlance.

Even newly discovered gas was an economic by-product of oil, to one degree or another, until the late 1950s. Only then did producers begin to employ sophisticated seismic surveys that gave more precise indications about the content of untapped reservoirs. Seismology can reveal the hydrocarbon content because oil and gas have different densities, meaning that sound waves pass through them at different speeds. This improved technology increased what producers referred to as *directionality* in exploration—they began to know in advance whether they would strike oil, gas, or a mix of both. Therefore, the economic subsidization of new, nonassociated gas by oil began to taper off rapidly in the early 1960s. Indeed, this point was raised repeatedly by producers in the FPC's *Permian* proceedings between 1960 and 1965.[27]

Clearly, price's function as an incentive can only influence future behavior, not actions taken in the past, and the original, giant gas fields—the Panhandle in Texas, the Hugoton in Kansas and Oklahoma, the Monroe in Louisiana, and several others—had been discovered in the search for oil decades before *Phillips*. Again, the FPC's two-tiered rates recognized this, as the lower rate was applied to nonassociated gas wells that began producing before 1961.

The last lofty economic precept that must be applied to the upstream US oil and gas industry in this era was that a dollar today is worth more than a dollar tomorrow. Businesses are not like well-mannered children, who devour the marshmallows in their bowl of Lucky Charms only after finishing the perfunctory task of eating the cereal. Businesses eat the marshmallows first and suffer through the cereal later, and even then only if they must. For oil and gas producers, this precept translates into drilling your best wells, in your best fields, first. All else equal, this means that the next barrel is always harder to get at than the last, and that, over the long-term, prices should rise. (Of course, all else is never equal, as we shall come to see.)

This far into this book, it must feel that the US oil and gas industry was already fully mature, but in fact it was still quite young. Many hundreds of thousands of good wells in near-virgin acreage had yet to be drilled when the Supreme Court issued its *Phillips* decision. By 1954 the United States had produced 50 billion barrels of oil since the first well was drilled in Pennsylvania in 1859. By 1972 that number had doubled, meaning that as much oil was produced in the eighteen years after *Phillips* as the ninety-five years preceding it. As of the end of 2019, the United States had produced a total of 230 billion barrels.

The corresponding numbers for gas are not as straightforward, since they do not include the amount of gas flared or vented,[28] but they tell the same story, and to an even greater degree. The cumulative amount of gas marketed by 1954 was just 121 trillion cubic feet. The number for 2019 is over 1,400 trillion cubic feet.

As the *Phillips* decision was issued, then, producers had barely begun to develop the United States' vast reserves of oil and gas.

In sum, gas was readily available at very low prices before the 1950s and at relatively low prices in the 1950s and early 1960s, thanks largely to the fact that oil production was rising steadily. It was only in the late 1960s, when demand began to outstrip reserves additions, that a markedly higher price was needed to incentivize producers to go and find new gas, for gas's sake, and with far fewer marshmallows in the mix.

The theory is underpinned by the historical record. Recall that the Tennessee Gas Pipeline was built expressly to find a market for the gas that the Chicago Corporation was producing on its oil properties in South Texas. Its CEO, Gardiner Symonds, had been a vice president at the Chicago Corporation before being dispatched to run Tennessee Gas. It sold its gas to customers in Appalachia for a bit less than twenty cents per thousand cubic feet when it began deliveries in 1944,[29] which was a bit below the prevailing price of oil in energy-equivalent terms at the time.

Oil prices were rising by the time Texas Eastern Transmission (TETCO) began deliveries in 1947. It purchased most of its gas from the oilman H. L. Hunt and United Gas—a large pipeline system that produced, gathered, and transported gas in Louisiana—and sold it to consumers for around twenty-three cents per thousand cubic feet. This was about the same price that Transco, which bought its gas from fifty-four separate Gulf Coast oil and gas producers, charged its New York City customers when they began receiving gas in 1950.[30] Another pipeline that we have not yet documented, Texas Gas Transmission, began moving gas from the oilfields of East Texas and southern Louisiana to Memphis, Tennessee, and cities in Indiana and Kentucky in 1949 for between fourteen and thirty-seven cents.[31]

These were delivered prices, which included the cost of transport. The prices received by producers were much lower. Texas Gas Transmission paid its producers between 11 and 22 cents,[32] which was on the high side

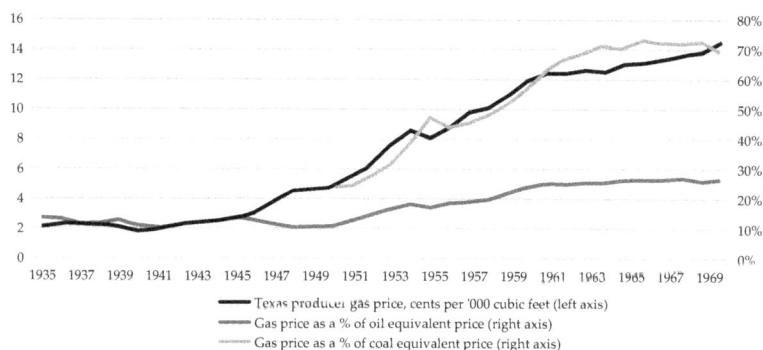

**FIGURE 7.5:** Average price of gas received by Texas producers versus energy-equivalent Texas oil and national coal prices, 1935–1970

at the time, as the average price received by a Texas producer for his gas in 1949 was just 4.6 cents.[33] Prices in the field were certainly rising, as pipelines connecting reserves with markets gave the gas greater value, but they remained well below the energy-equivalent prices of both oil and coal, the two main competing fuels.

This, combined with the fast economic growth throughout the 1950s and early 1960s, sent gas demand skyrocketing. There was thus a growing chasm between the importance of gas to producers and to consumers. From a producer's point of view, gas was still second prize. He was happy to sell it, and cheaply, so long as he was finding it in the hunt for oil. But gas had become a strategically important—indeed, an indispensable—energy resource for consumers.

In hindsight, this should have caused policymakers more concern than it did. If oil exploration and production dropped, for whatever reason, gas prices would need to rise or gas reserves additions would fall in lockstep. Equally, if oil prices rose and gas prices did not, gas demand would rise rapidly, as consumers switched to the cheaper fuel.

## THE AGE OF THE ELEPHANTS

The idea of a shortage of either oil or gas in the 1950s and early 1960s would have sounded ridiculous. This was an era of abundance: energy shortages were a thing of the past. In fact, overabundance would better characterize the global oil market during this era, which saw the emergence of meaningful offshore production volumes in the Gulf of Mexico and the rise of the Middle East from a middleweight producer to the largest and most important oil province in the world.

In the Gulf of Mexico, producers had been experimenting with drilling offshore since the late 1930s, but the technology of the day did not allow them to leave the cradle of land completely. Long piers made of wooden pilings with train tracks laid atop them could be extended for a mile or more out into the Gulf, but they were monumentally expensive, and anyway quickly overcome when hit by large waves from the all-too-common hurricane. As the industry developed, engineering advances allowed producers to drill and produce in open—albeit shallow—waters by 1947, but widespread development was soon hampered by legal issues.[34]

In question was who had the right to grant leases offshore: the federal government or the states? It will come as no shock that either side thought itself to hold the authority. President Truman attempted to settle the question shortly after taking office in September 1945, when he declared that the federal government wielded control over the entirety of the continental shelf. Louisiana, Texas, and California disagreed vehemently, arguing that their historical boundaries extended several miles offshore. Texas was especially dismayed, and had perhaps the most legally sound claim on offshore territory, having established and defended a maritime boundary of three leagues, or 10.35 miles, when it was an independent republic from 1836 to 1845. When the Republic of Texas was negotiating its annexation by the United States in 1845, President James Polk wrote to Sam Houston, formerly the president of Texas and an important pro-annexation advocate, that "you may have no apprehensions in regard to your boundary. Texas once a part of the Union and we will maintain all your rights of territory, and we will not suffer them to be sacrificed."[35]

The pressing matter at hand when Polk wrote those words was whether the United States would defend Texas's southern border, along the Rio Grande, from Mexican aggression, not its seaward boundaries. Indeed, war with Mexico did erupt in the wake of Texas's annexation, and the three-league seaward boundary was upheld specifically in the 1848 Treaty of Guadalupe Hidalgo,[36] and in several later court rulings.[37]

A century later, however, the US government under President Truman had decided that it held paramount claim to all submerged lands. It brought suit against California in 1946, and Texas and Louisiana in 1948. The states were furious, and continued to auction offshore leases until the Supreme Court sided with the federal government in 1950, bringing the nascent offshore industry to a standstill.

The tidelands issue—which was how it was referred to at the time, despite the fact that the vast majority of the acreage in question fell well outside of tidal areas—became prominent in Congress and in the 1952 presidential campaign. That year, Truman's last in office, Congress passed a bill that would have reverted authority over the submerged lands back to the states within their historical boundaries—three leagues for Texas and Florida, and three miles for the other states. But Truman vetoed it. In a parting shot, despite it being almost a foregone conclusion that

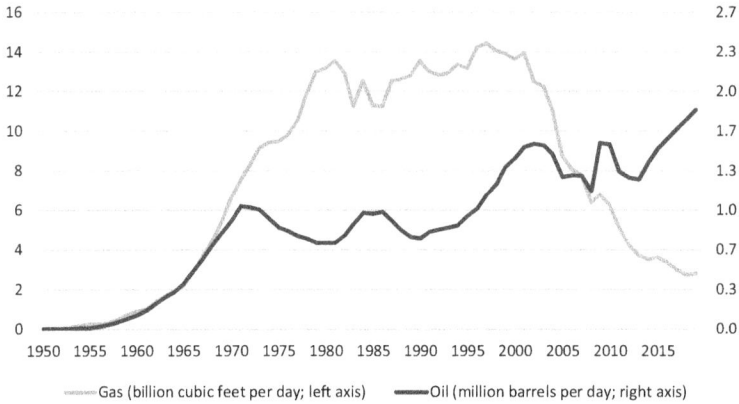

**FIGURE 7.6:** US offshore oil and gas production, 1950–2019

pro-business and pro-states-rights Dwight Eisenhower would cede authority back to the states, Truman issued an executive order just four days before leaving office, declaring offshore oil lands to be held as a petroleum reserve for the navy.

It did not take very long for the Congress to send the Tidelands Act to Eisenhower. The new bill, substantially similar to the one Truman had vetoed, would have been made law in March 1953, except for some filibustering that prevented it from reaching the president's desk until May. Eisenhower signed it immediately, and with gusto, saying: "I deplore and I will always resist federal encroachment upon rights and affairs of the states."[38] With that, the states assumed authority within their historic boundaries and the federal government was given title to all lands farther offshore.

Leasing and production ramped up almost immediately, with Louisiana quickly rising to prominence. As the offshore industry matured and became incredibly technically advanced—indeed, to such a degree that those in the offshore industry began to look down on their primitive, onshore brethren—oil production rose steadily from just 5,000 barrels per day in 1953 to 1.2 million in 1970. And, since gas was found alongside it, gas volumes rose from just 70 million to 8.3 billion cubic feet per day over the same period. Since it can be said that no one was looking for gas when they went offshore, this represented an enormous addition of economically subsidized gas to the nation's total supplies.

The Gulf of Mexico quickly became the pride of the US oil and gas industry. But while a million barrels per day was and is nothing to sneer at, the US Gulf Coast, then and now, pales in comparison to the Middle East.

In 1950 the Middle East (including North Africa) produced just 1.8 million barrels of oil per day, far below the US figure of 5.4 million, of which 2.2 million came from Texas alone. By 1965, however, the Middle East had leapfrogged the United States: production was up to 10.3 million barrels a day, versus 7.8 million in the United States, with Texas not having grown by much (Louisiana that made up the bulk of the growth, from its prolific offshore leases). It did not stop there: Middle East production would hit over 20 million barrels a day by 1971, and 25 million by 1976.[39]

This was a boon for the global economy, and changed almost every aspect of life in the West. Transportation was democratized, with many families buying one, then a second car, and air travel becoming ubiquitous (passenger miles doubled between 1954 and 1962, and doubled again by 1967).[40] Air conditioning allowed backwaters in America's South and Southwest to grow to become major cities (the author writes this passage in Houston, Texas, in August, without sweat on his brow). Plastics both replaced more expensive materials and allowed entirely new products to be created. Indeed, it would take another book to document how thoroughly cheap oil and gas changed society in the postwar era.

For producers in the United States, however, every new giant, or elephant, field in the Middle East spelled competition. Despite how rapidly demand was growing, both in the United States and globally, reserves were being added even more quickly. And economics favored the Middle Eastern fields, which could profitably produce oil at around twenty cents per barrel at a time when US producers were getting close to three dollars.[41] The United States would stand no chance in a battle for market share.

The vertically integrated, internationally active "majors" were unfazed by competition from the elephants overseas because they owned them. Before the nationalization of most Middle Eastern oil reserves in the early 1970s, these majors—Standard Oil of New Jersey (later Exxon, now Exxon-Mobil), Socony-Vacuum (later Mobil, now ExxonMobil; Socony stood for Standard Oil of New York), Standard Oil of California (now Chevron), Texaco (now Chevron), and Gulf Oil (now . . . Chevron), as well as British Petroleum and Royal Dutch Shell—actually owned Middle Eastern oil through "concession" agreements, and split profits evenly under the universally accepted fifty-fifty rule. The majors did all the work, and the governments of the host countries collected rents.

The majors did have extensive US operations, but any revenue lost at home due to a market share battle would be offset by gains from abroad. But for the independents in Texas, Oklahoma, Louisiana, and elsewhere in the United States, there was no such offset: it was a matter of life or death, in which death seemed the more likely outcome. And while these independent oilmen could do nothing to change the economics, they could band together to form a powerful political lobby and change the rules of the game in their favor.

Talk of oil import quotas began as early as 1949, when a Dallas geologist (perhaps unsurprisingly) named Tex Willis wrote in a letter to his senator, Lyndon Johnson, that there was "no sense in bankrupting every independent oil man in Texas for a few Arabian princes and because . . . Standard Oil of New Jersey claims they need the money."[42] Johnson was sympathetic, but President Truman would have none of it. The next year, Truman would veto the Harris-Kerr bill, which would have exempted independent producers from gas price regulation under the NGA, arguing that it might have endangered consumers by not holding prices to reasonable levels. In short, Truman did not count oil and gas producers among his constituency.

Imports accounted for only 8 percent of domestic crude oil consumption when Tex wrote his letter to Johnson. This number would grow as the Middle East came into its own, and a decade later, in 1959, the United States was importing 12 percent of its oil.[43] While not a tectonic shift, national security began to enter the argument in favor of import quotas after the Suez Crisis in 1956.

Egypt, under British rule since 1882, had formally gained its independence in 1922, but the British continued to occupy the country and install rulers sympathetic to the empire until the Egyptian Revolution of 1952. The revolution was led by a group called the Association of Free Officers, directed by Lieutenant Colonel Gamal Abdel Nasser. Nasser did not believe that his rank merited the ultimate position of power so he selected a general, Muhammad Naguib, to become the first president of the Republic of Egypt. The power-sharing arrangement did not last long, though, and Nasser successfully marginalized Naguib over the next year. By 1954 Nasser was the undisputed dictator of Egypt.

By this time, most British troops had left Egypt, but Britain retained a military base at the Suez Canal. The British government was the largest single shareholder in the profitable Suez Canal Company, which had operated the canal since it was built in 1869. The ninety-nine-year lease was therefore due to expire in 1968, after which time ownership would pass to Egypt.

Nasser and his fellow revolutionaries had come to power on a dual platform of Egyptian nationalism and pan-Arab unity. The canal was an unacceptable vestige of Egypt's, and the region's, colonial past, and a potential source of substantial revenues to boot. The British had grudgingly agreed to remove all troops from Suez by late 1955, but this was not enough. Nasser unilaterally nationalized the canal in July 1956.

After several months of non-action, events began to unfold quickly in October 1956. Unbeknownst either to Nasser or President Eisenhower—Israel, Britain, and France, who was another large stakeholder in the Suez Canal Company—had formulated a plan to invade Egypt, with Israel taking control of the Sinai Peninsula in a ground operation and Britain and France landing paratroopers about the canal to retake control. The operation went off, but with a major hitch: Israeli ground forces quickly overran the unprepared Egyptians and successfully took Sinai, but there was an unplanned gap of several days between this land action and the colonial powers' paratrooper campaign. During those crucial days, Nasser managed to scuttle several large ships in the canal, rendering it impassable to all tanker traffic.

At that point, three-quarters of Europe's oil passed through the canal. Europe had only several weeks of excess inventory and it was November: wintertime peak oil demand was just around the corner.[44] So began the first global oil shock.

Eisenhower was furious, most of all with British prime minister Anthony Eden, whom he had told repeatedly to avoid military action. Rather than come to the aid of the British and the French by allowing the US majors to institute an emergency supply program and calling on domestic US producers to increase output, the Eisenhower administration refused to lend a hand until Britain and France had removed all troops from Egypt. With extreme expediency, they acceded to Eisenhower's demands, and the emergency supply program was instituted, sparing Europeans from what would have been a very uncomfortable winter.

Normalcy returned when the canal was reopened in April 1957, but the Suez Crisis had shown how important Middle Eastern oil had become to the West and how vulnerable it could be to the closure of a choke point like Suez.

The Suez Crisis was an isolated geopolitical event. The constraint was not the supply of oil, but transport, or "deliverability," and Suez did not in any way change the fundamental oversupply situation on global markets, which would only accelerate over the next decade. What it did do was allow US politicians to veil their support for import quotas in a national security context. The irony cannot be overstated, since the Texas Railroad Commission, which above all represented the interests of independent, domestic oilmen, refused to increase its production allowance during the crisis.[45] But then, historical fact has always had an undersized influence on political discourse.

By 1958 talk of establishing oil import quotas in the United States had gathered strength. Lyndon Johnson was now Senate majority leader, and Sam Rayburn was Speaker of the House. Both were in favor of quotas. Even more important than having a Texan atop both houses of Congress was the broader reality that this reflected—namely, that the fundamental balance of political power had begun to shift to the Southwest, especially toward Texas. In the words of the historian Robert Caro: "When senators returned to Washington after the 1952 election, there was a new awareness on the north side of the Capitol. There was a vast source of campaign funds down in Texas, and the conduit to it—the only conduit to it for most non-Texas senators, their only access to this money they might need badly one day—was Lyndon Johnson."[46]

Eisenhower remained opposed to the quotas in principle, but the pro-quota current was so strong that he felt a veto would be overturned. In March 1959 he signed into law mandatory oil import quotas, only days after remarking that such measures were "in conflict with the basic requirement of the United States to promote increased trade in the world."[47]

Imports leveled off at around one million barrels per day, and US producers continued to receive their $3 per barrel, a full dollar above the Arabian posted price, which itself was a bit above the going market price for resale.[48] Insulated from further foreign competition, US producers were left to compete amongst themselves for their share of the rapidly growing US market. Prices remained stable.

Had market forces been left to play themselves out, free from government intervention, US production of both oil and gas would have been squeezed by the flood of cheap imports from the Middle East. Instead, it grew by 2.4 million barrels a day from 1959 to 1972, an increase of 34 percent. Gas production almost doubled over the same period, from 33 billion cubic feet per day to 62 billion.

Events in the domestic and global oil markets thus contributed to domestic gas supply growth, which helped stave off gas shortages.

# CHAPTER 8

# THE END OF ABUNDANCE

In the earliest days of the industry, oil and gas fields were found via decidedly unscientific methods. Aboveground seeps were, of course, tell-tale signs of hydrocarbons that lay underfoot, but producers very quickly drilled out these obvious prospects. Creekology, prospecting alongside creeks and rivers, was a common early method of finding oil. While the creekologists often had no understanding of geology, the practice was nonetheless quite successful, as creeks were more common atop anticlines—underground rock formations exhibiting an upside-down $U$ shape, which acted as natural traps for oil and gas. Divining rods, also known as witching sticks, were also used by early wildcatters, less successfully, and it is said that one particularly adventurous soul chose drilling locations by riding his horse at full gallop until his hat fell from atop his head, presumably onto a large oil deposit.

By the mid-1920s, the emerging science of geophysics yielded methods that quickly came to replace these more primitive techniques. Gravity measurements could detect large changes in the density of underlying rock formations and were used extensively to locate the many salt domes in the Gulf Coast region that often hold oil and gas deposits. (Spindletop produced from a salt dome, although it was discovered because of noticeable above-ground gas seeps.) Seismology overtook gravity measurements and quickly became the dominant method of reconnaissance, as it could detect smaller variations in underground rock density. Seismic crews, called doodlebuggers because of the strange equipment they carried, grew from a handful in the late 1920s to more than two hundred before the onset of World War II, and were responsible for the marked increase in reserves additions over that period.

Scarcity breeds innovation, and as the large, easy-to-find fields were discovered and developed, technology needed to improve. In 1930 a single US seismic crew could expect to find fourteen million barrels of oil each year; by 1954 this number had dropped to under two million barrels.[1]

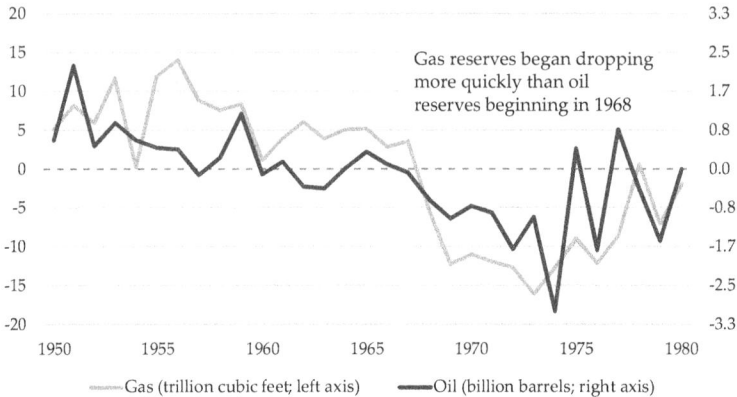

FIGURE 8.1: Year-on-year change in US proven oil and gas reserves, 1950–1980

While the advances in geophysics that occurred in the 1950s and 1960s are highly technical, they can be broken down into broadly two categories: (1) those that improved the accuracy of the data being gathered, and (2) those that allowed for better interpretation of the data after it had been gathered.

On the first front, more sensitive recording systems allowed exploration companies not only to see an underground reservoir, but indicated what type of fluids that reservoir contained—oil, water, or gas—and could sometimes even show the various contact points, where water met oil and where oil met gas. This is because the three fluids have different densities, which means that sound waves propagate through them at different velocities (more slowly for gas than for oil or water). Companies also began placing measurement equipment down into wells before they began full-on commercial development. A variety of different so-called well logging methods were developed to detect oil or gas, some using simple technology, some requiring a degree in physics even to comprehend.

The vast amounts of data that these physics-based exploration methods produced were originally inscribed onto paper or photographic plates, which meant that interpretation was tedious and prone to error, even if the data were accurate. In the 1950s, seismic data began to be recorded on magnetic tape, which allowed for playback and the development of bolt-on technologies that refined the data, correcting for things like small elevation changes and slightly less-than-straight lines between the geophones that detected the minute vibrations in the earth.[2] By the late 1960s, most data were being both recorded and processed digitally, using more powerful computers.

While these advances helped producers find more oil and gas, it also allowed them to distinguish more clearly between oil and gas before beginning expensive drilling operations. Somewhat counterintuitively, this technological advancement actually reduced gas supplies, as it enabled drillers to focus on developing more profitable oil-prone reservoirs.

But with demand continuing to grow rapidly, at nearly 6 percent per year over the 1960s, the pipelines needed more gas. If independent producers had no reason to look for gas, then certainly pipeline affiliates did.

Unfortunately, a series of court and FPC decisions in the early 1960s discouraged pipeline companies from ramping up their own exploration and production activities. In response to a suit brought by California utilities against El Paso Natural Gas, which sought to earn a higher return on its production properties than for its pipeline assets, a court ruled in 1960 that pipeline-affiliated producers were not entitled to book the same federal tax subsidies as independent producers. As regulated enterprises, shielded from the risks facing independents, these subsidies had to be passed on to consumers in the form of lower rates.[3] Upon appeal, the FPC further dug in its heels, ruling that all pipeline company assets were to be considered part of the same rate base, qualifying for the normal, cost-based rate of return of 6 percent.[4] This created a double standard, as the FPC used a 12 percent rate of return when it calculated rates that independents would receive for gas at the wellhead.

The pipelines felt that a 6 percent return did not justify riskier upstream investments, and so greatly reduced investments in production affiliates. Unsurprisingly, pipeline-owned reserves fell over the late 1960s.[5]

Adding to interstate pipelines' woes, producers in the 1960s preferred dedicating their gas reserves to intrastate pipelines. Even if prices were lower, which they usually were, until around 1970, there was substantially less regulatory uncertainty in intrastate markets. And the dedication decision was a final one: once reserves were committed to interstate commerce, they were committed for the life of the field.

In sum, the surplus of gas reserves that had been built up over the preceding decades was drawn down throughout the 1960s. By 1968 scarcity was apparent, and the FPC scrambled to institute new policies to encourage increases in supply. These new policies did help, but not quickly enough. The domestic gas shortage had thus begun before OPEC began an oil embargo in October 1973 that roiled global markets and, indeed, the entire postwar geopolitical order. But the embargo greatly exacerbated gas shortages in the United States, turning a crisis into a catastrophe.

The embargo caused oil prices to quadruple, which both increased demand for gas and reduced supplies on the interstate market. Demand rose nationwide as the physical shortage of oil compelled consumers to use more gas. In the unregulated intrastate markets, the increase in demand translated into a large price increase. Regulated interstate prices did not move,

Southwest                              Rest of United States

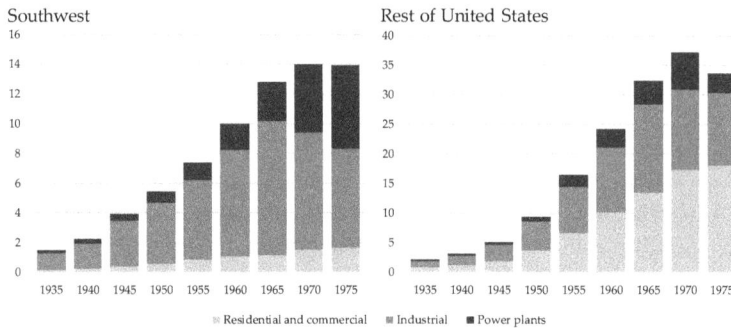

**FIGURE 8.2:** Natural gas consumption by end use, producing Southwest versus rest of United States (billion cubic feet per day)

however, making gas far cheaper than oil on a relative basis, which provided an additional, economic, motivation to switch to gas. This combination meant that interstate pipelines were unable to attract almost any new supplies after the oil shock, causing curtailments of firm supplies by interstate pipelines to skyrocket.

The embargo also brought energy policy to the very top of the political agenda, especially as the United States had only just ended its involvement in the Vietnam War, which had dominated the national conversation for several years. Gas market reform became a part of broader energy market reforms, and indeed, part of a deregulation drive that touched many of America's largest industries.

After the oil shock, the FPC focused on rationing scarce gas supplies, while lawmakers began in earnest the long and painful process of deregulation.

## THE GIANT SUCKING SOUND OF RESERVES DRAWDOWNS

As I have detailed the rise of natural gas from a byproduct commodity, essentially regional in nature, to a strategic and nationally important fuel, I have focused mostly on the build-out of the interstate gas pipeline network, designed to drain the supergiant fields of the Southwest by connecting them with large markets in the Midwest and Northeast. Equally, however, there was tremendous consumption growth within the Southwest throughout the 1940s, 1950s, and 1960s, but of a different character.

Whereas demand in the highly populated and generally colder consuming regions was split evenly between residential and commercial use—for heating—and heavy industry, in the Southwest, industry reigned supreme. The Gulf Coast had grown to become the preeminent refinery and petrochemical complex in the world, and by 1970, refineries in Texas, Louisiana,

**TABLE 8.1: NEW GAS DEDICATED FROM THE PERMIAN BASIN**

| Year | Billion cubic feet | To interstate pipelines (%) | To intrastate pipelines (%) | Interstate price (cents/mcf) | Intrastate price (cents/mcf) |
|------|------|------|------|------|------|
| 1966 | 178.0 | 84 | 16 | 17.6 | 17.0 |
| 1967 | 77.2 | 78 | 22 | 17.8 | 16.3 |
| 1968 | 156.1 | 13 | 87 | 16.8 | 17.2 |
| 1969 | 175.8 | 17 | 83 | 16.8 | 18.4 |
| 1970* | 113.4 | 9 | 91 | 17.5 | 20.2 |

*\* First half of 1970*

Source: US Department of the Interior, Office of Economic Analysis, "Deregulation of Natural Gas Prices, Final Environmental Impact Statement" (Washington, DC: Government Printing Office, 1974), 1–23; and Richard H. K. Vietor, *Energy Policy in America since 1945* (New York: Cambridge University Press, 1984), 159.

and Oklahoma processed 60 percent of all the crude in the nation, domestic and imported.[6] (The only other state that stood anywhere close was California, which processed 10 percent.) Moreover, whereas the rest of the country produced nearly all its electricity from coal or hydropower until the nationwide nuclear build-out of the 1970s, which was in no small part occasioned by the very oil and gas shortages that I am now discussing, the producing states generated a substantial portion from gas-fired generators. Indeed, in 1975, 95 percent of electricity produced in Texas and Oklahoma came from natural gas.[7]

The increasing scale of intrastate markets in Texas, Louisiana, and Oklahoma gave producers a viable alternative to dedicating their reserves to interstate transport. And while intrastate prices throughout the 1960s remained close to or even below interstate prices,[8] producers still preferred to dedicate their reserves to intrastate pipelines whenever possible for three reasons. First, once reserves were dedicated to interstate commerce, they could not be undedicated. Thus, if market dynamics changed after the original, twenty-year contract expired, such that the producer wanted to sign a new contract with an intrastate pipeline, he could not: his reserves were dedicated to interstate commerce for the life of the field. (Legally, it was the acreage that was dedicated, not the contract volume.) The second reason producers preferred to sell intrastate was that FPC regulations could overrule and invalidate the language in the original contract. Rates could be, and were, changed retroactively. As a result of rate investigations between 1962 and 1968, for example, the FPC ordered producers to refund the pipelines a total of $133 million, or about $1 billion in 2020 dollars, because the initial rates had been found to be too high.[9] In intrastate commerce, the sanctity of producer contracts was not so brazenly breached. Last, intrastate

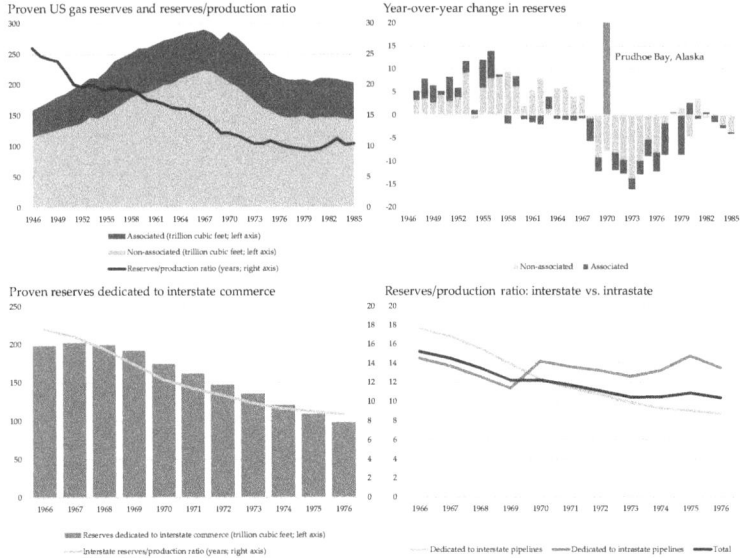

**FIGURE 8.3:** Gas reserves data

dedications had smaller upfront reserve requirements, usually fifteen years, versus twenty for interstate dedications,[10] meaning that producers could get more bang for their buck.

When intrastate prices did rise above interstate prices, which began in the Permian Basin in 1968 but was the case for most producing region by the early 1970s, producers almost totally ceased dedications to interstate pipelines. The 1973 OPEC embargo greatly widened the gap between interstate and intrastate prices, straining the already precarious situation.

Thus, while too-low prices and advancements in exploration techniques were contributing to a slowdown in gas reserves additions nationwide, the reserves drawdown was even more pronounced in the interstate market. The reserves-to-production ratio for interstate pipelines dropped from eighteen years in 1966 to just nine years in 1976, whereas the corresponding number for intrastate pipelines held steady at around fourteen years.

Scarcity can be said to have begun in earnest in 1968, even if signs of it were visible some years earlier. It was for the first time in 1968 when more gas was produced than found, nationally, resulting in a net decline in total reserves, and when LDCs found themselves being denied new supply contracts with the interstate pipelines. Economy-wide inflation was also on the rise, with the consumer price index having risen 3.5 percent in 1967 and 3.6 percent in 1968, after averaging just 1.4 percent for the prior eight years. The average cost per gas well had thus risen by 62 percent from 1960 to 1971, while the average wellhead price rose only 30 percent.[11]

All this serves to blunt the shock value of what otherwise would have been a policy turnaround as monumental as when Samuel Insull called for more regulation in electricity distribution seventy years earlier. In December 1968 the president of the American Gas Association (AGA), which represented some three hundred LDCs and transmission companies, wrote the FPC a letter, urging it to "take a new look" at how it arrived at area price levels, "To insure that the prices resulting will in fact be sufficient to occasion the necessary exploration, development and devotion of gas to interstate commerce."[12] The same LDCs that had fought tooth and nail for rock-bottom prices now joined together with the pipelines and the producers to advocate for higher wellhead prices. There was still substantial dissent within the ranks, of course, but the overall tone was now in favor of price increases to stimulate new drilling.

## DEALING WITH SCARCITY

General agreement on the need for higher prices may have been a prerequisite for bold FPC action, but, as all parties concerned found out over the next several years, the FPC certification process was structurally ill-suited to anything but incremental change. Too many parties could intervene in each proceeding and cite too many decades of precedents that new policies would inevitably contravene. While the AGA was now in favor of higher wellhead prices, its members were not uniform in their support. Many small, municipal LDCs and consuming state public utility commissions continued to oppose vociferously any and all price increases. Talk of shortage led these groups to call for more, rather than less, regulation. Their baseline view was that the shortage was a farce—there was plenty of gas in the ground, but producers were withholding it in wait of higher prices. If the FPC took a firm line, they argued, producers would give up hope on future rate hikes and bring the gas to market. And if they were wrong, and there really was a shortage, it was because the United States had simply run out of gas. In this case, it would do no good to increase prices just so that producers could earn a few more years of high profits without bringing any more gas to market.[13]

It also did not help that the shortages cropped up just as the Supreme Court upheld the FPC's area rate methodology in *Permian*, in 1968. The fairly clean and administratively practical area rate system was supposed to usher in decades of stability. But throughout the fourteen years between *Phillips* and the Supreme Court's affirmation of *Permian*, the US oil and gas freight train had ground to a near standstill.

Again, it is important to realize that while policy played an important role from a very high level, it was geology that led to the gas shortages of the 1970s. Returning to the Lucky Charms metaphor, the marshmallows had all been eaten, and it would prove difficult to sustain an economy as large and as energy intense as the United States on cereal. Said another way, had there been no regulations whatsoever, the higher gas prices that would have materialized in the 1970s would probably have translated first into more demand reductions, rather than supply additions, chiefly because coal would have eaten away at gas's share of the industrial fuel market. In extractive industries like oil and gas, demand generally responds more quickly than supply.

Richard Nixon assumed the presidency in January 1969 and appointed John Nassikas as his FPC chairman. Nassikas endeavored to embody the bold change that the supply/demand situation required, saying shortly after his appointment: "What we need is a Nassikas round of gas rate increases."[14] Raise rates he did, but Nassikas also found himself stymied by the limited wiggle room that the FPC was afforded post-*Phillips*. Nassikas did not have the legal or procedural mandate to abandon cost-based regulation, much as he wanted to, and many of the notable "wins" his commission did score were challenged in court and overturned. Their efficacy is questionable. Certainly, they succeeded in bringing some new gas to market, but not enough, or quickly enough, to prevent curtailments. Just as importantly, nothing was done to reduce demand. In fact, the FPC's policy actually shielded consumers from higher prices as they materialized by allowing pipelines to "roll in" these higher-priced supplies such that consumers only paid the weighted average, blunting any demand-side response.

The next few pages illustrate one theme above all others: that by the 1970s the situation had devolved into total chaos, and that, while all parties agreed in principle what needed to be done, it took several years of serious domestic gas shortages and a sea change in the global energy landscape to muster the political will to deregulate wellhead prices.

The FPC started its campaign to bring about new supply cautiously, and its first major policy change was upheld by the courts. In October 1969 it issued an order allowing pipeline-affiliated production companies to receive area rates, rather than having production costs added to their overall rate base, where they earned only 6 percent, thus rescinding the decision it had made in a case against El Paso Natural Gas in 1962.[15] This encouraged pipelines to restart their production arms, which several did. But three other, bolder programs were shot down just a few years after being promulgated.

In 1970 the FPC began an experimental program that allowed interstate pipeline companies to make interest-free loans to producers and include those loans in their rate bases, effectively charging consumers for upstream exploration and development activity. Whatever discoveries resulted from the loans would be dedicated to interstate commerce, with the pipeline that

TABLE 8.2: FPC AREA RATE INCREASES, 1970–1973

| Producing area | Date | Previous ceiling | New ceiling |
|---|---|---|---|
| Hugoton-Anadarko | Sept. 1970 | 17.0 | 20.0 |
| Appalachia | Oct. 1970 | 28.0 | 24.0-34.0 |
| Texas Gulf Coast | May 1971 | 17.0 | 23.1 |
| Southern Louisiana | July 1971 | 20.0 | 26.0 |
| Other Southwest | Oct. 1971 | 20.5 | 21.9–23.5 |
| Permian Basin | Aug. 1973 | 16.5 | 32.4 |

Source: Richard H. K. Vietor, *Energy Policy in America since 1945* (New York: Cambridge University Press, 1984), 281.

advanced the loans having first call on the reserves. The New York Public Service Commission (NYPSC) cried foul at what became known as the Advance Payment Program,[16] but in 1972 the circuit court ruled that the FPC had the authority to attempt this method in order to alleviate the shortages that were cropping up.[17] By 1975, however, after $3.3 billion had been spent and the NYPSC brought a second suit, the court had changed its tune, saying that the experimental program had become "institutionalized," which was never the idea.[18]

At about the same time, the FPC tried to exempt small producers from rate regulation altogether. The small producer exemption would apply to companies producing less than ten billion cubic feet annually,[19] which at the time would have exempted all but seventy of the country's 4,700 producers, but who collectively accounted for only around 15 percent of total gas output.[20] The exemption was opposed by both consumers, including the NYPSC, and large producers, who wanted to win the war and not the battle. Large producers continued to press for full wellhead price deregulation, and the issue was gaining support in Congress. Nevertheless, the FPC issued the small producer exemption in 1971, only to see it overturned in 1974 by the Supreme Court, which said: "The NGA makes unlawful all rates which are not just and reasonable, and does not say a little unlawfulness is permitted."[21]

The last failed experiment was the so-called optional program,[22] promulgated in August 1972, which allowed producers to apply for a rate higher than the prevailing area rate and then commence service at that rate while the FPC considered the application. If the applied-for rate was denied by the FPC, the producer was granted the option of abandoning service.[23] This program violated the in-line standard set in *CATCO*, and was meant only as a stopgap to take away the frightening finality of making a life-of-field dedication to interstate commerce while the FPC worked to bring interstate prices to parity with intrastate prices.[24]

While it was experimenting with novel programs to bring more gas to market, the FPC was also busy issuing new area rate opinions, all of which moved prices higher. Six such opinions were issued during Nassikas's tenure as FPC chairman, and another round of increases would no doubt have been implemented had he not made the decision in 1974 to abandon area rates altogether and institute a national price. After fourteen years of searching for a workable model, between *Phillips* in 1954 and the Supreme Court's decision to uphold *Permian* in 1968, that model lasted for only a few years.

Desperate times call for desperate measures, and it is worth stepping back to quantify just how bad things had gotten in the early 1970s. Curtailments of gas deliveries on interstate pipelines had begun in 1970, but were mostly limited to industrial customers that held so-called interruptible contracts with pipelines.[25] Interruptible customers purchased supplies at a substantial discount to firm customers because they had alternative sources of fuel or were willing to shut down during times of peak gas demand. So, while unwelcome, curtailments to interruptible customers were part of the bargain, and certainly not catastrophic in the eyes of lawmakers or the general public, who was served by LDCs that held firm contracts.

Beginning in 1972, however, the pipelines had run out of interruptible contracts to interrupt and instituted the first round of large-scale curtailments to customers with firm supply agreements. Only about 5 percent of firm requirements needed to be curtailed that winter, after some acrobatics moving gas from pipes that were long gas to those that were short, but the cuts served as proof positive that energy shortages had become a reality.

The FPC now took on the additional responsibility of rationing scarce gas, which it did in January 1973 by creating eight priority-of-service categories. Interruptible consumers occupied the lowest three slots on this totem pole, followed by firm commitments to large industrial consumers for whom the gas was not crucial, then large industrial consumers for whom gas was crucial, then smaller industrial consumers, and businesses and residences at the very top.

While rationing worked, it was not a solution to the larger problem, which was that consumers had no economic reason to use less gas, while producers had no economic reason to produce more of it. The interstate market lacked a price signal.

Unregulated intrastate prices had begun to pull away from area rates even before OPEC issued an oil embargo in October 1973, turning a regional gas shortage into a national energy crisis. This event is often referred to as the "first oil shock," and was the Arab world's response to the United States supplying armaments to Israel in the immediate aftermath of a surprise attack by Egypt and Syria in the Yom Kippur War. The "second shock" would come in 1979, when Iranian production dropped sharply after the revolution that unseated the shah and elevated the Islamic government under Ayatollah Khomeini. In reality, though, the 1973 embargo was actually OPEC's third—the cartel had stopped sales to the United States and the West during the Suez Crisis of 1956 and the Six-Day War between Egypt, Syria, and Israel in 1967. Unlike the prior two actions, however, which were more headaches than hemorrhage, the 1973 embargo led to an almost immediate quadrupling of global oil prices.

The first and most important difference between the embargo in 1973 and the prior two was that Texas, which had long served as the world's "swing" source of supply, had run out of spare capacity. When in 1967 OPEC stopped sales to countries friendly to Israel, the Texas Railroad Commission responded by increasing allowable statewide oil production, causing US output to surge by almost a million barrels per day. Several other producing countries brought on about the same amount, which rendered the "oil weapon" little more than a logistical problem, just had been the case years earlier during the Suez Crisis.

Despite the massive ramp up in Texas oil output following the 1967 embargo, Railroad Commission officials were still disappointed. Their figures showed that two million barrels per day of reserve capacity existed, versus the one million they got. A few years later, in 1971, they conducted an official test to find out exactly how much spare oil production capacity existed, and were extremely disconcerted when the answer turned out to be a paltry two hundred thousand barrels per day.[26] A few months later, in March 1972, the Railroad Commission issued a general order removing all production quotas statewide.[27] The United States was no longer the world's swing supplier of oil.

As demand continued to grow and US production began declining, the share of oil imports rose quickly. After remaining at 12 percent of total US consumption since Suez in 1956 through 1970—the result of import quotas—in 1973 they jumped to 26 percent.[28]

The second chief difference between 1973 and the earlier embargoes was that the balance of power had tilted toward the host governments of OPEC and away from the Western companies that produced oil within their borders. The large integrated "majors" that had dominated the global oil business through the 1950s—the so-called seven sisters—had been joined by a number of "international independents," which made operating in a unified manner more difficult. Realizing this, host governments in the Gulf and North Africa demanded larger and larger concessions, with some nationalizing or partly nationalizing their reserves.

The 1973 oil embargo marked the end of the old order and the emergence of oil as a diplomatic weapon, wielded by OPEC. Secretary of State

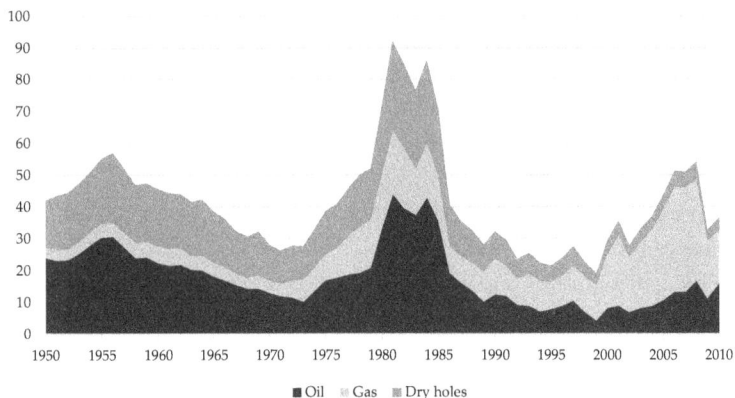

**FIGURE 8.4**: Oil and gas wells drilled, 1950–2010 (thousand wells per year)

Henry Kissinger remarked that the event "altered irrevocably the world as it had grown up in the postwar period."[29] The resultant shortages became so severe, so quickly, that hours-long lines formed outside of those service stations that were lucky enough to have gasoline in their tanks. For the first time, Americans confronted energy scarcity.

As oil prices rallied, natural gas instantly became the cheaper fuel. Where prices could adjust, bringing the two, substitutable fuels closer to party, they did—on the intrastate market. Meanwhile, the area rate increases for interstate gas achieved under the Nassikas FPC now looked wholly inadequate. With consumers clamoring for more gas, interstate pipelines unable to purchase it, and intrastate prices rising with oil, the FPC was stuck playing catch-up.

Nassikas's decision in June 1974 to do away with area rates and create a national price—FPC Order 699—was such a step. The new price of forty-two cents per thousand cubic feet applied to all wells dedicated to interstate commerce after January 1, 1972. The FPC held a rehearing almost immediately and increased this rate to fifty-two cents, adding that old gas would also be eligible to receive the national price after their original contracts expired.[30] The magnitude of this increase was unprecedented—the average wellhead price had remained under seventeen cents throughout the 1960s and had only just crossed the twenty cent threshold the year before the FPC instituted this new rate.

Given enough time, fifty-two cents probably would have elicited a good deal of new supply. Indeed, the number of exploratory gas wells began skyrocketing as early as 1972 after more than a decade of decline and stagnation. But the time lag between increased drilling activity and increased production was measured in years. Just as the freight train of US gas

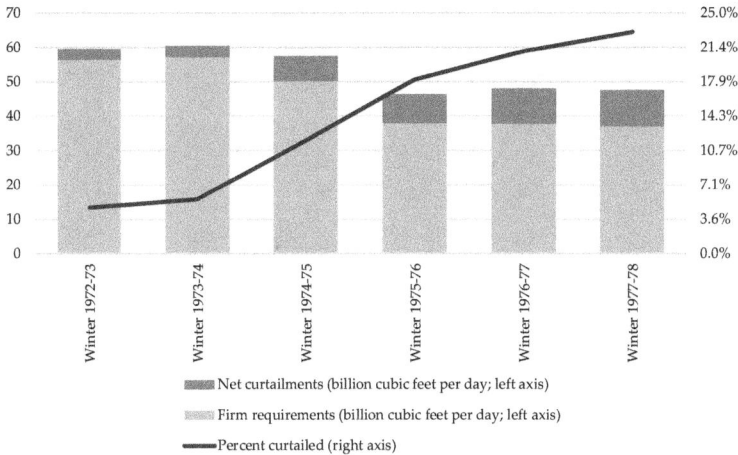

**FIGURE 8.5:** Firm interstate delivery requirements, net curtailments, and percent of firm requirement not delivered, winters of 1972–1978

production took fourteen years to stop, it would take several years to restart. Reserves only began to grow in 1978, so when interstate curtailments got worse over the winters of 1974–1975 and 1975–1976 and the gulf between oil-linked intrastate prices and the regulated interstate price continued to widen, the FPC jacked up prices again in July 1976, this time to $1.42.

This was the beginning of the end for FPC wellhead price regulation, which had lost all coherence. The 1976 price increase to $1.42 was both enormous and arbitrary, as, in formulating it, the FPC for the first time to into account various non-cost factors. This final decision "took regulatory adjustment about as far as possible without legislative intervention."[31]

Jimmy Carter assumed the presidency on January 20, 1977, during one of the coldest winters on record and at the apex of the gas crisis. Less than a week into his term he gave a televised address to the nation, saying: "Half the pipelines of our country have already curtailed shipments to the major industrial users. Four thousand plants are now closed. Four hundred thousand people have been laid off because of natural gas shortages. Shipments to homes have been curtailed by two major pipelines. And many other homeowners are now threatened with that same prospect, and the forecast for the rest of the winter is for continuing extreme cold."[32] That January, Pennsylvania Governor Milton Shapp ordered all elementary and secondary schools closed and for state buildings to set their thermostats to 62 degrees.[33] An almost unthinkable one-fifth of firm requirements were curtailed nationwide in the winter of 1976–1977, with the worst-affected pipes delivering only half of promised supplies. The following winter, though not as cold, saw similar volumes of curtailments. The industry had hit rock bottom.

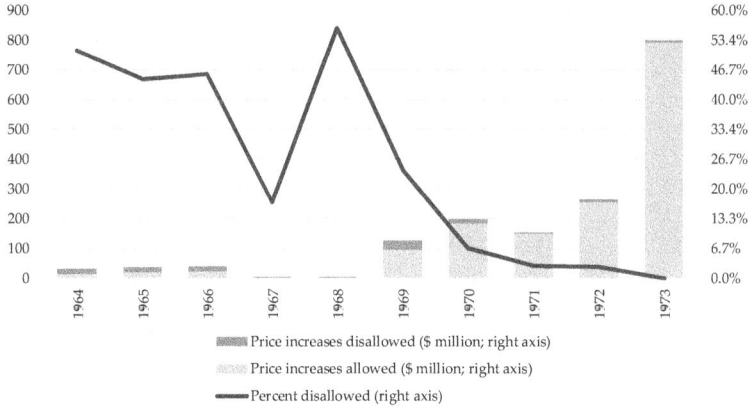

**FIGURE 8.6:** Producer rate increases allowed versus disallowed

Higher prices, increasing curtailments, and continued declines in gas reserves was bound to lead to consumers pushing back on the FPC. Throughout the 1970s, they did so with increasing intensity, as the agency struggled to get producers to commit more gas to interstate commerce using the carrot of higher prices. The national average wellhead price was seventeen cents as the decade began. By 1974 the FPC had raised it to fifty-two cents. Two years later, it was $1.42. The FPC also began rubber-stamping price increases filed by producers, disallowing only 0.3 percent of all rate hikes sought in 1973, versus around 50 percent in the 1960s. The use of so-called emergency gas purchase programs, where industrial consumers bypassed regulation altogether and paid whatever the market would bear, for up to 180 days, also grew.

While the higher prices were causing a massive ramp-up in drilling activity—more than 10,000 new gas wells were drilled and completed in 1977, up from 3,500 before Nassikas became chairman of the FPC—consumers can be forgiven for questioning the validity of an approach that left them paying more money for less gas, year after year. Paying, essentially, for the privilege of shivering through the winter and watching as their industries migrated south.

Clearly, the surge in global oil prices was out of the FPC's control. But rather than convincing consumer groups that the agency's large gas rate increases were justified, the oil crisis and the widening gap between well-supplied intrastate markets and undersupplied interstate markets led them to call for even more regulation. It also allowed them to couch their arguments in sinister tones: *We cannot let the domestic gas market be held hostage to the whims of a Middle Eastern cartel!* This was effective messaging, though it of course omitted mention of the fact that the gas crisis had been brewing

for two decades, since *Phillips*. The fact that the 1973 oil crisis began, somewhat coincidentally, just as domestic gas shortages attained emergency status, bolstered consumers' calls for deepening regulation.

Consumer advocates began calling for a three-pronged regulatory response to the shortages as early as 1974. First, gas prices should be set by
Congress, and any future price increases should be clearly defined. The
FPC, they argued, had been captured by the industry it was formed to regulate. This first stratagem was designed to coax more gas to market using
the stick rather than the carrot—producers would have nothing to gain by
withholding reserves if they knew prices would not rise.

The next two sought to redistribute physical gas supplies from producing states to consuming states. Congress would grant the FPC the power
to prioritize use categories, meaning that less gas would be used in power plants and industrial facilities, which could switch to coal more easily
than smaller consumers. Of course, the majority of gas burned in power
plants and large industrial facilities was in the producing Southwest, served
via intrastate pipelines, which meant that the bulk of conversions to coal
would occur there, freeing up more gas for use in the consumer states. The
last prong of the consumer-favored regulatory response was to extend FPC
price control to intrastate markets, taking the regulatory gap argument to
its logical denouement.[34]

## SELECTIVE MEMORY IN THE SOUTHWEST

Producer states cried foul as the federal government considered coming in
and tampering with their markets. Understandably so, as their market was
working—the governor of Texas was not shutting down public schools for
weeks on end, and the Southwest had become a manufacturing powerhouse,
in both petrochemicals and other energy-intense industries. But neither were
intrastate markets immune from fiascos, as the following episode proves.

Coastal States Gas Corporation started as a gathering company, in the
truest and most literal sense of the term. South Texas was home to many
large gas fields, which the major pipelines had developed during the 1940s
and 1950s. It was home to an even greater number of small gas fields—
too small for the majors to bother with. One man who knew these small
fields as well as anyone was Oscar Wyatt, who had spent the past few years
working as a salesman for the Reed Roller Bit Company. Whenever Wyatt
visited a drilling site, he would learn as much as possible about which wells
were good performers and which were dusters. He even learned how to read
drilling logs, and would rifle through these records whenever the drilling
supervisor left him alone on site.

Wyatt and a partner started their own production company in the early 1950s, but he had grander plans than being co-owner of an also-ran

exploration and production company (E&P). He thought there were millions to be made linking up the small fields of South Texas, which were flaring away their gas, then aggregating the volumes, and finally signing large sales contracts to the major pipeline companies. In 1955 Wyatt split with his partner and formed Coastal States Oil and Gas. His was a surefire strategy: he would approach small landowners without producing wells on their property and offer them a better deal for their gas than anyone else could. He did this by turning pipeline companies' own monopolistic practices against them.

Interstate pipelines were required to have twenty years of gas supply, which meant they produced around one million cubic feet per day per eight billion cubic feet of reserves. But this was an FPC regulation, not one imposed by the Texas Railroad Commission. An intrastate gathering system was not subject to FPC regulation, meaning it did not face reserve requirements. Wyatt offered to take one million cubic feet a day per two billion cubic feet of reserves—a rate four times faster than the majors. This gave producers more money, more quickly. And it made the major producers and pipelines livid because the gas was often being drained from their reservoirs, in yet another case of milkshake drinking.

Angry as they were with Wyatt, pipelines could do nothing to stop him. The rule of capture said that whomsoever pulled the gas from the ground was its owner, regardless of whether it came from beneath another landowner's property. And while their enmity toward Wyatt never faded, the more that Coastal grew, the more that major interstate pipelines found themselves needing to do business with Wyatt because of how they had structured their supply contracts. Most pipelines had signed contracts with most-favored-nation clauses, which guaranteed their producers the highest price paid to any other producer in the field. If Tennessee Gas Pipeline had a contract with Phillips for ten cents, but then signed another contract in the same field with Exxon for fifteen cents, the most-favored-nation clause would be triggered, and Phillips would begin receiving fifteen cents. As wellhead prices rose in the late 1950s, every new contract that a pipeline signed with a producer triggered a field-wide price increase. Coastal, however, was not a producer, it was a gatherer—if Tennessee Gas signed a contract with Coastal, it would not trigger most-favored-nations clauses. This is how Coastal was able to outbid the major pipelines and then resell the gas to them, essentially acting as a gas broker. The situation was such that whenever a major pipeline found itself bidding against Coastal, it would let Coastal win.

This was legal, if not duplicitous. But it was in Wyatt's nature to straddle, and often cross, the lines of legality. He tampered with meters so that he would be paid for more gas than he actually sold. He set up tanks to inject propane close to measuring stations to fool his customers into thinking he was delivering gas with a higher heating value than was actually the case. He even tried to redefine the long-accepted geographical regions of the state of Texas, to get the highest possible rates for his gas after the FPC began its area rate system. He would do anything to bring more money in the door.

In that light, it seems surprising that Coastal was the low bidder to supply the cities of San Antonio, Austin, and Wyatt's native Corpus Christi with gas for twenty years starting in the early 1960s. This was the next incarnation of Coastal, as a public utility, regulated by the Railroad Commission. It kept its gathering business, of course, and renamed it Lo-Vaca Gathering Company, a portmanteau of two counties in South Texas: Live Oak and Lavaca.

Wyatt's basic strategy was a simple "heads I win, tails you lose." Coastal would bid low to ensure that it got the long-term contracts with the cities, and it even signed provisions promising never to go back to the Railroad Commission for a rate increase. If the prices that it guaranteed the cities were too low to earn a profit, however, it would simply renege on that provision and file for a rate increase. After all, once Coastal became a public utility, it became subject to public utility regulations that guaranteed it a rate of return on prudently incurred costs, regardless of however imprudent a contract it had signed. What is most damning about the whole episode is that the cities who signed with Coastal did not realize that they could not simply contract away the Railroad Commission's regulatory authority, even after their incumbent pipeline suppliers and other competitors of Coastal told them point blank that the provision prohibiting Coastal from filing for a rate increase was "Mickey Mouse language." The cities' decisions to choose Coastal was based in equal parts on a lack of foresight, greed, corruption, and naïveté.

Problems began to emerge just a few years after the contracts were signed. Coastal curtailed deliveries to San Antonio in January 1968, even before shortages became acute on the interstate market. Let that sink in: the citizens of south Texas were left cold in the winter for lack of gas many years before the citizens of Ohio suffered the same fate. Things normalized after the cold blew through, however, and in April 1971 Wyatt wrote reassuringly to the San Antonio City Council that Coastal had secured substantial volumes in the Permian Basin for between sixteen and twenty cents. What he failed to mention is that the West Texas gas would never travel to San Antonio.

When extremely cold temperatures descended on south and central Texas again in late 1972, curtailments began once more. This time, the cold did not let up: even as power plants switched over to fuel oil, which they burned "as fast as it could be hauled in by truck," Coastal could not come up with enough gas to serve the remainder of its customers. Despite imposing sixty-five days of curtailments on its customers that winter, it maintained all the while that it was having "mechanical difficulties." The truth was that it had run out of gas. Sure enough, just after temperatures normalized, in March 1973, Coastal went to the Railroad Commission and asked for a rate increase.

Where was the gas from West Texas that Coastal had purchased in 1971 for its customers in San Antonio? In Dallas, where Coastal had no utility customers, and thus no obligation to serve. How did it end up there? Wyatt's defense was straightforward: it had no other place to go, since the only pipeline from West Texas to San Antonio—also owned by Coastal—was already full. This was a truthful lie. Yes, that pipeline was full, but it was full of gas that Coastal was transporting on behalf of another company, one who had simply offered to pay Wyatt more than his contract with San Antonio. After having built the pipeline to serve its utility market in San Antonio, Coastal had sold its capacity to the highest bidder. It was still acting like a gas broker, not a utility, and Texas law did not force it to.

In fact, after its first west-to-east pipeline from the Permian Basin became full, Coastal had built a second pipeline, which seems logical enough. Except, because it really did not care about its obligation to provide gas to its existing customers but rather to grow at all costs, it did not run the second line into San Antonio but rather into Dallas, where Lone Star Gas had a monopoly. Wyatt had been salivating over acquiring Lone Star for years.

Wyatt gave a second excuse to officials from San Antonio and Austin, which had also been heavily curtailed. It is relatively unimportant, except for the fact that involves a contract structure that would become extremely important in the years to come. San Antonio demanded to know why Wyatt was selling to a market where he had no utility customers. Even if the first line was full, meaning that San Antonio could not have received the volumes that winter, the volumes that were sold to Dallas belonged to San Antonio—the reserves were dedicated, even if they could not be used in a given moment. There was, after all, still another ten years left on the contract.

Wyatt defended selling his West Texas reserves to markets near Dallas because he had signed take-or-pay contracts with producers. A take-or-pay clause stipulates that the purchaser take the physical gas or else pay for it anyway. It is a form of insurance for producers, insulating them against shifting market dynamics by providing a revenue guarantee. Wyatt himself was one of the pioneers of the provision, which he had used as a sweetener in negotiations with small producers while he was building Coastal as a gathering company. In the 1940s and 1950s gas producers regularly had their production prorated by the pipeline companies when demand was low. From a pipeline company's point of view, this was cheaper than building underground storage, which would allow them to inject gas when demand was low and withdraw when it was high. From a producer's point of view, it was lost revenues, pure and simple, all the more so if gas prorating meant that oil production had to be shut in, too.

The take-or-pay provisions plus the fact that the original pipeline, from West Texas to San Antonio, was full meant that Wyatt had no choice but to sell the gas somewhere else. He could have told the producers to keep the gas in the ground and paid for it anyway, leaving the reserves for San Antonio in years to come, but leaving money on the table was not Oscar Wyatt's way. In sum, Wyatt had not technically lied when he said that there was no way that Coastal could transport more Permian gas to San Antonio in 1972. He had just bastardized the truth.

It wasn't that Coastal made no efforts to increase its reserves, but that it was always willing to sell newly acquired reserves for a quick buck, despite having fixed price contracts for another ten years. Wyatt was a trader at heart, and he had faith that he could buy more gas later and still make a profit. This was nothing more than a massive gamble that wellhead prices would not rise quickly. Heads, I win. If prices did rise, however, he could always go to the Railroad Commission for a rate hike. Tails, you lose.

The coin had come up tails during the frigid winter of 1972–1973, and Coastal went to the commission for a bailout. For years, San Antonio and Austin had demanded data on reserves and deliverability, but Wyatt had stonewalled them. Now that the matter had moved to adjudication, the truth came out, and it was much worse than the cities had expected. In 1971 the company had claimed to own reserves of 11.3 trillion cubic feet. In 1973 that number had somehow shrank to only 9.4 trillion, of which only a small fraction—3.7 trillion—was available for its utility customers. The rest had been sold to other companies. The 3.7 trillion cubic feet only represented five years of supply, despite Coastal still having around ten years left on most of its contracts. The statistics were even more bleak when expressed in terms of daily deliverability: Coastal's customers consumed around 2.4 billion cubic feet on a peak day, but the company could only deliver 1.4 billion cubic feet, and it was going to lose 300 million of that number the following November, as it had committed another chunk to the Dallas market.

Here was a problem created wholly within the confines of the state of Texas, by a native Texan, being adjudicated by the Texas Railroad Commission, in Austin. And yet somehow cries rang out of federal overreach and Yankee imperialism. Fears that a bankruptcy trial would end up in Delaware or New York prompted shouts that Texas gas would "go up Yankee smokestacks." Coastal seemed conveniently to forget that it was, at that very moment, selling 200 million cubic feet per day to interstate pipelines.

Bills had gone up for consumers in San Antonio and Austin. They were about to go up for everyone. The Arab Oil Embargo would begin in October 1973 and usher in an era of energy scarcity. Despite Wyatt's almost total lack of scruples, Coastal's average cost of gas was only 22.75 cents in 1973, even after he had sold off several trillion feet of reserves. His average sales price was 24.1 cents, barely any margin. By 1975, however, new gas was selling on the unregulated intrastate market for around $2, and the national ceiling price for interstate gas had been increased to $1.42.

In this light, Coastal seems like the lesser villain. Its behavior, while reprobate, cost consumers mere dimes. OPEC cost them dollars. So why tell it?

For one, it is a good yarn, and Oscar Wyatt remained a potent force in the US natural gas industry for the next twenty years. For another, it illustrates that sclerosis was not specific to the interstate gas industry—of all the places to suffer crippling curtailments of wintertime gas deliveries, South Texas is probably the last that comes to mind. In fact, the story of Coastal in the 1960s and 1970s is perhaps the strongest case for ensuring that pipelines secure twenty years of reserves, as was required of interstates.

What really went wrong in the case of Coastal—which is the same thing that was going wrong at the national level—is that producers, consumers, and policymakers all thought that they could insulate themselves from market forces. Before it signed with Coastal, the city of San Antonio had the option to renew its contract with United Gas, which had supplied it for decades, and, as an interstate, had ample reserves. But United was not willing to guarantee a fixed rate for twenty years like Coastal had. It promised a fixed price for the first few years, whereafter the contract would revert to market prices. Houston Natural Gas, a large and highly reliable Texas intrastate system, also competed for the San Antonio contract, and offered similar terms to United. When it found out that Coastal had offered a fixed price for the full twenty years, one of its executives issued a prescient warning to city officials: "Anyone offering to supply your entire requirements at fixed prices must be gambling on his ability to buy or discover gas in the future as your requirements increase. In view of the great uncertainties as to what the future cost of gas will be and the large volumes involved, no responsible supplier can afford to assume the risk that the cost of gas in the field may approach or rise above the prices quoted for delivery to the City." United and Houston Natural Gas lost. San Antonio opted for the cheapest offer: Coastal's. It did not realize that market realities would, one way or another, dictate the terms of service.

Easy as it is to demonize Oscar Wyatt, that would miss the point. Paul Burka, who chronicled this entire episode in a highly researched article in a 1975 issue of *Texas Monthly*, summarized the situation well: "Coastal did not conceive and execute a master plan; rather, it took advantage of circumstances that never should have been allowed to exist. . . . In the end, the cities bought twenty-year gas supplies with little more caution than if they were buying two years' worth of ball-point pens."[35] The same criticism could be applied to the FPC, which was, as Burka wrote, using up its supplies of ballpoint pens by increasing area rates, then doing away with them and creating a national rate, then increasing that, and all the while signing off on whatever rate increases producers asked for. The capriciousness with which billions of dollars in future costs and revenues were being shuffled around was astounding, no matter what state you lived in.

# CHAPTER 9

## FROM SCARCITY TO SURPLUS

The histories generally remember Ronald Reagan as the great deregulator, but it was during Jimmy Carter's four years as president that most of the landmark deregulation legislation was written and signed into law. Carter deregulated airlines (the Airline Deregulation Act of 1978), the trucking industry (the Motor Carrier Act of 1980), the railroad industry (the Staggers Rail Act of 1980), banks (the Monetary Control Act of 1980), and began the breakup of Ma Bell, which held a near absolute monopoly on telephony in the United States. He also deregulated vast swathes of the energy industry, with the five statutes passed as part of the National Energy Act of 1978.

These statutes began the gradual deregulation of the electric utility industry, instituted the gas(oline) guzzler tax, which encouraged Americans to use more fuel-efficient vehicles, limited the amount of natural gas that could be burned at industrial boilers, and instituted a completely new regime for wellhead gas prices.

I will of course focus on the natural gas–related statutes, but it is important to keep in mind that gas market reform was a part of a broader program of reforms, which were of paramount political importance. While the wording of the legislation varied by industry, the underlying philosophy was the same: reduce price and entry controls to increase competition and allow the free market to allocate resources more efficiently. The massive US regulatory apparatus, borne of the New Deal in the mid-1930s, was largely disassembled by the early 1980s.

Though only a part of the broader agenda, the drive among lawmakers to fix America's broken energy system was visceral, more so than for other industries. Since its founding, the United States had been a land of abundance and of locomotion. The 1973 oil shock and the ensuing gasoline lines were more than a simple inconvenience—they had wounded the national psyche. Congressmen, senators, and President Carter had campaigned on promises to fix the energy system. To this end, in 1977 they established a new, Cabinet–level federal department, the Department of Energy (DOE), which assumed responsibility for coordinating a national energy policy and managing the nation's nuclear industry, both its arsenal and its reactors. The Federal Power Commission (FPC) was brought under the DOE for organizational purposes and renamed the Federal Energy Regulatory Commission (FERC), but it remained an independent commission made up of five presidential appointees. For all intents and purposes, FERC differs from the FPC in name only.[1]

The legislation that most affected the gas industry was passed in November 1978 and called the Natural Gas Policy Act (NGPA). It took the job of setting wellhead prices out of FERC's hands, instead establishing a complicated schedule of incentive-based pricing and providing for complete deregulation by 1985. Some gas was deregulated earlier, in 1979, and prices paid for these "high-cost" volumes served as a bellwether for market conditions. Just as important as the outright price level was the fact that the NGPA extended price controls to the intrastate markets, to the ire of producers and consumers in those states, who had for several years been paying higher prices to keep the industry alive and subsidizing interstate consumers in the process.

The NGPA's price schedule was designed to ensure an orderly transition to a deregulated market—it was regulated deregulation. To do this, it relied explicitly on a forecast of oil prices, as oil was gas's main competitor in the industrial fuels market. But the oil market did not cooperate with the NGPA's schedule, as the Iranian Revolution of 1979 led to a shock that saw prices more than double. Gas pipelines bid aggressively to bolster their reserves after a decade of drawdowns, but found themselves paying through the nose for both high-cost domestic supplies and imports, which were rising in lockstep with oil. The high prices led to a bonanza in the US exploration and production industry, spawning a gilded class of oilmen that included some larger-than-life characters, such as T. Boone Pickens.

While wildcatters reveled, policymakers fretted that higher oil prices would cause natural gas prices to fly up once full deregulation kicked in, in 1985. But once again, events on the global stage turned domestic assumptions upside down. Oil prices began dropping in 1982, and had tanked by 1986, beginning a more than decade-long slump. The pipelines, who had signed long-term gas contracts at prices that were barely justifiable when oil peaked at $40, found themselves caught long as the cost of a barrel dropped back down to $15 and industrial consumers switched fuels. Even pipelines' core consumers—residential and commercial users behind LDC city gates—had reacted to the shortages of the 1970s by installing heating systems that used electricity or oil. In just a few years, the US gas market had transitioned from shortages so severe that they literally left consumers shivering through the winter to a surplus so large that multi-billion-dollar pipelines were teetering on the verge of bankruptcy.

As early as 1983, Band-Aid solutions were conjured to try to allow pipelines to offload the gas for which they had paid far too much. These solutions created the first spot markets for gas, and began the inexorable transformation of the pipeline business model. For the five decades since PUHCA separated LDCs from vertically integrated utility holding companies, pipelines had operated as energy merchants, matching and optimizing a portfolio of producer supply contracts with end-user sales contracts. By the mid-1980s it was clear that they would need to cease these merchant functions and become contract carriers, providing transportation and storage services only, and letting consumers deal directly with producers. A pipeline executive at Natural Gas Pipeline of America, the grand dame that Samuel Insull had built to serve Chicago, summed up policymakers' attitudes toward pipelines well when he said in 1983: "What the Commission is telling us is 'Get out of the business. You have done an abysmal job in managing it to date, so get the hell out and we'll let the producers in. You guys just go about your business of transporting gas.'"[2]

## THE NGPA AND ITS MANY FLAVORS OF GAS

As the Carter administration formulated the rules that would remake the domestic gas market, there was bitter acrimony over how to solve a problem borne of regulatory failure. Producers demanded deregulation. Consumers were split by geography. Consumers in consumer states, who received their gas via interstate pipelines, demanded deeper regulation, including extending price controls to intrastate markets and mandating that large industrial users convert from gas to coal. They wanted an even playing field with consumers in producer states, who had access to ample gas supplies.

Consumers in the producer states of Texas, Louisiana, and Oklahoma rightfully resented this view. For one thing, they had been paying higher intrastate prices for several years, keeping the industry alive and thus defraying costs for the consuming states. This was the rate argument, and was expressed in cents per thousand cubic feet. For another, converting their industries to use coal instead of gas would be a massive undertaking. At the time, about 95 percent of the electricity generated in these states came from natural gas-fired boilers. This was the scale argument, and was expressed in billions of dollars.

Pipeline companies mostly sided with producers, but were less strident: they only required parity between intra- and interstate markets, and for enough reserves nationwide to meet their contractual commitments. Coal interests also became involved in the political debate, and sided with the consumer-state approach, given that industrial conversions would greatly benefit the coal industry. To throw yet another wrench into the works,

Democratic president Jimmy Carter, who counted consumer states as his largest constituency, had won the presidency because he carried both Texas and Louisiana, which was due in no small part to campaigning on a deregulation platform.

From this polarized situation emerged polarized legislation. The House of Representatives, where states are represented proportionally to their population, issued proposals favorable to consumer states, whose representatives were joined by those from coal-producing Appalachia. The Senate, where each state has two votes, was more amenable to producer-state concerns. The Carter administration, despite having won office on a promise to deregulate, switched sides and favored a consumer-oriented approach.

Under normal circumstances, the various sides would have been too far apart to come to agreement. But circumstances were far from normal. Natural gas market reform was a part of a much broader package of deregulation, but a key part. Politicians who had run on the promise of fixing America's energy system engaged in unprecedented political jockeying to push through the Natural Gas Policy Act (NGPA), with "tangible quid pro quos"[3] promised in return for votes. Lame duck congressmen were offered judgeships and executive appointments, and pork was doled out on a grand scale. One senator from Idaho, who had no particular interest in natural gas regulation one way or another, was promised a $400 million nuclear reactor research project for his state. Intimidation tactics were also employed, with the administration reminding executives from the banking, steel, automobile, and textile industries—all of which still operated in regulated environments—of ongoing tariff negotiations, Securities and Exchange Committee proceedings, and, of course, upcoming deregulation legislation whose wording might depend on their cooperation.[4]

Producers would have been a harder bunch to cajole, especially independents. They had less to gain and far more to lose by supporting the NGPA. A principal feature of the proposed legislation was to extend federal regulation into the only place where they had gotten a fair shake: the intrastate market. But they had reason to fear that the FERC bogeyman would get them one way or another, and even though the legislation was far from perfect, it did call for a small initial price increase and for eventual deregulation.

Producers' fears that they would be better off grudgingly accepting the new legislation were justified by several recent developments. In May 1978 the Supreme Court upheld a FERC ruling that gas, once dedicated to an interstate pipeline, could not be transferred to an intrastate pipeline under any circumstances, even if the owners changed or went bankrupt. The ruling applied retroactively, as far back as could be proven. This case, *California v. Southland Royalty*, meant that FERC had jurisdiction not only over natural gas companies but over the actual reserves of gas, once they had been dedicated to interstate commerce. There was worry that "hungry pipelines" would dig through their archives to find old contracts and then compel producers to dedicate gas that they had been selling on the intrastate market, even if those contracts predated the producer's very existence. The *Southland* decision meant that a hypothetical firm, only established in 1960 and that had been selling gas for $3 on the intrastate market, might be forced to sell its gas on the interstate market at half that price because the reserve's prior owner had signed a contract with an interstate pipeline all the way back in 1930.

Moreover, if the pipelines could produce any such evidence, producers knew they would be dealing with a FERC that was unusually eager to prosecute. In 1977 and 1978 the commission submitted as many potential violations to the Justice Department as it had in the forty years prior, the vast majority of them cases of "unlawful diversion" from the interstate market. Each violation carried the potential of a very large fine.

Perhaps most troubling to producers was FERC's statement in 1978 that it would begin applying a prudent investment standard to producers. This would have amounted to a wholesale cancellation of the Natural Gas Act's gathering-and-production exemption, and seen FERC rule retroactively on whether certain drilling techniques and other physical processes, long regulated by the states, had been economically justified.[5]

Dismayed by what looked to be substantial expansions of FERC's authority under the old NGA, producers toned down their opposition to the NGPA. In November 1978 President Carter signed the bill into law, and although the final version did contain some of the language in the Senate bill that was more favorable to producer states, it was closer in both substance and spirit to the consumer-oriented legislation put forth by the House and the Carter administration.

At its most basic, the NGPA replaced cost-based regulation with central planning, where the government used blunt but powerful controls to bring the market into equilibrium. Once equilibrium had been reached, and enough new supplies were being brought online to meet a reduced demand level, the government would relinquish its control and let the free market take over. As such, it reflected tacit admission even among consumers that the utility compact could not be suitably applied to natural gas production.

For the utility compact to function, there must be give and take. Promotional regulation must be combined with protectional regulation: the utility accepts a regulated rate of return based on costs only because it is shielded from competition. While it is thus suitable for pipelines and local distribution companies, utility regulation was always a bad deal for natural

**TABLE 9.1: NGPA PRICE SCHEDULE, FROM TITLE I**

| Gas category | Section # | Price as of April 20, 1977 | Monthly price increase | Price as of October 1981 | Deregulation date | Description |
|---|---|---|---|---|---|---|
| New gas from new reservoirs | 102 | 1.75 | Inflation + 3.5% | 2.91 | 1/1/1985 | New onshore reservoirs (began production after 4/20/1977) |
| | | | | | 1/1/1985 | New onshore wells (completed after 4/20/1977) from old reservoirs but at least 2.5 miles from nearest existing well |
| | | | | | 1/1/1985 | New federal offshore leases effective after 4/20/1977 |
| | | | | | Never | Reservoirs discovered after 7/27/1976 on old federal offshore leases |
| New onshore wells from old reservoirs | 103 | 1.75 | Inflation only | 2.51 | | Drilling started after 2/19/1977 not within a proration unit. |
| | | | | | 1/1/1985 | Deeper than 5,000 feet |
| | | | | | 7/1/1987 | Shallower than 5,000 feet |
| Interstate gas dedicated before NGPA (11/9/1978) | 104 | Just and reasonable | Inflation only | 2.08 | Never | Post-1974 gas |
| | | | | | Never | 1973-1974 biennium gas |
| | | | | 1.76 | | Small producers |
| | | | | 1.35 | | Large producers |
| | | | | | Never | Replacement contracts or recompletion gas |
| | | | | 0.99 | | Small producers |
| | | | | 0.76 | | Large producers |
| | | | | | Never | Flowing gas |
| | | | | 0.50 | | Small producers |
| | | | | 0.42 | | Large producers |
| | | | | | Never | Certain Permian Basin gas |
| | | | | 0.59 | | Small producers |
| | | | | 0.52 | | Large producers |
| | | | | | Never | Certain Rocky Mountain gas |
| | | | | 0.59 | | Small producers |
| | | | | 0.50 | | Large producers |
| | | | | | Never | Certain Appalachian gas |
| | | | | 0.47 | | Small producers |
| | | | | 0.44 | | Large producers |
| | | | | 0.26 | Never | Minimum rate gas |

| | | | | | |
|---|---|---|---|---|---|
| Intrastate gas dedicated before NGPL (11/9/1978) | 105 | Lower of contract as of 11/9/1978 or Section 102 price | | 1/1/1985 | If contract price was more than $2.08 on 11/9/1978 |
| | | Higher of contract as of 11/9/1978 or Section 102 price | | 1/1/1985 | If contract price was less than $2.08 on 11/9/1978 |
| Rollover contracts | 106 | Higher of just and reasonable as of rollover date plus inflation and 54 cents | Inflation only | 0.77 | Never | Interstate |
| | | Higher of contract as of rollover date plus inflation and $1 | | 1.43 | 1/1/1985 | Intrastate |
| High-cost gas | 107 | Section 102 price or higher incentive price, as indicated two columns to the right | As per relevant section, or market based | Market | 11/1/1979 | Gas from wells deeper than 15,000 feet drilled after 2/19/1977 |
| | | | | Market | 11/1/1979 | Gas from shale, coal seams, and geopressured brine |
| | | | | 2x Section 103 price | Never | Tight gas |
| | | | | Section 109 price | Never | Intrastate Section 105 gas from wells with qualified production enhancement work having been done. |
| Stripper well natural gas | 108 | 2.09 | Inflation + 3.5% | 3.12 | Never | Nonassociated gas produced at a rate less than 60,000 cubic feet per day over 90-day period |
| Other | 109 | 1.45 | Inflation | 2.08 | Never | Prudhoe Bay gas, or otherwise not covered |

gas producers, who were not offered any protection from competition. They gave much and got nothing back.

The NGPA removed FERC from producer rate regulation altogether, as it contained a schedule of wellhead prices promulgated by the Congress and signed into law by President Carter. This schedule was title I of the act. Title II defined what consumers paid. Both were extremely complicated, so much so that the third-ranking Republican in the House, Congressman John Anderson, said of the NGPA: "This bill is nothing but a masterpiece of confusion. It is so murky in its utter complexity, other than being a law-yers' and accountants' relief act, that it does nothing to advance the energy interests of this country."[6]

This is only somewhat hyperbolic. I thus boil down titles I and II of the NGPA to their essence. (The act had five titles, but only the first two warrant discussion.)

Title I created eight categories and thirty subcategories of gas with dif-ferent starting prices, price escalators, and deregulation dates. These cate-gories can be grouped into three basic "flavors" of gas: old, new, and high-cost. Old gas, from wells drilled prior to passage of the NGPA, remained more or less at their previous, regulated prices, until the wells ran dry, and provided pipelines with a "cushion" of low-cost supplies through the transi-tion to a fully deregulated market. New gas—be it interstate or intrastate—from wells completed after the NGPA was passed, was set at an initial price of $1.75. The rate at which this price escalated depended on whether the gas was from a previously discovered reservoir, in which cases the price would rise in line with inflation, or from a newly discovered field, in which case it rose at a faster rate, to encourage exploratory drilling. Price controls for new gas would expire in 1985.

Volumes from old gas and new gas wells represented the bulk of na-tional production, but the bulk of the problems that plagued the industry through the 1980s came from the third flavor: high-cost gas. This was gas from more expensive formations—such as coal seams, shale beds, tight res-ervoirs, and, most importantly, deep gas—from wells drilled to more than fifteen thousand feet. The price for high-cost gas was deregulated just one year after the NGPA passed, and thus represented the first true market price since area rates were instituted, a decade or so earlier. The dedicated reader can study Table 9.1, which lays out a simplified version of title I's rate schedule. The more aloof can be stupefied by the fact that Table 9.1 can, in good faith, be described as simplified.

The $1.75 price that applied to most new gas, initially, was not as high as the energy-equivalent price of oil, but because the NGPA brought the intrastate market under regulation, there was no longer any threat of con-sumers there paying more and "stealing" interstate gas. It was also higher

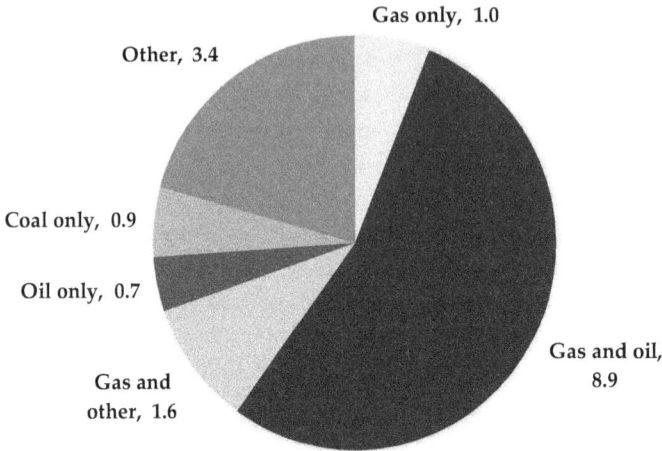

**FIGURE 9.1:** Fuel consumption at large industrial boilers by boiler fuel design, 1979 (billion cubic feet per day of gas equivalent)

than the prevailing interstate price of $1.42, set by FERC in July 1976, which was positive for those producers who sold a large portion of their output to interstate pipelines.

Title I of the NGPA accomplished two of the act's goals: (1) leveling the playing field between inter- and intrastate markets, and (2) increasing supplies. But the NGPA also created a situation where pipelines would be incentivized to bid up prices for unregulated gas to extraordinary levels because they were explicitly allowed to pass these costs on to their consumers, most of whom were captive LDCs who could not readily switch fuels. This was especially so for those pipelines with large cushions of old gas, as the relatively small volume of new, high-cost supplies would be "rolled in" and averaged with a much larger volume of cheap supplies. Cognizant of such a possibility, the NGPA's authors formulated title II, which implemented what was known as "incremental pricing."

Title II split consumers into two groups: (1) large industrial users and (2) smaller users, which included all residential and commercial consumers (LDCs). The large industrial users paid the "incremental" price, which was set, in a very roundabout way, at the prevailing price of fuel oil. This higher price paid by industrial users served two functions. First, it shielded residential and commercial consumers from higher prices during the phased deregulation. Second, it kept pipelines from bidding up prices to unreasonable levels because the industrials would not be willing to pay prices higher for gas than for fuel oil, since they had the ability to switch fuels.

In theory, the incremental pricing provision of title II allowed residential and commercial customers to continued receiving a subsidy without creating incentives for pipelines to overpay for gas. And it is true that most industrial boilers did have fuel-switching capabilities. About half of the fuel consumed in the industrial sector at the time was consumed by boilers that could switch between gas and fuel oil, while only 6 percent of consumption was from boilers that had no option but to fire gas.

Title II did prevent many contracts from being signed above the oil-equivalent price. Many, but not all. LDCs still received rolled-in rates, where the price was equal to the weighted-average cost of gas. This allowed pipelines to sign some contracts with producers above oil parity because the weighted-average price was cheaper than oil, and because the NGPA allowed the pipelines to pass through the purchased gas costs automatically, without customers having a say-so. This practice was most common among pipelines with low-cost cushions, who outbid their higher-cost brethren.

More importantly, title II did nothing to prevent gas pipelines from chasing gas prices up close to oil parity, and then signing ten- or twenty-year contracts with take-or-pay terms, in which the pipeline guaranteed that it would pay for the volumes regardless of whether they were actually taken. High-cost contracts with high take-or-pay percentages, often for more than 90 percent of the annual contract volume, left pipelines exposed both to falling oil prices and falling gas demand.

As I detail over the next few pages how pipeline companies' behavior in the years following the NGPA led to a massive gas bubble, which in turn led them into severe financial distress, it is tempting to write them off as impatient, imprudent, incompetent, short-sighted, and downright greedy. Some of these accusations doubtless have merit, but historical context reveals that, more than anything, the pipelines were caught off-guard by difficult-to-predict market dynamics and led into a morass by managers that had, for their entire careers, been coddled by regulations that protected them from financial risk due to such fluctuations in the market.

At a very basic level, interstate pipelines had just experienced a decade of severe shortages, where they were forced to curtail deliveries to their core customer base. Moreover, total demand for gas in the United States had grown rapidly for forty years straight. It is only natural that, in these conditions and with their experience, pipelines were tremendously eager to sign new supply contracts with producers and get back to the business of delivering ever-growing amounts of gas.

Fanning the flames was the fact that the 1979 oil crisis started just months after the NGPA was passed, which led to an almost tripling of the oil price, from $13 per barrel to as high as $40. Gas once again became a much cheaper fuel, by comparison, and the bidding war began.

## THE 1979 OIL SHOCK AND THE MAKINGS OF A BUBBLE

The 1979 oil crisis was very different from the 1973 oil crisis. In 1973 OPEC placed an embargo on exports to the United States and other nations that had supported Israel in the Yom Kippur War. The ensuing price spike was due to a physical shortage of oil on the markets, and coordinated price hikes among OPEC members.

The 1979 crisis, on the other hand, went through several stages and was driven as much by panic as it was by physical shortages. OPEC also functioned much less cohesively over the course of the 1979 crisis, which continued throughout 1980. Two of its largest and most influential states, Iran and Iraq, went to war with each other. Meanwhile, Saudi Arabia, the single largest and most powerful member of the cartel, urged moderation against the will of many smaller members, who wanted simply to reap the highest prices possible.

The crisis began in Iran, then the world's second-largest oil exporter, where strikes and demonstrations against the US-backed shah had intensified throughout 1978. By December all but a few of the expatriate workers who tended the oilfields were evacuated, and exports dropped to nothing. There were even shortages domestically. In January 1979 the shah left the country for a vacation, never to return, and his last and most ardent supporters were defeated the following month, elevating Ayatollah Khomeini to the post of supreme leader. The Islamic Republic of Iran was officially born in April.

The Iranian Revolution thus marked the beginning of the 1979 oil crisis, even though the drop in Iranian output was mostly offset by increased production from Saudi Arabia, which favored long-term stability over short-term profits. So, while the revolution did not actually lead to a major drop in global supplies, it did incite panic in a market that was still quite rigid and lacked the flexibility to cope with short-term shocks. The global seaborne oil market was dominated by bilateral agreements—the spot market was still quite small. Even though supplies were available, somewhere, many companies did not have access to them. Under normal circumstances, companies would have been willing to trade amongst each other. However, the psychology of panic meant this did not occur. Instead, all the oil majors looked frantically for supplies and exacerbated the bidding war by purchasing volumes in excess of demand, increasing storage inventories in the fear that the barrels might simply be unavailable in the future. Seeing opportunity amid the fear, OPEC raised prices several times, against the wishes of Saudi Arabia.

Fear built as 1979 progressed. In early November a band of Iranians took hostage sixty US embassy workers, beginning a 444-day crisis that

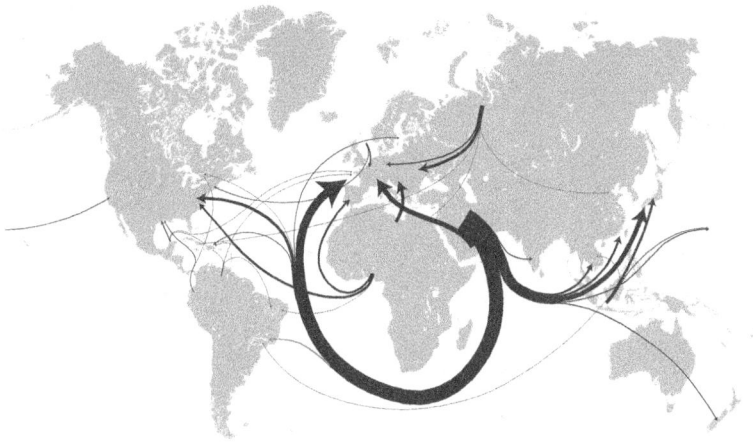

**FIGURE 9.2:** International crude oil trade, 1980 (thousands of barrels per day)

served as a major embarrassment to the Carter administration. A few weeks later, a group of armed fundamentalists seized the Great Mosque in Mecca, Islam's holiest site. Both this and another protest in eastern Saudi Arabia, close to the oilfields, were successfully defused by the Saudi government, but they gave credence to the notion that the whole region was being swept up by a wave of violent Islamic fundamentalism. With Islamists casting America as "the Great Satan" and the rest of the Western world in only slightly less spiteful terms, such a movement could only be bad for the free flow of oil out of the Persian Gulf, which was by far the most important trade route in the world.

Saudi Arabia, as OPEC's largest producer and exporter, seemed to have the most to gain from rapidly escalating prices. But it also sat on by far the world's largest reserves, and its oil minister, Ahmed Zaki Yamani, took the long view. While the run-up in prices was bringing the Saudi government a torrent of profits, Yamani advocated strongly for moderation, fearing that high prices would lead both to long-term demand destruction and to new, more expensive sources of supply coming online and pushing OPEC barrels out of the market.

While Yamani, the most important man in OPEC, had the ear of all his colleagues, even he was powerless to stem another run-up in prices in September 1980, after war broke out between Iran and Iraq. Production fell in both countries, and Iran cut off almost all of Iraq's export routes, shaving another two and a half million barrels off a sixty-million-barrel per-day market. Another round of panic buying ensued. This time, not only did buyers hold on to the large inventories that they had amassed the year prior, they actually began to build up floating storage, holding oil on tankers.

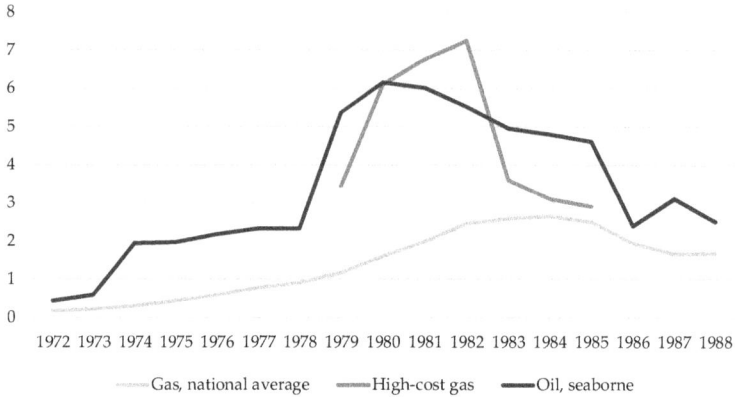

**FIGURE 9.3:** Oil versus gas prices ($/MMBtu)

This was extremely expensive, as the hefty daily charter rates had to be paid even while the ships sat moored in harbor. By the end of 1980, the oil price was $35 per barrel, with spot cargoes selling for upward of $40.

A brief aside on energy measurements helps translate the effect of the global oil crisis on the domestic US gas market. From now on, I refer to natural gas prices in dollars per million British thermal units, or MMBtu, instead of dollars per thousand cubic feet. A Btu is a measure of energy (the energy required to heat one pound of water by one degree Fahrenheit), while a cubic foot is a measure of volume. Conveniently, one thousand cubic feet of gas contains almost exactly one MMBtu of energy—the terms are used almost interchangeably. One barrel of oil, which is also a volumetric measurement, equal to forty-two gallons, contains about six MMBtu. This means that, in the wake of the 1979 crisis, the price of oil was just under $6/MMBtu ($35 per barrel, divided by six MMBtu per barrel). Given that oil was the principle competitor to natural gas, both LDCs and large industrial users in the United States were happy to pay anything less than that for their gas.

US gas pipelines were happy to oblige, paying huge sums to secure supplies of high-cost gas, especially for their LDC customers, who received rolled-in pricing. In mid-1981 the average price paid for high-cost gas was $6/MMBtu, and had risen above $7 a year later. Some pipelines were paying upward of $9. Thus, while only representing 6 percent of total supplies in 1982, the high-cost gas that the NGPA had deregulated accounted for 20 percent of total costs, nationally, and for upward of 30 percent at several pipelines.[7] One executive at a production company recalled: "It was like a kid in a candy store. For a long time his mother kept him at home; then he went and had a terrible shopping spree."[8]

If pipelines were like kids in a candy store, producers were the ones selling the candy, and were delighted for the business. The run-up in both oil and gas prices in 1979 and 1980 led to record levels of drilling in the United States: 92,000 wells were drilled in 1981, versus 46,000 in 1977, before the passage of the NGPA.[9] For gas producers, the real run-up only applied to wells qualifying as high-cost—new gas prices were still capped at $1.75, plus whatever escalators had been applied. This pushed producers to drill deeper and into tighter formations than ever before, as the two most important subcategories of high-cost gas were deep gas,—from wells drilled to a depth of more than fifteen thousand feet, and tight gas—from reservoirs with very low permeability.

Ironically, the NGPA did not terminate but rather extended a situation that had characterized the gas market throughout the 1970s, in which lower-cost supplies were made unavailable because their prices were held down at artificial levels. To illustrate this concept, imagine two gas fields, Alpha and Beta, owned by the same producer, Charlie. The cost of production at Alpha is $4, while the cost at Beta is $7. Alpha qualifies as new gas under the NGPA, and Charlie can therefore only receive the regulated price of $1.75. Because this is below the cost, Alpha will not be drilled. However, field Beta qualifies as high-cost gas, meaning that Charlie can receive whatever price the pipelines bid. If the market needs more gas, the pipes will bid up to the cost of production at Beta, $7, rather than the much lower cost of production at Alpha, $4. Regulations make the cheaper supplies unavailable.

## PIPELINE COMPANIES' BIDDING WARS

It arouses little controversy to say that the 1954 *Phillips* decision was not good for the long-term health of the US gas industry. The most fundamental criticism of the ruling is that it applied monopoly utility regulation to a sector that was characterized by fierce competition and low barriers to entry: the upstream. That criticism is certainly correct. However, there is at least one enlightened explanation for why the Supreme Court saw the logic in regulating producer prices, and that is to prevent pipelines from engaging in bidding wars, which would lead to prices skyrocketing above costs.

In most markets, buyers face sellers directly, meaning that prices will rise or fall until equilibrium is reached. In the natural gas market, however, both buyers and sellers faced pipelines. In allowing interstate gas pipelines to function as energy merchants, rather than common carriers, the 1938 Natural Gas Act incentivized competing pipeline companies to bid up the price they paid to producers, without regard for the public interest. This is because, under cost-based regulation, the only way to grow profits is to grow market share. Had they been regulated as common carriers, pipeline companies could only have competed for producers' business by offering a cheaper or otherwise superior transportation service. But as merchants, they competed by offering a higher price for the commodity itself.

Seen from this angle, in issuing *Phillips*, the Supreme Court did not seek to regulate producers per se, but how pipeline companies transacted with them. Whether producers were operating monopolistically was beside the point: *Phillips* was not meant to prevent producers from gouging consumers because producers never faced consumers. Instead, the court issued *Phillips* to prevent pipeline companies from throwing money at producers, against the long-term public interest, in order to gain market share.

It could be said, then, that the root cause of the US gas industry's sclerosis was not producer regulation but rather that the pipeline companies had not been regulated as common carriers all the way back in the 1930s.

As *Phillips* was issued in 1954, the conditions for a bidding war seemed ripe. It was the height of the postwar boom, and gas demand was rising at such a rapid pace that obtaining the requisite twenty years of supply was becoming increasingly difficult. Access to supply became a key differentiator between those pipelines that would win an FPC certificate and those that would not, and prices had been rising steadily. Still, if *Phillips* is to be interpreted as a measure to prevent an imminent pipeline bidding war, it was probably premature by several years. In 1954 wellhead prices had only risen to around ten cents—even less in Texas and Oklahoma, and below the oil equivalent price everywhere—and the new pipelines built or substantially expanded during the era sourced gas from different production provinces.[10] While this is not proof positive that pipelines were not overpaying for gas supplies, it is at least indicative that there were still "many fish in the sea."

By the late 1960s, however, nearly all of the expansive petroleum provinces of the United States had been tapped, and the pipeline network, while not looking especially different on paper than it did in 1954, was much larger and more interconnected. There is no doubt that by the 1970s pipelines would have begun to bid aggressively for new supplies. In fact, that is just what they did, on intrastate markets. *Phillips* and the FPC's area rates prevented interstate pipeline companies from entering the fray and competing with the intrastates for domestic supplies, but this did not stop them from looking elsewhere. As I illustrate over the next few pages, neither *Phillips* nor title II of the NGPA, which had large industrial consumers pay the fuel oil equivalent price for gas, was truly effective in preventing bidding wars. In fact, the NGPA merely changed the geography of the bidding war that had begun nearly a decade earlier.

Before the NGPA, there were four higher-cost alternatives to lower-cost, but regulated, domestic gas. The most reasonable was pipeline imports from Canada, which were indexed to the price of fuel oil. These rose from 1.5 billion cubic feet per day in 1967, before shortages began, to almost 3 billion

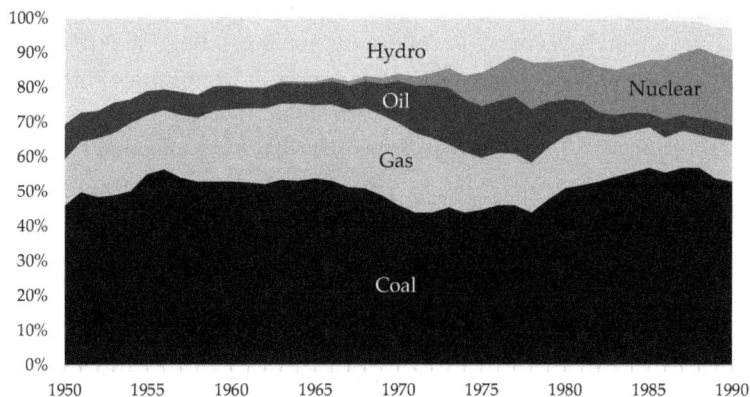

**FIGURE 9.4:** US electricity generation by fuel type

by 1973.[11] In 1973, however, the Canadian government began husbanding its gas supplies, disallowing any new exports contracts. The ban on new exports was not lifted until 1981, meaning that this supplement could not realize its full potential.[12] There were also various projects discussed to import up to 2 billion cubic feet per day from Mexico, but negotiations eventually collapsed.

Far less reasonable than pipeline imports from either of the United States' neighbors was the idea of developing the natural gas stranded in Alaska's North Slope. Prudhoe Bay, by far the largest oil field in all of North America, was discovered in 1968. In addition to ten billion barrels of oil, representing an incredible one-third of total US reserves at the time, the field was estimated to contain twenty-six trillion cubic feet of gas, which added up to a less astounding but still impressive 10 percent of the US total. Despite the prolific size of the resource, its location—some three thousand miles from the border between Canada and the Lower 48 states, as well as the extremely harsh Arctic conditions—meant that the project was too expensive to have ever gotten off the ground. Those twenty-six trillion cubic feet remain beneath the permafrost, continually reinjected to maintain reservoir pressure as the oil is sent to a terminal in Valdez, some eight hundred miles south.

Coal, the supply of which the United States has always had in excess of conceivable demand, also came back into the limelight in the 1970s. For one, many industrial facilities converted from gas to coal for fuel. The share of electricity generated from gas dropped from a high of 24 percent in 1970 to just 10 percent by 1986, nationally, while coal's share increased from 46 percent to 55 percent. But consumers were not the only ones eyeing the nation's vast coal reserves. Natural gas pipeline companies began investigating and investing in so-called synthetic gas facilities to convert coal into a more modern version of manufactured gas. But unlike the manufactured gasworks of yesteryear, which were built by utilities within their service territories and used coal that arrived on railcars, the syngas facilities contemplated in the 1970s were envisioned as mega complexes located close to mines that would transport gas to consumers through the existing, long-distance natural gas pipeline network.

While several syngas projects were considered, only one was built: a $2 billion facility in North Dakota, which produced just 125 million cubic feet per day. Soon after it came online, in 1984, it was deemed uneconomical, and the partnership that owned it defaulted on the $1.5 billion loan that the DOE had guaranteed in order to get the project off the ground. The grand-sounding syngas experiment was, by all accounts, a failure.

The last high-cost supplement to domestic gas in the 1970s was waterborne imports of liquefied natural gas (LNG). Activity in the LNG space clearly showed that a bidding war between interstate pipelines existed, regulations be damned, and foreshadowed the events of the early 1980s, after the NGPA was passed.

The most economical way to transport natural gas is via a pipeline. This is not only because it is gaseous at normal conditions—it is not so difficult to create an airtight vessel—but because its energy density is much lower than for a liquid fuel like oil. Indeed, oil is about twenty times more energy-dense than natural gas. A pipeline solves this problem by having a large diameter (more empty space per pound of steel wall) and by packing the gas in at high pressure, usually fifty to one hundred times atmospheric pressure.

This is not to say that gas transport cannot be compartmentalized. Small volumes of natural gas can be compressed to two hundred or even three hundred times atmospheric pressure and put on trucks, trains, or boats. But these extremely high pressures of three thousand to five thousand pounds per square inch require specially designed, and very expensive, tanks.

A more elegant and economical solution is to cool the gas to such a point that it condenses into a liquid, a process called liquefaction. For methane, the principal component of natural gas, liquefaction occurs at negative 260 degrees Fahrenheit and shrinks the product by a factor of about six hundred, making it that much more energy dense. Best of all, the LNG does not need to be held under pressure to keep it in its liquid state. Special cryogenic, thermos-like tanks are needed to prevent too much of the LNG from boiling off, but these are much less expensive than vessels designed for extremely high pressures.

The commercial origins of LNG can be traced back to the United States—the first liquefaction plant was built in West Virginia in 1917, and the first transoceanic LNG cargo departed Louisiana for England in 1959—but the global export market can truly be said to have begun in 1964, when Algeria sent its first tanker to the United Kingdom.

Algeria has always been one of OPEC's lesser producers of oil, but is home to several world-class gas fields in the Sahara Desert. As can be imagined, this location posed a problem for Algeria's state-owned oil and gas company, Sonatrach. Without a large enough domestic economy to absorb the massive supplies, and geographically isolated in northwest Africa, the only practical way for Sonatrach to market its gas was via LNG. Nor was Algeria the only remote country with a small domestic economic and large gas reserves. Indonesia began exporting LNG in 1977, and overtook Algeria as the largest exporter in the 1980s, before losing its title to Qatar, a small peninsula that juts out from Saudi Arabia into the Persian Gulf, and which shares the world's largest gas field with Iran.

Gas in Algeria, Indonesia, Malaysia, Brunei, Qatar, Australia, and many other countries was stranded, without a market, much in the same way that the giant fields in the southwestern United States were stranded before large-diameter, high-pressure pipeline technology was developed in the 1920s. And, just like in the United States, the initial development of these resources was extremely costly and involved a great deal of risk. Unlike US holding companies in the 1920s, LNG exporters did not have the ability to integrate vertically, from wells in Algeria to burner tips in Boston, and thus manage risk. The best they could do was sign long-term contracts that afforded as much financial protection as was feasible.

One such system of protection was imposing take-or-pay obligations on the buyer, meaning that, even if he it did not or could not take possession of the LNG, it would have to pay for it. This was, of course, not unique to LNG: Oscar Wyatt's Coastal States Gas Corporation had been signing contracts with take-or-pay provisions with domestic producers since the 1950s, and major pipelines had followed suit, so they were quite comfortable making these concessions to their LNG suppliers. Their willingness to do so was underpinned by two conditions that they regarded as constants. First, that gas demand would continue to grow, which it had, consistently, since World War II. This made the question of whether the pipeline would actually take the contracted-for volume moot—of course it would. Second, that its volumetric commitments to producers would be matched by volumetric commitments from consumers. Pipeline companies had, since their inception, imposed minimum bills on their customers. Take-or-pay obligations on the costs side were matched by minimum bills on the revenues side—risk was symmetrical.

With the shortage in full swing in the late 1970s, several pipeline companies drew up plans for LNG import, or regasification, terminals. In the end, four were actually built, making LNG the most popular of the pre–NGPA high-cost supplements. All were supplied by Sonatrach. Technically, each of the projects was successful. Economically, they were all disastrous.

In 1971 the first terminal was built and operated by a company called Distrigas, in Boston, and served local LDCs. Distrigas's terminal was quite small, meant to serve peak winter demand in New England. As such, its contract volume was for only sixteen billion cubic feet per year, or forty-four million cubic feet per day.

The three others were much larger. El Paso Natural Gas, at that point the nation's largest pipeline company, built two. One on the Chesapeake Bay in Cove Point, Maryland, to serve the mid-Atlantic and Appalachia; the second on Elba Island, near Savannah, Georgia, to serve the Southeast. Between them, El Paso Natural Gas's two terminals had contracts to import 365 billion cubic feet per year, or one billion cubic feet per day. The fourth was built by Panhandle Eastern, to feed its Trunkline system with 450 million cubic feet per day of gas at Lake Charles, Louisiana.

Despite their different geographies, customer bases, and sizes, all four projects suffered identical fates.

Trunkline LNG received its first cargo three years later than planned, in September 1982, after the gas shortage had turned into a glut, and after its budget had more than doubled to $567 million, versus the original estimate of $240 million. Its contract with Sonatrach called for 3.3 trillion cubic feet of deliveries over twenty years. In the end, the project took just 4 percent of that number, and shut down after fifteen months.[13] Trunkline LNG was the last terminal built in the United States until the mid-2000s.

El Paso Natural Gas's two terminals were mothballed in 1980, even before Trunkline LNG received its first cargo. The company wrote off the $365 million it had invested in the project after it could not come to agreeable pricing terms with Sonatrach.

Distrigas hobbled through the first part of the 1980s, but went bankrupt in 1985 after its customers would not take delivery of gas, saddling it with asymmetric take-or-pay obligations.[14]

From a spectator's seat, the issue appeared to be about pricing. Algeria was, in every case, demanding an oil-equivalent price, with the country's energy minister quipping, "A therm is a therm"[15] (a therm is equal to one hundred thousand British thermal units, and is often used in small customer billing). As he spoke these words, in September 1980, war had just broken out between Iran and Iraq, and the oil price had never been higher. The US pipeline importers thus found their customers revolting against paying such a high price for gas. Still, FERC had approved each of the LNG projects and the pricing provisions in their contracts: the price should have been the price.

But by the time the large LNG facilities actually came online, the NGPA had been passed, and the rush for high-cost domestic gas supplies was on. Politically, passing such high costs on to consumers for foreign-sourced gas was simply unpalatable. Thus, while pipelines all over the country were paying upward of $6 for domestically produced high-cost gas in 1980, the Department of Energy told Sonatrach on El Paso Natural Gas's behalf that it could not accept a price higher than $5.[16]

In reality, the issue was not about pricing so much as it was about whose side of which contracts the government would choose to abrogate. A foreign, state-owned oil and gas corporation was easy pickings, but officials would soon be forced to make more difficult decisions, where all the actors were domestic.

The oil crisis began to dissipate in early 1981, as buyers realized that they were as responsible for the price increases as was any actual shortage of oil. They slowed their purchasing programs and began to draw down the massive inventories that they had built over the preceding two years. More importantly for the long run, the high prices of the 1970s had encouraged new production from the North Sea and Mexico, which was rising steadily.

The contracts for high-cost gas that US pipeline companies had signed looked reasonable in the late 1970s and 1980, when oil prices were at record highs. They began to look too high in 1981, higher still in 1982, and downright ludicrous in 1983. Market forces thus dictated that losses needed to be taken. The US government had a large say over who would take them. By and large, it protected consumers and left the pipelines and producers to duke it out, without much intervention. The pipelines, in turn, abrogated thousands of contracts with producers, and used what in most cases were their larger and more sophisticated legal experts to force producers to agree to more realistic price levels.

The managers at the helm of pipeline companies in the 1970s can be accused of grandiosity and ego, eager as they were to push through large, technically complicated projects. Mostly, however, they were operating as efficiently as possible within the confines of a distorted regulatory framework. Panhandle's 1979 annual report to shareholders said of Trunkline LNG: "The company's commitment to LNG is based upon one paramount consideration, namely, that the gas reserves involved cannot be duplicated from domestic sources."[17] In hindsight, this was patently incorrect, but it reflected the prevailing wisdom of the time. Less than two years earlier, a government report had predicted that between 5 percent and 15 percent of total US natural gas consumption would eventually be served via LNG, which would have equated to a whopping 2.7–8.2 billion cubic feet per day.[18] This would have necessitated several more import terminals, and would have cast Panhandle Eastern, El Paso Natural Gas, and Distrigas as savvy and farsighted leaders.

Instead, imports petered out at just seven hundred million cubic feet per day in 1979 before falling steadily in the years after, as a glut of domestically produced gas developed.[19] This was financially ruinous for the owners of the LNG terminals that were built, and, ironically, put the pipeline companies who had tried but failed to build several others in a stronger competitive position. The change in policy rewarded them for their failures.

The historian Richard H. K. Vietor sums up the pre–NGPA high-cost supplements concisely: "These projects, which seemed reasonable at the time, now stand as monuments to the perverse consequences of regulation."[20]

## HIGH-COST GAS AND LOW-FLEX CONTRACTS

Just under a year after the NGPA was signed into law, on November 1, 1979, high-cost domestic gas was deregulated. The oil crisis was in full swing, with prices rising, and pipelines were in shopping mode. Some had just finished, or were still in the process of building, extremely large and expensive facilities to import LNG or gasify coal. The luckier ones had either not trod down this path, or had stumbled early enough to avoid the painful endgame. Both types drooled over high-cost gas.

This may have seemed odd, because prices quickly rose above $6, with some contracts signed at higher than $9. Why were pipelines so excited about paying so much, when they balked at similar pricing demands from Algeria?

For one, there was more regulatory certainty. The NGPA was law, passed by Congress and signed by the president. FERC had a long history of waffling and changing its mind, but pipelines could be certain that they could buy NGPA–qualifying high-cost gas at market prices, and then roll those costs in with the rest of their supply portfolio and pass it on to LDCs. For another, high-cost gas contracts represented bite-sized chunks that had an almost invisible effect on the weighted-average cost of gas in their much larger overall supply portfolios. Whereas Trunkline LNG was a 450-million-cubic-feet per-day dollop of very expensive gas, a contract for 20 million signed with an Oklahoma producer at $8/MMBtu would pass almost unnoticed into the pipeline's portfolio.

One such Oklahoma producer was the GHK Company, founded by Robert Hefner III—the H in GHK—who is regarded as the father of deep gas.[21] It was Hefner who advocated most ardently for the deregulation of deep gas, and was one of the first to break with his peers in the exploration and production industry by supporting the NGPA. In fact, at the nadir of the bitter and lengthy campaign to make the NGPA law, when that

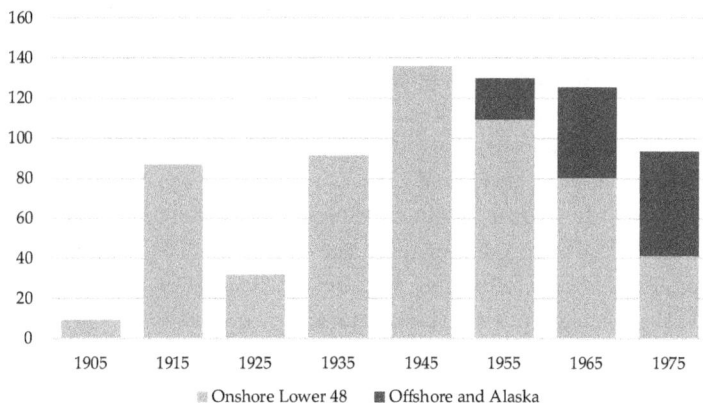

FIGURE 9.5: US natural gas discoveries over ten-year intervals (trillion cubic feet)

prospect looked most improbable, Hefner received an offer from James Schlesinger, whom President Carter had appointed as the first secretary of energy. Schlesinger had little faith in the prospects of deregulating all new gas, but told Hefner that he could convince Carter to sign an executive order exempting deep gas from price regulation. Hefner balked, stating that such an order could be reversed by the next president. As ever in the energy industry, large sums of capital would need to be spent, and amortized over decades of sales. It was too much to risk on an executive order, which could easily be canceled by the next administration. Hefner continued to press the case for a permanent solution, and the NGPA was still the best hope.

It was also Hefner who suggested 15,000 feet as the definition of "deep" gas. This was not a stratigraphic distinction—stratigraphy varies greatly both within and among petroleum basins—or even an operational one. It was, in fact, almost totally arbitrary. An industry magazine to which Hefner subscribed split out some of its statistics by well depth, and >15,000 feet was the deepest category.

Throughout the late 1970s and into the mid-1980s, most of the major producers shared the conviction that the United States was simply running out of oil and gas. Even as a gas surplus emerged in 1981 and 1982, they did not change their stories: the glut was temporary, and there were simply not enough fields of large enough size to stem the inexorable production declines.

There were good data to back this up. Giant fields were being found with less and less frequency, and when they were, they were offshore—or, in the case of Prudhoe Bay and several others—in Alaska. Nor were the majors perpetrating some great scam on the American people, telling them oil and gas supplies were running out while secretly withholding billions of barrels and trillions of cubic feet from the market. They truly believed that the United States was tapped out, and put their money where their mouths were. While the majors had been doing business internationally for some decades already, it was during the 1970s that they largely exited the business of finding and producing oil and gas in the continental United States, abandoning their onshore E&P roots and instead ploughing their capital into challenging offshore prospects in the Gulf Coast or North Sea, or into parts of the world where security and property rights were poorly established. Along the way, the majors made themselves into large bureaucracies with centralized decision-making processes—a business model that was well-suited to managing risk, but made them inept and unimaginative wildcatters.

Robert Hefner III saw the United States through a completely different lens. He did not have to find one hundred million barrels of oil and one hundred billion cubic feet of gas every year just to keep his reserves flat and his investors from revolting. And he knew from experience that the majors had always hunted for oil, ignoring dry gas prospects. Before founding GHK, Hefner worked at the economic analysis division of Phillips, and even here, at the nation's largest gas producer, he was barred from assigning any economic value to gas in his calculations: it was, economically speaking, purely incidental to oil. When Hefner left Phillips in 1959, he knew well where to find gas. Whether he could make any money extracting it was less certain.

The Anadarko Basin of western Oklahoma, southwestern Kansas, and the Texas Panhandle does not look much like a basin. On the surface it is flat, sometimes astoundingly so. Belowground, it does not have a typical bowl-like shape but rather gets progressively deeper as one moves south, and then slopes sharply back upward. In western Oklahoma up to forty thousand feet of sediment overlies the basin's "basement rock," with both oil- and gas-bearing rock formations stacked atop one another like pancakes. It was here, from a base in Elk City, that Hefner and GHK went in search of deep gas in the 1960s.

Hefner soon found out that the relationship between depth and cost is not linear—drilling six times as deep could cost thirty times more. His wells generally cost upward of $6 million, and sometimes as much as $20 million, in an era where wells being drilled into shallower, more conventional formations cost around $200,000. The intense pressures and extreme heat of ultra-deep reservoirs required much more robust well designs. For example, the metal tubing through which gas flows from the reservoir up to the wellhead had to withstand pressures in excess of twenty thousand pounds per square inch, which GHK learned after the normal tubing it had

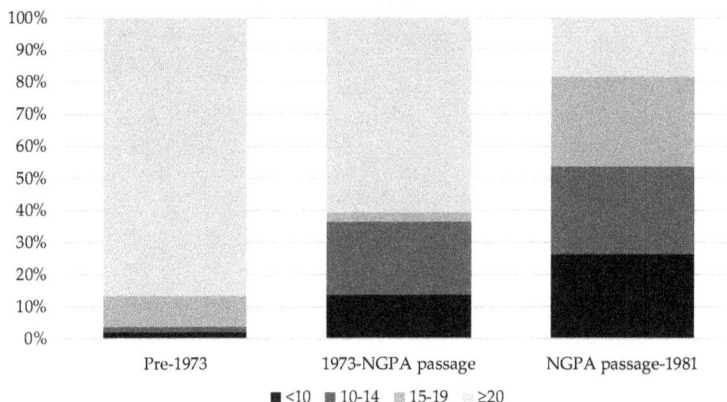

**FIGURE 9.6:** Length of pipeline/producer contract terms (years)

used in one well simply split apart at the joints. The company also found out that run-of-the-mill drilling mud was unable to fulfill two of its main functions: lubricating the drill bit and preventing blowouts while drilling was still under way. (It normally accomplishes this latter task because the dense mud forms a column thousands of feet tall, whose great weight overcomes the propensity of the gas to flow up-hole.) Even after the kinks were ironed out and the well was drilled and completed successfully, the company found that there were no gauges made that could measure the extreme pressure of gas flowing at the wellhead. Hefner needed to have an oilfield services firm build him custom gauges in its Tulsa facility.

But if the costs of deep gas wells astounded, then so did the production levels. The same tremendous pressure that made drilling so difficult also meant that many deep gas wells flowed 8–10 million cubic feet per day during their first year of production, at a time where rates at conventional wells were measured in the low hundreds of thousands. Deep gas proved to be a prolific resource, and as GHK and others ramped up activity, 15,000 feet quickly became commonplace. The deepest well that GHK drilled reached 31,441 feet and from 1974 until 1979 held the title of deepest hole in the world, when scientists from the Soviet Union drilled one that eventually bottomed out at more than 40,000 feet.

The 15,000-foot cutoff created a perverse incentive for some audacious producers holding acreage in hilly topography. It was sometimes practical to make a 14,000-foot reservoir qualify as "deep" by drilling from atop a mountain, rather than from a valley floor. The prospect of getting triple or quadruple the price was well worth the additional cost, with one executive quipping: "We would have drilled from a zeppelin if we could've found one."[22] Entertaining as the thought is, drilling from mountaintops was not

**TABLE 9.2: MAJOR PRICE PROVISIONS IN NATURAL GAS CONTRACTS**

| | | Quantity (quadrillion Btu) | Price ($/MMBtu) | Most-favored nation clause (%) | Oil parity (%) | Market-out clause (%) | Maximum price (%) |
|---|---|---|---|---|---|---|---|
| Interstate | Pre-NGPA | 4.13 | 1.01 | 92 | 12 | 6 | 8 |
| | Post-NGPA | 2.72 | 2.53 | 87 | 28 | 19 | 8 |
| Intrastate | Pre-NGPA | 3.37 | 1.64 | 66 | 1 | 14 | 8 |
| | Post-NGPA | 0.96 | 2.19 | 45 | 3 | 27 | 15 |
| Total | Pre-NGPA | 7.50 | 1.29 | 80 | 7 | 10 | 8 |
| | Post-NGPA | 3.68 | 2.44 | 76 | 21 | 21 | 10 |

Source: US Department of Energy, Energy Information Administration, "Natural Gas Producer/Purchaser Contracts and Their Potential Impact on the Natural Gas Market," DOE/EIA -0330, June 1982, 54.

a widespread practice, largely because most of the petroleum provinces of Oklahoma, Texas, and Louisiana are located beneath table-flat plains. It certainly played no role in the glut-driven crisis of the 1980s.

What did play a large role was contracting practices, which had changed little over the previous decades. In this regard, the experience of GHK serves as a near-perfect microcosm for the industry at large.

GHK's largest single contract was with El Paso Natural Gas, who was the biggest buyer of high-cost gas in Oklahoma. In most ways, this contract was perfectly normal. It had a term of twenty years, which was fairly standard (although contract lengths did begin to shorten after the passage of the NGPA).

It also had take-or-pay provisions, which were almost universal. Lastly, it contained a most-favored-nation clause, which meant that GHK was guaranteed to receive the highest price that El Paso Gas paid any other producer in the same area. This was also a very common practice, especially in the interstate market, where around 90 percent of contracts included a most-favored-nation clause. In fact, the majority of contracts contained three-way most-favored-nation clauses, which required the pipeline to pay the producer the highest price that any producer received in the area, even if it was from a different pipeline. In other words, if GHK were selling gas to El Paso Gas for $5, but then Panhandle Eastern signed a new contract with Phillips for $7, this would trigger the most-favored-nation clause in GHK's contract, and it too would start receiving $7 from El Paso Gas.

Again, all this was perfectly typical. What set GHK's contract apart from others was its extremely favorable pricing terms: Hefner had negotiated for GHK to receive the higher of the fuel oil equivalent price or $3.50/

MMBtu, from a 1979 base year, plus a 6 percent per-year escalator (which, after twenty years, would lead to a price of $11.22).

This one contract illustrates explains almost singlehandedly how the bidding war of 1978–1982 led directly to the so-called take-or-pay crisis, which bankrupted pipelines and producers alike. A single contract like the one between GHK and El Paso Gas—with a high floor price and a ceiling at the oil equivalent price—would trigger most-favored-nation clauses across the entire producing area. The pipelines had essentially, and without realizing it, turned hundreds of small, high-cost gas contracts into one enormous contract, with the weakest negotiator among them having determined the field price.

It was this utter lack of foresight in contracting practices that convinced regulators that pipelines were unsuited to act as merchants. But again, context makes the pipelines seem less clueless than shell-shocked: demand had grown for four decades straight, and the majors were exiting the United States because there was no more left. They had been using these types of contracts for years without incident, and in fact these long-term, rigid practices had helped stave off the shortages of the 1970s (though they also contributed to its severity when it inevitably arrived). Producers must also be held to account. The days of monopolistic pipelines paying them peanuts, if not outright stealing the resources from under their feet, had long since passed. Both most-favored-nation clauses and take-or-pay clauses were much-needed protections when pipelines did most of their business with affiliated producers, but by 1980 there were more than fifty pipelines that received gas from the field in Louisiana, and an even greater number in Texas.

As shortage gave way to surplus, few in the industry thought it was anything more than a blip. Those who believed the bubble was more than a short-term phenomenon, like Hefner, were mostly ignored.

## "DRIVE 75 AND FREEZE A YANKEE"

Producers did not like the fact that the NGPA had extended the federal government's reach into intrastate gas markets, but they had accepted this intrusion in exchange for higher interstate prices, a schedule for eventual deregulation, and because they were fearful of the consequences of not accepting it. As it turned out, the NGPA took effect just ahead of the 1979–1980 oil shock, which led to a drilling bonanza that made countless Texans, Louisianians, and Oklahomans into millionaires.

The NGPA ended up being a fine deal for producers, but a terrible deal for consumers in producer states. The act not only took away the advantage of more ample supplies that intrastate pipeline companies enjoyed over their interstate brethren, it actually turned it into a marked disadvantage.

Intrastate pipelines had maintained healthy reserves throughout the shortages of the 1970s because they paid higher, unregulated prices to producers. Interstate pipelines were prohibited from paying more than the low, regulated price, and thus could not compete. Accordingly, their reserves shrank. All the while, interstate consumers were enjoying a subsidy on the back on intrastate consumers. Most were unaware of it, and certainly did not see themselves as lucky, given the shortages they were enduring, but the higher intrastate prices were keeping many gas producers in business, which benefited everyone and prevented the devastating shortages of the late 1970s from happening even sooner. The profits producers earned from sales to intrastate pipelines allowed them to continue taking losses on interstate sales. Ample supplies, then, were not a privilege that the intrastates enjoyed by virtue of being geographically close to the fields, but because they paid for it.

After the NGPA became law, the intrastates could no longer bid above the interstates for regulated gas. Leveling the playing field was, of course, one of the main goals of the act, so this was regarded by many as a success. But in fact the NGPA did more than level the field—it tipped it deeply in favor of the interstate pipelines, and hence in favor of consumer states.

The first reason was because of the cushion of old gas that the pipeline companies began with after the passage of the NGPA. With some exceptions, the NGPA kept old gas prices at their original contract levels, escalated by the inflation rate. This alone would have left the intrastates at a disadvantage because they had been paying higher prices before the NGPA: their cushions were made up of more expensive gas. It was made worse when Congress, in a bid to court producers, deepened the disadvantage by allowing producers of old (i.e., pre–NGPA) intrastate gas to receive the "new gas" price, $1.75, escalated by inflation plus an additional 3.5 percent. Hence, while many interstates continued to receive substantial volumes of sub-$1 gas into the 1980s, intrastates had no such cushion.

The second reason followed from the first. The lack of cushion meant that the intrastates could not viably compete in the post–NGPA bidding war because they could not absorb the high-cost gas to the same extent as the interstates.

Finally, even if they would have been outbid, the intrastates were explicitly prohibited from bidding on gas from the federally controlled Outer Continental Shelf (OCS), which represented an important source of supply during this period. (Between 1979 and 1982, OCS wells brought online added nearly ten billion cubic feet per day to overall supplies.)[23]

Rumor has it that the first "Drive 75 and Freeze a Yankee" bumper stickers turned up in Wyoming in 1974 in response to the National Maximum Speed Law, passed by Congress in January of that year. This law, part of the Emergency Highway Energy Conservation Act, set a national speed limit of fifty-five miles per hour, taking the decision out of states' hands. The speed limit, along with several other measures, was meant to reduce oil consumption in the wake of the 1973 oil crisis.

Wherever the bumper stickers originated, by the mid-1970s they were widespread in producing states, and in 1978 a minor Texas–based folk group recorded the song "Freeze a Yankee." The song never hit the top forty, but its lyrics spoke to the mood in producing states with regards to Washington, DC, which was viewed as being dominated by northeastern interests.

> Us Texans love our Cadillacs,
> Big Continentals and Pontiacs.
> We're gonna' keep all the gas we can make,
> And let them Yankees shiver and shake.

> Well them Yankees say they need our oil
> And they gotta have gasoline,
> But don't you put no refineries way up north,
> They wanna' keep their air real clean.

As a result of the NGPA, the trend that had prevailed in the late 1960s and continued through 1978 abruptly reversed—reserves began to decline at intrastate pipes, while growing at interstates. From 1977 to 1980 total reserves dedicated to intrastate pipelines dropped by thirteen trillion cubic feet, or 20 percent, while interstate reserves grew by six trillion, or 8 percent.

In a final insult to consumers in producer states, the Carter administration had passed the Power Plant and Industrial Fuel Use Act (FUA) as part of the 1978 package of energy reforms. The FUA prohibited any new power plant from using either oil or gas, and mandated that even all existing gas-fired plants be converted to coal by 1990. For Texas, Louisiana, and Oklahoma, whose grids were almost entirely powered by gas generators, this amounted to cruel and unusual punishment.

It was not that producer states wanted so badly to remain 100 percent gas-dependent. Their electric utilities were well aware of market dynamics, and had begun taking active steps to diversify their generation mixes as gas prices rose in the mid-1970s. Around twenty coal-fired power plants had either already been built or were under construction in these three states by the time that the FUA passed, which accounted for one-third of their total installed capacity base of seventy-five gigawatts—a substantial amount. These facts muddy the narrative that some like to recount, of the federal government forcing producer states to give up their local, clean-burning gas only to import dirty coal. In fact, Oklahoma Gas & Electric, that state's largest electric utility, ran weekly ads in the *Oklahoman* in 1975 and 1976 promoting using coal from a mine 1,100 miles away, in Wyoming, to be delivered to its new Muskogee plant, which began operations in 1977. An op-ed in the *Oklahoman* from the period—written a full three years before the FUA—hailed the state for going ahead and building the Muskogee coal plant and the Black Fox nuclear power plant in spite of inaction from Washington.[24]

What rightfully irked producer states was that, with the FUA, Washington was demanding that they scrap their existing units by 1990, at which point many would still have had ten or twenty years of useful economic life. It was largely because of this conversion provision that producer-state politicians railed against the FUA, citing studies suggesting that compliance would cause electric rates to triple or even quadruple.

The Department of Energy recognized the impracticality of conversion, especially among smaller utilities, which did not have the financial wherewithal to raise the capital that would be required. "Forcing the utility to completely convert to coal in the near term will create financial difficulties that the company may not be able to meet successfully. Similarly, a rapid conversion schedule would present almost insurmountable environmental difficulties."[25]

I began this chapter by pointing out that most of the landmark deregulation legislation so often associated with Ronald Reagan was actually enacted under Jimmy Carter. However, as we have seen in natural gas, and which was equally true in the other industries that Carter deregulated—telecoms, airlines, trucking, and banks—Carter had a very heavy-handed style of deregulation. In fact, the Carter era could be characterized as an era of highly regulated deregulation.

When Ronald Reagan became president in January 1981, he began an era of deregulated deregulation. In June his new energy secretary, James Edwards, proposed easing the FUA, and the conversion provision was eliminated in August. It would take another six years to repeal the FUA in its entirety, as the industry finally came to realize that there was no gas shortage.

The latter half of the 1970s and the entirety of the 1980s was coal's last hurrah in the United States. While a good deal of the coal capacity brought online over this period certainly would have been built even in the absence of regulations, regulatory distortions were responsible for the marked drop in gas's share of the US electricity market. Between 1960 and 1974 almost 100 gigawatts of gas-fired generation were built. Over the next fifteen years, that number dropped to less than fifty. Coal build, meanwhile, was virtually unchanged over the two periods, at 130 gigawatts.

Texas, Oklahoma, and Louisiana                    Rest of US

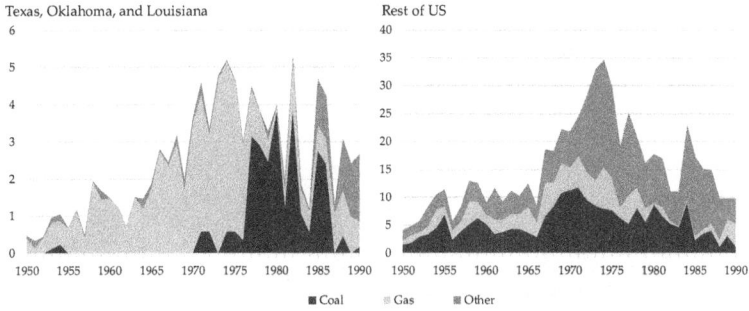

**FIGURE 9.7:** New electric capacity by year, gas versus coal versus all other fuels (mega-watts)

The rapid increase in production and simultaneous drop in demand could mean only one thing: a glut was around the corner. Beginning in late 1982, the United States experienced a drawn-out oversupply of gas. The era is known as the "gas bubble."

# CHAPTER 10

## THE GAS BUBBLE AND THE END
## OF MERCHANT SERVICE

In early 1981, consensus among economic forecasters was for the oil price to continue to rise, from $40 per barrel to over $100. In hindsight, this prediction was wrong not only in magnitude but also in direction, as prices actually fell. Given the circumstances, though, the forecast seemed reasonable. "Death to America" was the clarion call from across the world's largest oil-producing region, the Middle East. Those drillers with the most money and experience—the majors—had largely exited North America in their search for more hydrocarbons. And, even in the gas-rich Southwest, electric utilities were in the midst of large-scale conversions to coal.

It serves as bibliographic irony, then, that the data used to examine US gas pipelines' purchasing behavior during their shopping spree was collected by the government in order to quantify the risk of a fly-up in prices, which was expected after the complete deregulation scheduled for 1985 under the NGPA. Lawmakers feared that gas prices might jump from the regulated price, $2.50–$3.00/MMBtu, to the oil-equivalent price, which, at the $100 per barrel that was forecast, would be almost $17/MMBtu. The irony is that we now use the data for the exact opposite purpose: to quantify pipeline companies' take-or-pay obligations to producers, which made them vulnerable as demand fell dramatically and they were unable to market their high-cost gas.

The price of high-cost gas was linked to the price of oil. It was only natural that total US gas demand began to fall as prices ran up in the aftermath of the 1979 oil crisis. It is a basic economic principle that consumers demand less of something the more it costs. But as the oil crisis waned and prices fell, US gas demand did not recover, which led to a protracted situation of oversupply—the gas bubble.

Ill-advised regulation—namely, the Power Plant and Industrial Fuel Use Act of 1978 (FUA), passed alongside the NGPA as part of President Carter's energy reforms—played a role. While it was not responsible for the large-scale coal and nuclear build-out that electric utilities undertook in the late 1970s, which was a response to (admittedly distorted) market conditions, it prevented them from switching back to gas as the glut emerged in the early 1980s. Gas demand had also begun to fall in the all-important residential and commercial sector, after forty years of rapid and competition-free growth, as real estate developers installed electric-powered heating in new homes and offices. Since homeowners are unlikely to switch their heating systems, decisions made in the 1970s had effects that lasted for several decades. Last, an intangible but undeniable conservation mind-set led overall energy consumption—of oil, gas, and electricity—to drop beginning in 1981, proving out the fears that Saudi oil minister Ahmed Zaki Yamani had voiced a year and a half earlier, but that had fallen on deaf ears.

The broader US economy was also at its weakest since the Great Depression, with the unemployment rate reaching a devastating 10.8 percent by the end of 1981. Much of this was due to extremely high interest rates, which Federal Reserve chairman Paul Volcker had raised in order to combat the inflation that the oil crisis had ignited. The energy-intensive manufacturing sector was one of the hardest hit.

The primary culprit for why gas demand did not recover as the oil crisis abated was that gas prices did not fall, initially, and then did not fall by enough when they finally did. Pipelines, who had been so eager to secure new supply, had signed onerous take-or-pay and must-take provisions with producers. This led to an extremely perverse situation where the drop in demand actually caused their weighted average cost of gas to rise, as pipelines were forced to shut in cheaper, cushion gas, which they had contracted for decades earlier, and hence had less stringent take terms. As take-or-pay liabilities spiraled upward, pipelines passed the costs onto consumers through their minimum bills, meaning that delivered gas prices did not fall in lockstep with oil.

The sum total of all this was that from 1979 to 1986, total US gas demand fell by 20 percent—or by an incredible eleven billion cubic feet per day. Over the same period, interstate pipelines accumulated tens of billions of dollars in take-or-pay liabilities.

The situation was untenable, and the laissez faire–inclined Reagan administration was not about to bail out pipelines at the consumer's expense. Reagan's FERC canceled all minimum bills for LDCs in 1983, leaving pipelines with one-sided commitments to producers, with whom they were left to negotiate directly. It might be said that the last, truly merchant service that pipeline companies performed, where they took on direct and substantial market risk, was in renegotiating billions of dollars of supply contracts with producers. Indeed, their business model from around 1983 through around 1995 was not that of an energy trader but of an aggressive law firm. In this, they were fairly successful: when all was said and done, the pipelines recovered a bit more than seventy cents on the dollar of what their contracts said they owed producers.[1]

Completely by 1992, but for the most part by 1988, interstate pipelines transformed themselves into common carriers,[2] providing transportation and storage to producers and consumers, who would deal with one another directly, or through a new type of company: energy marketers. Though the latter term may be unfamiliar to some, almost everyone can name at least one such energy marketer: Enron.

## THE PARTY AND THE HANGOVER

When the NGPA deregulated deep gas in 1979, "It was as though Washington, DC, had pointed its finger at western Oklahoma and singled out its residents to become rich."[3]

"The glamour is in the Deep Anadarko," a Dallas geologist remarked in 1981, at the height of the deep gas frenzy. Said another: "Every prospect available is being sold. I don't think there's ever been a time when people would drill with so little justification."[4] Robert Hefner III's GHK Company had started out in the deep basin almost alone, but now found itself surrounded by other drillers hoping to strike it rich beneath fifteen thousand feet.

To do this, of course, required money, and lots of it. This was more difficult in Oklahoma than in other parts of the country because, until 1983, Oklahoma laws prohibited banks from having more than one branch. This meant there were hundreds of community banks, none of which became anywhere near large enough to finance a drilling boom that would eventually soak up between $10 billion and $15 billion.[5]

So begins the story of an unassuming, Oklahoma City–based lender named after the strip mall in which it was located: Penn Square Bank. The small bank, next-door neighbor to Shelly's Tall Girl Shop,[6] could not have been confused for Goldman Sachs. But Penn Square grew to account for such an outsized role in energy lending that Federal Reserve chairman Paul Volcker himself needed to get involved in its affairs when boom gave way to bust.

If there is a central figure in the Penn Square story, it is Bill Patterson, who was the head of the bank's energy lending division. He is also the most colorful.

Patterson's attitude toward lending can be summarized by an exchange he had with a young man, fresh out of college, who wanted to set up his own drilling operation. Nervously, the young man said to the group of bankers before him that it would be nice to have $500,000, but that he could probably make do with $100,000. Patterson leaned over the table and said: "Son, you can't do shit with five-hundred thousand. How about five million? What could you do with that?"[7]

Patterson's path to success was his embrace of a lending practice that allowed small Oklahoma lenders, like Penn Square, to play in the proverbial big leagues: correspondent lending. Local banks had relationships with drilling companies and expertise in energy lending. Meanwhile, large, national banks desperately needed new loan opportunities. Their deposits were mushrooming, which was largely because high oil prices had generated a massive inflow of petrodollars from the Middle East and other countries. Meanwhile, their traditional client base of large US corporations had circumvented the banks by issuing short-term commercial paper directly to the market, which shrank their loan books. Some had responded by investing large sums in developing countries in the late 1970s, but these had proven very risky, as defaults or steep haircuts led to substantial write-downs.

All this meant that US banks were eager to plough billions into nontraditional sectors of the domestic market. They found a perfect sink in Oklahoma, which was home to an upstream oil and gas industry that had a nearly bottomless appetite for capital (and risk), and where large, entrenched lenders were conspicuously absent.

The solution that connected Bill Patterson and his tiny, shopping-mall bank to national behemoths was for Penn Square to originate loans to oil and gas companies, whom they knew well, Oklahoma City being a fairly small and tight-knit conurbation, and then sell those loans downstream, to the big banks. Once sold, Penn Square would continue to participate by servicing the loan—making sure the interest was being paid and keeping tabs on the borrower's activities—but the loan itself was no longer on Penn Square's balance sheet. Correspondent lending a la Penn Square was, effectively, selling other peoples' money. Patterson excelled at this task, and Penn Square quickly became the largest correspondent, or upstream lender, in the country.

Part of the textbook definition of economic bubbles is that they are notoriously difficult to identify until after they have burst. Certainly, the market in the years 1979–1981 had features that made the Oklahoma drilling frenzy look sustainable, and even likely to grow. There was unrest in the Middle East, the majors continued to predict that US production was in terminal decline, and gas was rapidly losing share to coal in the electric power sector. Hefner and his acolytes, which had grown from virtually zero to the entire population of western Oklahoma, assured anyone who would listen that trillions of cubic feet of gas lay buried deep in the Anadarko. And at the new, unregulated deep gas prices of $5, $7, or even $9, the wells would pay for themselves in just a few years, or even months, despite their staggering costs. Equally, though, the behavior of Penn Square, and Bill

Patterson in particular, should have tipped off more sober observers of the leaner times that lay ahead.

Three downstream, money-center banks accounted for most activity in this era: Chicago–based Continental Illinois, New York–based Chase Manhattan, and Seattle-based Seafirst. When Seafirst threw its annual Christmas bash in December 1981, Patterson arrived on a bank-owned private jet "filled with good lookin' young dollies." So full of women was Patterson's plane, that the other Penn Square loan officers in attendance were unable to secure seats, so they did the "sensible" thing: they hired another private jet.[8] Over the next several months, as business continued to boom, Penn Square bought another jet, which allowed Patterson to issue what became fairly frequent exclamations of "Let's rodeo!," which meant an impromptu trip to Las Vegas. On these flights, he grew infamous for wearing Mickey Mouse ears and guzzling champagne out of his cowboy boots, which became something of a tradition among energy bankers of the day.[9] Around this time, Penn Square also began pursuing historically less-than-stable investments, purchasing a large number of racehorses.

Patterson disputed some of these accounts. For example, he insists that the champagne was in a loafer, not a cowboy boot.[10] But the numbers spoke louder than the anecdotes.

In 1977 Penn Square had just $62 million in assets, before it began its foray into correspondent energy lending. By 198 it had originated more than $2 billion in loans. Continental Illinois, which was the seventh largest bank in the United States at the time and enjoyed a reputation as an extremely conservative lender, took on more than $1 billion alone,[11] while Seafirst, the thirteenth largest, took on $400 million.[12]

Oil prices peaked in 1980 and began to creep downward in late 1981. But the creep was steady, and much more pronounced in real terms than nominal, as inflation was running at almost 10 percent annually. In 2020 dollars, the oil price had remained at around $100 per barrel from February 1980 through June 1981, but dropped to $85 a year later, and to just $70 the year after that. As prices slid, the upstream banks began to swoon.

In February 1982 Continental Illinois prepared a long list of small loans that it wanted to sell back to Penn Square. These loans were riddled with documentation exceptions—such as missing signatures, dates, handwritten addenda, and, more often than not, a total lack of financial due diligence on the borrower. Mostly, Continental had bought these subpar loans in order to maintain and deepen its profitable relationship with Penn Square. It appears that, as of early 1982, the relationship continued to matter more than loan quality, since none of these problems loans moved off Continental's books. Continental did not press harder because, at this point, it had only grown worried about a portion of its portfolio. The bank remained constructive on the long-term health of the US upstream oil and gas industry, even if Penn Square hadn't dotted the i's and crossed the t's on a few loans. Oil prices were off their highs, but still at historically elevated levels, and most of the big loans that it had bought from Penn Square seemed in order.

Cracks in this façade appeared the very next month. In March 1982 Seafirst decided that it needed to shore up its balance sheet by selling off some of its massive energy loan portfolio, which by then totaled more than $1 billion, of which more than $400 million had come from Penn Square. Continental, still generally constructive on oil and gas lending, sent an executive to pore over Seafirst's portfolio, and gave him a remit to snatch up whatever high-quality loans that he could buy at a bargain. Instead, the executive returned to Chicago in shock, having confronted "a graveyard of deals we had turned down."[13] It became clear very quickly that the problem loans that Continental had taken on were not outliers but the average. Penn Square had been loaning money to anyone who had asked, and, indeed, to many who hadn't. The same collateral was often used for more than one loan. And always, the only conceivable way for borrowers to make interest payments was if oil and gas prices remained high. Alarm bells started ringing.

Hefner says he saw the writing on the wall when he went to the dry cleaners, where he had been going for fifteen years, and found it closed because the owner had gone into the oil business. "That's when I knew it was over," he recalled in an interview.[14] GHK began pulling money out of its Penn Square accounts in June 1982, just weeks before the bank was liquidated by the Federal Reserve.[15] Impeccable timing, it would seem.

In reality, Hefner's decision to withdraw funds from Penn Square was borne of fury, not of prescience. It came immediately after Bill Patterson had foisted onto Hefner seven newly built rigs for $30 million, assuring Hefner that he was only acting as a temporary warehouse until a permanent buyer could be found, which would not take long.[16] Bank examiners had set up shop in Penn Square's basement and began classifying the quality of its loans, and Patterson was in a mad scramble to keep things afloat. The seven rigs had been bought by a partnership using a loan made by Patterson, that Penn Square had then sold downstream to Chase Manhattan. The partnership had essentially gone bankrupt, and Chase was livid. Transferring the assets—and the liabilities—to a big name like Hefner appeased Chase, and kept up the appearance that Penn Square was in fine condition, since it was still having no trouble selling its loans downstream. How could the bank examiners assign a substandard classification to loans that downstream banks were eager to buy?

It goes without saying that no buyer for Hefner's rigs emerged. He had assumed $30 million in debt that he did not want, for rigs that he did not need and could not sell.

In context, that $30 million was a drop in the bucket. By all accounts, Hefner was a true believer in the potential of the Anadarko Basin, not a huckster out for a quick buck. But he was also, unarguably, a degenerate debtor, who used every new dollar borrowed to pay past dues, and had been doing so well before the NGPA deregulated deep gas and Penn Square went on a spree selling other people's money.

Things came to a head after Penn Square was liquidated on July 5, 1982. While oil prices had not yet plummeted, the prices that producers like GHK were receiving for their deep, high-cost gas had, dropping to just $5–$6 from as high as $9–$10. Worse still, even at these lower prices, the pipelines could not market the gas in the face of rapidly declining demand, and they began simply not to take the volumes for which they had contracted. When GHK inevitably became insolvent, Hefner's creditors estimated that he owed between $400 and $600 million, with some "convinced that the [true number] was unknowable."[17] It was too difficult to distinguish between the various entities that Hefner used to fund different pursuits—between what was personal and what was business-related. In November 1982 he and his creditors reached a restructuring deal, and GHK narrowly avoided bankruptcy.

Hefner fared much better than his peers. After the Penn Square collapse, the western Oklahoma economy seized up. At an August 1982 conference, a national bank examiner remarked: "Nobody's paying anybody—there's gridlock in the oil patch."[18] The year 1983 saw new records set each month for Chapter 11 filings in the Bankruptcy Court of the Western District of Oklahoma, where the Anadarko was. More than one hundred drillers, and more than one thousand small services companies that provided them with parts and labor, went out of business over little more than a year, and a former Oklahoma Energy Department director guaranteed that "for every company that has declared itself bankrupt, there are five others teetering on the brink."[19]

It took another year for the bad Penn Square loans to bring down the once-venerable Continental Illinois. In May 1984 depositors withdrew billions, and failure became both inevitable and imminent. In the case of Penn Square, the Federal Deposit Insurance Corporation (FDIC) had simply liquidated the bank and paid off all deposits of less than $100,000. Continental, however, was deemed too systemically important—the catchphrase would soon become "too big to fail"—so the FDIC kept it on life support, bailing it out. Continental remained the biggest US bank failure until the financial crisis of 2008. Seafirst avoided failure only because it was bought by BankAmerica for a paltry $400 million.

For his role in the crisis, Patterson was acquitted of multiple charges in Oklahoma City, but ended up serving two years in prison when charges were brought in Chicago, on Continental Illinois' turf. Continental's chief energy loan officer likewise served prison time. Hefner destroyed vast sums of capital without ever generating net income, but never ceased living a charmed life. He became a noted collector of Chinese art, tried to develop gas resources in China (and other countries), and wrote a book whose rough thesis is that humanity is destined to use gas because it is gaseous. We have evolved beyond the lesser physical states of matter, like solid fuels—wood and coal—and liquids, like oil. Hefner never stopped looking for gas, and in the late 1990s did find a large field in eastern Oklahoma called Potato Hills. He currently lives on a winery in Virginia.

While the Penn Square episode gives the narrative life, a top-down look gives a more objective quantification of just how bad the oil and gas bust of the 1980s was nationally. The best and easiest metric to measure overall oil and gas activity is the rig count, which is as important an indicator of economic health to Texas, Louisiana, and Oklahoma as the Dow Jones or S&P 500 is to the rest of the country.

The rig is the nucleus of the upstream; mother to the well, into and around which so many other resources are poured, and from which oil and gas flows to market. Rigs generally operate 24 hours a day, 364 days per year (the only day off is Christmas), and as such utilize their crews at about four times the intensity of a more typical nine-to-five job. That means that even a basic, onshore rig with six people on at any given time employs the equivalent of twenty-four or so, and that number is just to operate the rig itself. Before the rig can begin drilling, the drill pad must be cleared of trees and brush and leveled by a preparation crew, and a dirt access road must be built. While operating, the rig needs a steady supply of diesel, mud, and new casing to build the well. It wears through drill bits, and its engines eventually need replacing. Even after the rig's work is done and the well is drilled, it must be completed and tied in to a gathering system for oil, gas, or both. A fair number for direct, full-time employment associated with each drilling rig is closer to one hundred people.

Rig builders turned out 1,100 rigs in 1981, and more than another thousand in 1982, bringing the total number of available rigs in the United States close to 5,500 by the end of 1982. But then, suddenly, the rig count—those rigs being actively utilized—dropped precipitously, from around 4,500 down to just 2,000. The industry shed more than 200,000 jobs in a heartbeat. In Oklahoma, where deep gas had made the party the bubbliest, and the hangover the most nauseating, the rig count dropped from a high of 882 to just 250, which was of course why no buyer ever emerged for those seven rigs that Patterson had foisted on Hefner.

Hefner recalls that, just before the NGPA became law, Energy Secretary James Schlesinger had told him something extremely prescient. By deregulating deep gas while continuing to control the price of old and other new

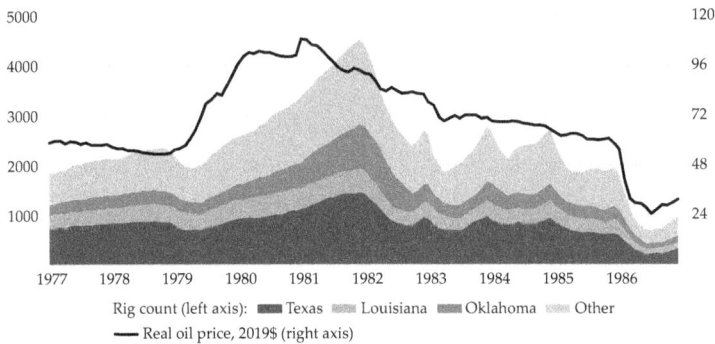

**FIGURE 10.1:** US rig count and real oil price, January 1977–December 1986

gas, "You will release macroeconomic forces on a microeconomy that will only create hyperinflation."[20] That is exactly what happened in the deep Anadarko Basin in western Oklahoma, and why the pain was felt more sharply there than in any other part of the country.

More sharply, at least, in 1982 and 1983. With oil prices remaining quite high, there was still work in the upstream, just not in the gassy Anadarko. Come mid-1984, after the failure of Penn Square and the bailout of Continental Illinois, the situation appeared to have stabilized. In reality, however, the global oil market was softer than prices led on. Saudi Arabia was going to great pains to keep the market in balance, reducing production from 10 million barrels per day in 1981 to just 3.5 by 1985. This might have been acceptable had there been solidarity within OPEC, but Saudi production cuts came as the cartel's lesser members were "cheating" and producing above their quotas. By 1986 the Saudis had had enough: they turned on the taps, flooding the market with almost 2 million barrels a day of oil, sending prices tumbling. This "good sweating" marked the true bottom, and made every oil field in Texas, Louisiana, Wyoming, or California just as empty as the deep Anadarko.

## TAKE-OR-PAY BECOMES A NATIONAL CRISIS

One of the most remarkable and practical achievements of modern science is the ability to forecast weather. Extremely complicated and continuously refined computational models yield almost eerily accurate results for up to a week out, and less reliable but still generally correct predictions for another week beyond. Of course, there is a natural bias to remember big misses, but short-term forecasts are right much more often than wrong.

The models lose accuracy farther out. There are simply too many variables that interact with one another in difficult-to-predict ways. Professional meteorologists will argue about model accuracy in the three- to four-week

Oil production (million barrels per day)

Real oil price (2019$)

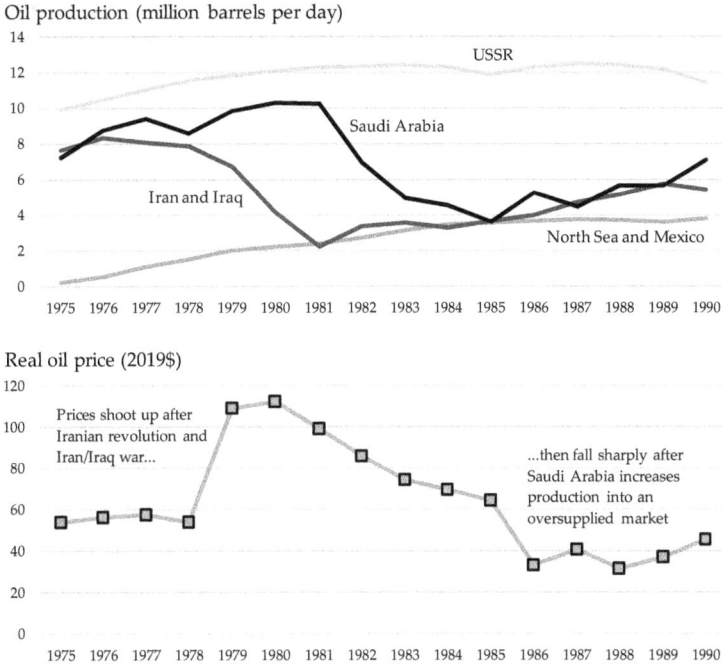

**FIGURE 10.2:** Oil production from select countries/regions and real oil price, 1975–1990

range (ad nauseam, if you allow them to), but very few put any faith in longer-term forecasts. Still, there are some climatic phenomena that can presage temperatures and precipitation for an entire season if they attain sufficient strength. Probably the most important, and certainly the most well-known, is El Niño—or, more formally, the El Niño-Southern Oscillation.

This phenomenon got its name hundreds of years ago in South America. West of the Andes Mountains, Chile and Peru contain some of the most arid terrain on the planet. But every few years, the skies open up, and wildflowers paint the deserts in Monet-like pastels. These periods of great abundance usually happen around Christmastime, which led the sixteenth-century Spanish-speaking colonists to term the event *El Niño*, after the baby Jesus.

El Niño is caused by temperature oscillations in the waters of the eastern equatorial Pacific Ocean, about 1,500 miles south of Hawaii, which have been occurring for at least 125,000 years, and probably for several million.[21] The reasons are unclear, and the oscillations are as-yet unpredictable, but knowing the trend of sea temperatures in the so-called Niño region often gives the most reliable indicator of winter weather in the United States. A strong El Niño, where sea temperatures are well above normal, translates into very mild winter temperatures in the Northeast and Midwest and above-average rainfall in California. A weak or neutral El Niño, or a La Niña, where Pacific sea temperatures are below normal, yields less certain winter outlooks.

By the fall of 1982, it was clear that the world was on track to experience one of the strongest El Niños since scientists began keeping records thirty years earlier. Meanwhile, the unemployment rate was running above 10 percent and the energy-intensive manufacturing sector was producing well below capacity.

Gas pipelines were in a precarious situation. They had only recently stopped signing expensive contracts with producers, and were required to take gas even if it was impossible to market. It is tempting to call the situation a perfect storm, but in fact storms were just what the pipes needed. The extremely mild winter of 1982–1983—the result of an extremely strong El Niño—was, from a pipeline's point of view, tortuous calm.

During the bidding wars for high-cost gas, which started in late 1979 and continued for about two years, pipelines competed with one another on both price terms and take terms. They quickly bid up prices to oil parity, but few were willing to bid much above that. That meant that the competitive dimension shifted to the take side of the contract. Take-or-pay clauses were already de rigueur, and provisions that obligated the pipelines to pay for 90 percent or more of the contract volume, regardless of whether they took the gas, became commonplace. Eager pipeline companies thus began offering producers an even better deal by including must-take provisions. These were much more stringent than take-or-pay, as they required the pipelines not only to pay for contract volumes but actually to take the physical volumes from the wells on schedule. When the dust settled, the vast majority of high-cost gas contracts included must-take provisions.

Surely, producers were in business to make money, and both types of provisions offered revenue assurance. Why then did they prefer must-take provisions over take-or-pay? The concepts of drainage and the rule of capture come into play once again.

Take-or-pay terms gave some flexibility to the pipeline company, allowing it to take less than the contracted-for volume in one year and to make up for the shortfall in later years. If the company had two contracts in a field, and it did not need the full volumes of both contracts—which could happen anytime demand fell, as in the case of a warm winter—it would make sense to take the full amount of the cheaper contract, and then make up for the shortfall in the more expensive contract in later years. But during that year, the cheaper producer would have done some milkshake drinking, since he was producing from the same reservoir as the more expensive producer. The more expensive producer would suffer irreparable harm, as it would lose reserves without producing any gas. This situation obviously cannot occur if the more expensive producer has a must-take provision. In fact, the reverse can occur, and did in 1982–1983.

Even before the warm winter, the pipeline companies found themselves long gas, required to take volumes that they could not market. As demand fell substantially short of expectations during the strong El Niño winter of 1982–1983, pipelines simply could not fit their full contract volumes into their systems—there was nowhere for the gas to go. Having signed so many must-take contracts with high-cost producers, they did the only thing they could do, and took the more expensive gas first, demanding that lower-cost producers shut in their wells.

So, perversely, just as the market was sending the signal for prices to drop, pipeline companies saw their weighted average cost of gas (WACOG) skyrocket. Even stranger, this was all happening as oil prices were dropping, bringing down the price of oil-indexed, high-cost gas. But the effect of must-take provisions greatly outweighed this drop in prices because of how the weights were adjusted. A simplified example is thus: In year one, the pipeline company purchases 90 percent of its gas at a low price of $2, and 10 percent at a high, oil-indexed price of $8, leading to a WACOG of $2.60. In year two, after the oil price has fallen, the high-cost gas only costs $6, but because of must-take provisions, the company must purchase 25 percent of its gas at $6 and the remaining 75 percent at $2, increasing the WACOG to $3. It was this effect that led to a factual, but economically nonsensical headline from a *Washington Post* article written in September 1982: "Natural Gas Surplus Leads to 20% Price Rise."

Pipeline companies had no choice, and simply walked away from their contracts with producers. Several enacted emergency gas-purchase programs that invalidated certain provisions of their contracts.

President Carter's regulated deregulation program was falling apart at the seams, and the Reagan-era FERC was anxious to do away with gas market regulation once and for all. The first step it took in this direction was to approve special marketing programs (SMPs), whereby producers sold gas directly to consumers and agreed to release that volume from the contract with the pipeline company. These programs, begun in 1982 and extended through 1984, marked the very beginning of a true spot market for natural gas, where producers and consumers faced each other directly and prices adjusted to real-time supply-and-demand dynamics. It also marked the beginning of what would turn out to be a fairly rapid change in pipelines companies' business models, from energy merchant to transportation provider.

SMPs were meant to provide a measure of relief to pipelines, as every cubic foot released from under contract reduced their take-or-pay obligations, which were ballooning into the billions of dollars. But, as ever in partial deregulation, perversities arose, as producers only released low-cost volumes, saddling pipeline companies once again with a higher WACOG. Essentially, producers used the program to sell 50-cent gas for the going market price of $2.50.

The next step was much bigger, and can truly be said to be the beginning of the end of merchant service.

Throughout the early stages of the take-or-pay crisis, starting in 1982 and lasting into the first half of 1984, pipeline companies  retained two vitally important hedges against the chaos: (1) minimum bills with their LDCs, and (2) the ability to automatically pass through gas costs. Minimum bills were, in essence, the mirror image of take-or-pay contracts with producers, and required LDCs to pay for a large percentage of their contracted volumes, regardless of whether they actually consumed the gas. They had been in use for decades, and, just like take-or-pay provisions, had aroused little controversy when demand was growing. The automatic pass-through of gas costs was included as part of the NGPA, called the Purchased Gas Adjustment (PGA). Lawmakers realized that the NGPA's many price schedules and the deregulation of high-cost gas would necessitate thousands of rate adjustments, and that it would have been wholly impractical to hold a lengthy FERC hearing for every such adjustment. Minimum bills and the PGA sheltered pipelines from bankruptcy-inducing losses during the early days of the gas bubble.

President Reagan's FERC saw minimum bills between LDCs and pipeline companies as the "primary culprit for the failure of price signals to transfer from the burner-tip to the wellhead."[22] Indeed, insulated as they were from the extremities of market swings, executives at pipeline companies "minimized phone calls, not WACOG,"[23] calling their largest producers and asking them to shut in low-cost gas because it was simply easier than arguing with many producers, and asking them to shut in their prized, high-cost wells. FERC recognized not only that pipeline companies were less driven to renegotiate contract terms with producers so long as minimum bills remained in place but also that producers would be less likely to budge: why should they make concessions if the pipelines could go on recouping their full costs from LDCs?

In May 1984 FERC issued its watershed Order 380, which did away with minimum bills and left pipelines with asymmetrical commitments to producers. The importance of this order cannot be overstated: an attorney with the American Gas Association wrote, years later, that it represented "the opening salvo in an assault on the existing regulatory and market regime."[24] It is not hyperbolic to say that once minimum bills were canceled, the transition of pipelines from energy merchants to transportation providers became inevitable.

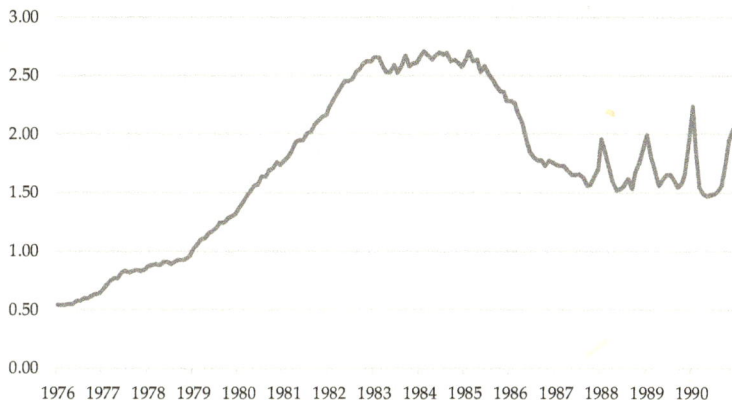

**FIGURE 10.3:** Average US wellhead gas price ($/MMBtu, nominal)

FERC followed up a few months later with another order that prohibited pipeline companies from passing through take-or-pay costs automatically, through the PGA mechanism. From the spring of 1985 onward, take-or-pay costs could only be recovered after a full proceeding, where they would be subject to challenge on the grounds of prudence.[25]

Also in 1985, all NGPA–qualifying new gas was deregulated, as had been scheduled. As should be obvious, prices did not fly up, as policymakers feared they would a few years earlier, but began a long decline. The average wellhead price in the United States peaked in February 1985 at $2.71 before falling to $2.28 by the end of the year, and just $1.75 another year later.

By 1985 President Reagan's FERC had taken the gloves off, exposing pipeline revenues fully to market forces. The timing was unfortunate, as national gas demand continued to drop in spite of the nascent economic recovery. For one thing, the FUA was preventing new gas power plants from coming online to sop up the excess supply. (Absent the Act, one might have expected new plants to have come online by 1985, a few years after the glut had emerged in 1982.) For another, the new jobs being created were concentrated in the services sector, rather than in the more energy-intense manufacturing sector. Home use continued to decline as well.

Pipeline companies began abrogating contracts in the spring of 1983, after the winter that wasn't. They invoked force majeure (i.e., Act of God clauses), mutual mistake (on the part of both pipelines and producers), and the greater good as justification, and, in the years that followed, began to resemble high-powered law firms as much as energy merchants. Tenneco, which owned the massive Tennessee Gas Pipeline, hired thirty or forty people straight out of law school to "pencil-whip" producer contracts, going over every letter and punctuation mark to find weaknesses. The "Midnight Crew," as they were referred to internally, were as revered for their hard work and long hours as they were reviled for stealing furniture after the rank-and-file left the office.[26]

Few things arouse such animal spirits among businesspeople in the United States as the sanctity of contract, serving as it does as the very basis for commerce. Producers were understandably furious with pipelines for violating this sanctity. Nevertheless, at least in retrospect, one can empathize with the pipeline companies. Can it truly be said that they entered freely into contracts with producers, when the market in which they operated was not a free market but one characterized by pervasive regulation?

Some state governments shared this empathy, or at least understood that the marketplace could not physically handle the amount of gas for which pipelines had contracted. In 1983 Oklahoma reduced allowable production to just one-quarter of each well's capacity in response to a petition from an intrastate pipeline, Oklahoma Natural Gas Company, that could no longer market the high-cost gas that it was mandated to purchase under its take-or-pay contracts.[27] This was the first use of a statewide production quota in more than a decade.

But this was the exception, not the rule. The courts were mostly not receptive to the plight of the pipeline companies: The language in the contracts was plain and simple, and it was the pipelines themselves who had written them. A Louisiana judge's opinion summarized the judiciary's response to the take-or-pay crisis well: "Shifting supply and demand and changing governmental regulations are normal factors considered in any business transaction. By the clear and unambiguous terms of the remainder of the contract, the plaintiff [producer] accepts the supply risk, the defendant [pipeline] accepts the market risk, and both parties adopt the risk of changed governmental regulations. . . . Modifications in governmental policy and regulation are economic facts of life and are not a basis for suspending a contract."[28]

In the short term, the solution was to let pipeline companies and producers duke it out and try to come to settlement agreements. In the long term, it was now clear that there was only one solution: to take pipelines out of the business of buying and selling gas.

## THE JOURNEY TOWARD OPEN ACCESS

By 1985 it was clear to most parties that pipelines were being taken out of the merchant business and turned into transportation providers. A free market, where buyers would transact directly with sellers with no price controls, was the destination. But the journey took another seven years to complete. Just before the turn of the decade, Senator Phil Gramm summed up the situation from his Texan perspective: "We deregulated natural gas, though I'm sorry to say that given the timetable for deregulation, if perestroika works in the Soviet Union, the Russians will deregulate natural gas before we do."[29]

The transition of the pipeline business away from merchant and toward transporter was not inevitable because the Reagan and Bush administrations were hellbent on achieving ideological purity—so much as it was the inevitable outgrowth of actions that FERC had already undertaken. The SMPs begun in 1982 had created a parallel market for natural gas, outside of pipelines' jurisdiction. Spot gas—sold by a producer, an independent marketer, or the pipeline itself—now competed with sales gas, sold exclusively by the pipeline under old-school, long-term contracts.

Had the market been booming, demand setting new records every year, spot volumes could have been sold over and above sales volumes. But the market was soft: customers were not taking their contract volumes from pipelines, pipelines were not taking their full contract volumes from producers, and producers were thus willing to offer short-term, spot gas at a lower price. Accordingly, in the market of the mid-1980s, spot gas was always and everywhere less expensive than sales gas.

This of course meant that consumers wanted it. But had FERC not issued Order 380, in May 1984, they would not have taken it, since they would still have needed to pay for their minimum volumes of sales gas. If, for example, an LDC had a contract with a minimum bill of 80 percent of contract volumes and it took only 60 percent, the cost of the next unit of sales gas was, economically, zero, because it was being paid for anyway. The LDC's bill would not drop if it substituted spot gas for sales gas—in fact, it would grow, as spot gas was cheaper, but not free. When FERC issued Order 380, voiding minimum bills on the grounds that it was not just and prudent for pipeline companies to charge customers for gas not actually taken, it incentivized those customers to seek out cheaper alternatives.

Pipeline companies did not like this situation, but they had no choice. And since they could no longer recover all their fixed costs from consumers via minimum bills, they too were incentivized to maximize throughput, and hence revenues. That meant transporting gas on behalf of spot sellers.

It was thus the confluence of Order 380 and SMPs that made the transition to transport-only service inevitable. But why so long a journey?

At a basic level, an entrenched and strategically important industry was being fundamentally altered. Relationships that had existed for decades were being broken, and new ones had to be forged. Producers knew nothing about consumers, as they had never faced them, and the same was true in reverse. Moreover, tens of billions of dollars per year of transactions were being relocated, along with billions more in potential future gains and losses. While this may seem like a simple shift in accounting, assessing market and price risk requires a different thought process than signing long-term deals, and assuming that risk can be a frightening experience. In such an undertaking, it is only natural to expect a multiyear transition, as the various parties sorted out the most equitable way to reshape business models and distribute risk.

The Reagan-era FERC seemed eager to get out of the business apportioning this risk—perhaps too eager. The courts took issue with several of the commission's rulings, remanding them in part or vacating them outright for exposing pipelines to undue hardship. A heated, and often downright venomous back-and-forth between producers, distributors, the pipeline companies, FERC, and the courts shaped the tumultuous, seven-year journey to mandatory, nondiscriminatory open-access transportation of gas.

The issue of discrimination among customers came up early on. The first spot gas was sold in late 1982 and 1983 to non-core customers, usually industrial facilities. One of the very first such transactions involved a small producer in Montana, whose wells were shut in because the pipeline could not take his gas, and two large fertilizer plants in Iowa, which had not operated for two years because the regulated price of gas had become so expensive that it was actually cheaper to import fertilizer from Peru and ship it up the Mississippi River. A young executive at InterNorth, which owned the Northern Natural Gas, managed to cobble a deal together at a much lower price, which saved the fertilizer plants $40 million in annual fuel costs. The plants awoke from their long slumber and began producing fertilizer for the massive Iowa agricultural market once again, consuming almost two hundred million cubic feet per day of gas in the process.[30]

SMPs at other pipelines likewise targeted industrial facilities, many of which had the ability to use alternative fuels and had switched over to fuel oil. An interesting phenomenon in the domestic oil refinery market made this switch even more stark than the prices of either crude oil or sales gas would have suggested. Spurred by the high prices throughout 1979–1982, production of heavy grades of crude surged from Mexico and Venezuela, which made their way into the United States. US refiners thus began producing a proportionally smaller share of valuable products, like gasoline and diesel, and a larger share of heavy products, like extremely viscous residual fuel oil. Cars, trucks, and airplanes could not use residual fuel oil, but many industrial boilers could, and found that they could get these unwanted heavy products for as much as a 40 percent discount to the prevailing price of crude oil, on an energy-equivalent basis. Refiners undertook a decade-long investment program that eventually allowed them to crack these heavier grades into lighter and more useful products, but in the interim, barges full of residual fuel oil trawled up and down the Mississippi, offering to sell the fuel at whatever price industrial boilers would pay. Gas demand among these facilities dropped to almost nothing.

Regardless of the market being served, all SMPs sought to address two, related problems: low demand and rising take-or-pay obligations. These had the compounding effects of creating a death spiral, where fixed costs were being spread among fewer and fewer consumers, necessitating higher rates, leading demand to drop even further, and causing an exodus among those users who could fire alternative fuels. Offering discounted spot gas to industrial facilities increased demand, allowing the fixed costs to be spread across a larger group.

While SMPs were thus a win-win from a net economic benefits point of view, they clearly discriminated against the pipelines' core LDC customers, who were not offered the opportunity to buy this cheaper gas, and could not switch to an alternative fuel. The captive core was being forced to pay for more fixed costs, per unit of gas actually taken, than industrial facilities. FERC did not contest the fact that SMPs were discriminatory, but made a greater good argument, noting that fixed costs would be even higher for core consumers in the absence of SMPs because of the death spiral effect. But this was not enough for the DC Circuit Court, which ruled that the programs were unacceptable, as they allowed pipelines to use their monopoly power over captive consumers to enhance their position in the competitive spot market.

Practically, this court ruling did not impact SMPs, which were experimental programs with near-term expiration dates, and which were left to "die a natural death."[31] However, the decision did shape future policy, as it made clear that pipelines would be required to offer access to transportation services on a nondiscriminatory basis to all customers.

As SMPs were expiring, in October 1985 FERC issued another landmark order, Order 436, which gave pipelines the option to become nondiscriminatory, open access-transportation providers. In return for offering open access to their systems to all shippers, FERC allowed transportation service to become self-implementing—pipelines and their shippers could transact at will, without certification hearings, for whatever volumes and over whatever term suited them. Pipeline transportation rates still fell under cost-based regulation subject to FERC review, but there was no longer the requirement to prove twenty years' supply, and the price of the gas itself was not subject to review. In the words of the DC Circuit Court, Order 436 envisaged "a complete restructuring of the natural gas industry."[32]

In issuing Order 436, FERC had learned its lesson from the DC Circuit Court ruling prohibiting discriminatory access, but now, by offering all customers the ability to contract for transportation-only service and purchase spot gas, the commission was taking away pipelines companies'

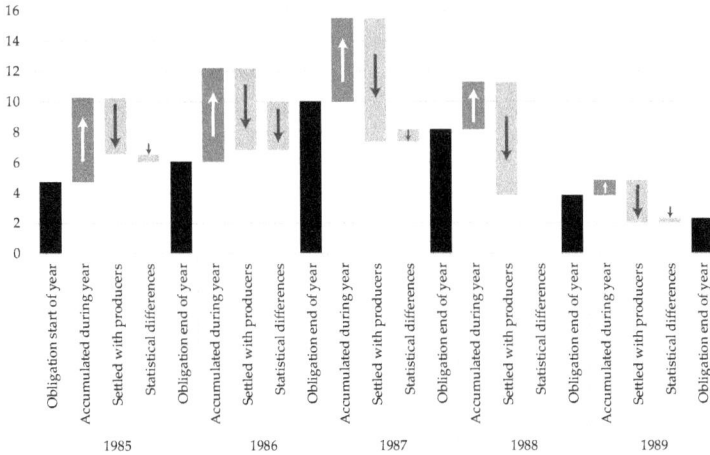

**FIGURE 10.4:** Interstate pipeline take-or-pay obligations, 1985–1989

ability to recover their take-or-pay obligations, and creating a multi-billion-dollar hole in their balance sheets.

This whole episode can get rather confusing, rather quickly, so let me quickly summarize how we got here.

Pipeline companies had signed high-priced contracts with take-or-pay or must-take terms in the wake of the NGPA and the 1979 oil crisis. As oil prices fell and the macroeconomic picture deteriorated, especially in the manufacturing sector, gas demand fell dramatically. With their markets shrinking, pipeline companies were forced to shut in large volumes of gas. They tried to shut in the high-cost volumes, but those were generally newly signed contracts with high take terms. Producers were sure to litigate—and had—when their most profitable wells were shut in, especially with the ink still wet on contracts that the pipeline companies themselves had written, so companies did the only thing they could: they shut in low-cost gas. Thus, sales gas was expensive gas.

Order 436 put pipeline companies in a tighter bind by expanding and greatly accelerating the process that had begun with SMPs, where pipeline companies saw their sales gas volumes displaced by transportation volumes. Obviously, the companies could not recover their take-or-pay liabilities through the price of spot gas, as they did not own that gas. It is tempting to think that they could recover the liabilities by imposing higher transportation fees, but there was a limit to this: if higher fees caused the spot gas to become too expensive to market, the whole exercise would be in vain. It would have been robbing Peter to pay Paul.

The market needed lower prices to clear, and that is what it got, on the spot market. As transportation volumes surged at the expense of sales volumes, take-or-pay liabilities skyrocketed to more than $10 billion mid-way through 1987,[33] and that was after several billion had already been spent to buy out contracts from more pliant producers—the low-hanging fruit—at around twenty cents on the dollar.[34] Pipelines—historically a sta-ble, even boring business with virtually guaranteed returns—now faced financial strain, which threatened widespread bankruptcies. Expectedly, interstate pipeline bond ratings dropped sharply from investment grade to junk after Order 436 was issued.[35]

On the whole, producers were somewhat conciliatory. There were bad actors, who used the perversity-laden situation to their advantage. In a gen-eralized example, a producer would contract directly with an end user on the spot market for a low price of something like $2, and use the pipeline for transportation, but then also demand that the pipeline company pay the full rate in its sales contract for gas not taken, say $5. The argument was that the pipeline company had not extinguished its liability just because the gas had found a home. Though obviously perverse, the general stance of the lower courts during this period was that the plain language of the sales contract should be followed, which gave the producers an edge in any eventual litigation.

But examples like these, of intransigent and litigious producers de-manding their full revenues as specified in their contracts, were outweighed by examples of producers who were more ready to deal and accept a steep haircut. The Montana producer that InterNorth matched up with the Iowa fertilizer plants was happy to accept a price of around $2.50, rather than the contract rate of $9, because he was not actually foregoing $9—he was foregoing no revenue at all, as his wells were shut in for lack of demand.

FERC initially took a hard line with respect to take-or-pay liabilities. Early on in the deregulation process, in 1982, it ruled that pipeline compa-nies could not include payments in their rate base for any contract where the minimum take was over 75 percent, which was the vast majority of them. In 1985 it went a step further, explicitly barring costs for gas not actually taken from being automatically passed on to consumers.[36]

This hard line was too much for the judiciary to bear. Once again, the DC Circuit Court stepped in. In June 1987 it vacated Order 436 on the grounds that it did not do enough to address the take-or-pay problem. And once again, FERC responded not by challenging the court's decision, but by issuing another rule: Order 500.

In its softened stance, the commission conceded that, while it had found some pipelines to have acted unwisely and imprudently in signing contracts with both high prices and high take terms, many others had simply been forced into signing contracts with such onerous terms to provide for their customers.[37] Order 500 stated that pipelines could condition transportation of producer-owned gas on the producer issuing volume-for-volume credits against take-or-pay liability. So, if the producer sold $2 spot gas to an LDC using the pipeline as a transporter, it would relinquish its claim to a $5 take-or-pay liability. This conditioning was clearly discriminatory, but FERC and the courts were willing to accept it as just discrimination.

More importantly, Order 500 ruled that pipeline companies could recoup an equitable portion of their take-or-pay liabilities from their customers by establishing a deficiency-based charge. That means that pipeline companies needed to absorb between 25 percent and 50 percent of their take-or-pay liability, but could pass the rest on to customers as a surcharge on sales gas.[38] If this seemed like backtracking, equivalent to reestablishing minimum bills, it was.

FERC also issued a large request for data as part of Order 500, in order to examine the take-or-pay issue in more detail, and perhaps impose a more definitive solution.

Virtually all parties concerned were displeased with the take-or-pay relief offered by Order 500, even some FERC commissioners. United Distribution Companies, an LDC lobbying group, wrote colorfully about FERC's approach: "The take-or-pay problem preys like a rogue elephant on the long-term stability of the gas market. But rather than shooting the elephant, the Commission proposes merely to weigh the beast."[39]

A parallel can be drawn between FERC's actions in the late 1980s and the FPC's actions forty years earlier, when it was being forced to make up rules to determine which producers were subject to rate regulation and which qualified for the production-and-gathering exemption. In both examples, the commission's judgment was hijacked by outside forces, be they the judiciary, Congress, or lobbying from the private sector. As the commission attempted to please everybody, the language of its regulations became increasingly convoluted. It comes as no surprise, then, that the courts threw out parts of Order 500; namely, the convoluted equitable-sharing feature.

As the rogue elephant quote evinces, most parties wanted FERC to act more boldly, and simply decide what had or had not been prudent on a case-by-case basis. After forty-some years of heavy-handed regulation, and another decade of even more traumatic regulated deregulation, the industry was almost childlike in its desire to be told what to do. Great men of business, from across the country, seemed unable to tolerate the turbulent transition to a free market.

But the commission, to its credit, did not reenter the fold of cost reallocation. After the courts threw out the solutions it had proposed to the take-or-pay problem in Order 500, it punted, leaving pipelines, producers, and consumers to renegotiate their contracts. The take-or-pay crisis diminished in magnitude over the latter years of the 1980s as settlements were reached, but many cases dragged on well into the 1990s. Twenty years after the take-or-pay crisis, an attorney from the American Gas Association wrote: "The exact numbers will never be known precisely, but it appears that several hundred billions of dollars of contractual exposure of pipelines were extinguished by expending approximately twenty or thirty billion dollars."[40]

## THE DEATH OF MERCHANT SERVICE

Order 436 led consumers to shift rapidly away from sales to transportation service. Within a year of the order, about half of the interstate pipeline companies had applied to become open-access transporters. Within another two years, virtually all had converted.

Order 436 did not, however, take pipeline companies completely out of the merchant business of buying and selling gas because it did not force LDCs to abandon their firm sales contracts. Many LDCs preferred the security of firm sales for at least a portion of their portfolio, as these contracts required pipelines to stand ready with gas on even peak days. As the amount of sales gas delivered to LDCs halved over the latter 1980s, on an annual average basis, peak-day deliveries were relatively unchanged. LDCs wanted to have their cake and it eat too—buying spot gas when it was cheap and available, but having a backstop of sales gas for the coldest days of winter. This loophole was forcing pipeline companies to remain in the least desirable portion of the merchant business: the firm, long-term market. By not completely unbundling the pipelines companies' sales functions from their transportation functions, Order 436 created an untenable long-term situation.[41]

Recognizing that this was unfair to the pipeline, FERC, as part of Order 500, outlined a program to "allocate the risks and costs associated with maintaining gas inventory to those customers actually deriving benefit from the maintenance of that inventory."[42] This program became known as gas inventory charges.

It is the correct response for the reader to think to himself: "Here we go again . . . " But worry not, for we will not get into the minutiae of this program. The point can be made simply: these gas inventory charges were nothing more than repackaged minimum bills—the same ones that FERC had voided with Order 380.[43] Customers, unsurprisingly, liked the free ride they were getting by using their firm sales contracts only when needed, and otherwise using the pipeline to transport cheaper gas. There were only two solutions. One was to engineer a complicated program to allocate costs, which in the late 1980s is what gas inventory charges were shaping up to be. Another was to take pipeline companies out of the merchant business once and for all. If they remained responsible for peak-day supplies, they would remain peak-day monopolies. For the market to function properly, and without distortion, LDCs needed to assume responsibility for securing their own gas supplies, including storage volumes for peak-day needs, while producers needed to balance their output to avoid cratering prices.

The industry reached its destination in August 1992, when FERC issued the last order that mattered: Order 636. It required pipeline companies to completely unbundle their sales from their transportation function—it was not, like Order 436, a voluntary program. And even though Order 636 did not force pipeline companies out of the merchant business, it gave them no incentive to stay in it. This is because Order 636 expressly prohibited pipeline companies from moving sales gas to customers under any conditions of service preferential to what they offered third parties. They could not thus stand ready to supply guaranteed peak-day gas, any more than could any unaffiliated supplier. After Order 636, there was no more distinction between sales gas and spot gas—there was only gas. Service was equal, whether you were a pipeline affiliate, a large LDC, a producer, or a tiny, start-up marketing firm.

## CONCLUSIONS FROM FOUR DECADES OF WRONG-HEADED REGULATION

With each new catastrophe of the 1970s and 1980s—gas curtailments, the Penn Square Bank fiasco, and the take-or-pay crisis—it is tempting to demonize those most obviously involved. The Supreme Court and the FPC, Oscar Wyatt and Bill Patterson, the ayatollah and the sheiks of Araby all deserve some blame. But rather than conclude the second part of this book with a desultory whodunit, I address the root cause of these crises.

There was a belief among regulators, regulated businesses, and unregulated businesses alike that market risk could be legislated or contracted away. That a producer, after signing a contract with a pipeline company, could go on drilling and forget about supply and demand. That a consumer could get however much gas he or she needed at low, fixed prices forever. That a bank could lend billions at 20 percent interest rates using only a rosy price outlook as collateral. There was a belief, in other words, that risk could be shifted entirely to "the other guy."

In economics, when supply and demand curves are plotted on a chart, the vertical axis is the price, and the horizontal axis is the quantity. In a free market, the point where the two curves meet is the equilibrium price and quantity. In a regulated market, it is possible to control one or the other—price or quantity—but you can never control both. This was the fallacy.

In *Phillips*, the Supreme Court did not insulate consumers from risk: it merely shifted the dimension of the risk that consumers bore from price to quantity. You must sell your gas at such a price, the high court ordered. But it could not compel producers to produce. The result—gas shortages in the regulated market—was a simple and predictable equilibration.

After several rounds of price hikes, culminating in the NGPA, the diktat became: "Go and get gas, at whatever price." But the US government could not compel consumers to consume. The result—a gas glut—was equally simple and predictable.

Contracts and regulations are capable of reallocating risk, but not extinguishing it. Perhaps a small Oklahoma producer, like Hefner's GHK, thought that its contract with the giant El Paso Natural Gas guaranteed it a high price and a ready market. In isolation, it probably would have. But when every producer had contracts with high prices and take-or-pay provisions, what was, on paper, the pipeline company's problem, became everyone's problem.

Before the restructuring of the 1980s and early 1990s, pipeline companies were like gas banks, where sales contracts with consumers were their assets and supply contracts with producers were their liabilities. In the take-or-pay crisis, consumers owed the bank billions for high-cost gas, but then the government, in FERC 380, which voided minimum bills, said they didn't have to pay. The whole system came undone.

With the gas banks' liabilities far exceeding their assets, they needed to be liquidated, just like Penn Square. This was the end of merchant service for pipeline companies—they would no longer act as gas banks. But this wasn't the end of the gas banking business altogether.

# PART III

# A (MOSTLY) FREE MARKET AND ITS DISCONTENTS, 1992–2020

It seems appropriate to evoke Francis Fukuyama and christen FERC Order 636 "the end of history" for the US gas market. The past twenty-five years has seen very little in the way of major regulatory change. Through cold snaps, El Niños, and hurricanes, the market has functioned without intervention. Even as renewed shortage gave way to the "shale gas miracle," which has left the domestic market swimming in cheap gas, the construct has not changed: prices are unregulated and pipelines transport gas on behalf of others.

It is perhaps a blunder to start the final section of this book by saying that history has ended. But just as Fukuyama did not imply that peace would reign forevermore, nor do I imply that Order 636 forever freed the gas market from turmoil. Rather, Order 636 marked the end of the transition, and the arrival to a permanent, sustainable, steady-state form—or, as Fukuyama wrote in 1989, as it became clear that the Soviet Union would collapse: "What we may be witnessing is not just the end of the Cold War, or the passing of a particular period of post-war history, but the end of history as such: that is, the end point of mankind's ideological evolution and the universalization of Western liberal democracy as the final form of human government."[1]

In politics, both globally and domestically, the arc of history has veered from Fukuyama's predicted straight path, whereas, applied to the US gas market, the "end of history" adage has held true.

# CHAPTER 11

## GAS BECOMES A COMMODITY

As pipelines got out of the business of buying and selling gas, a spot market quickly emerged. Prices were set for a month at a time, and responded freely to supply-and-demand dynamics. A new breed of market participants—gas traders—went to work breaking down the inefficiencies that had accumulated over the prior decades, buying gas where it was cheapest and selling it where it was most expensive, sometimes jumping across three or four pipelines to cash in on a wide spread.

This was all positive, but consumers began to fret at their sudden exposure to risks that before had sat with the pipelines. What if prices spiked? What if they could not get deliveries when they needed it? The market was making an abrupt U-turn, from operating solely under firm, long-term contracts to operating solely on interruptible, one-month spot sales. Balance was needed.

Enron answered the call. More than any other company, Enron invented the natural gas market as it exists today. Through a series of legitimate, and legitimately innovative mechanisms, it allowed producers and consumers to hedge price and volume risk to suit their particular needs. Enron essentially recreated the old pipeline portfolio model, rebuilding the vertical integration from wellhead to sales that was lost in the 1980s, but with a twist. In order to operate with maximum flexibility, and avoid the long-term obligations that had crippled merchant pipelines in the past, Enron and a handful of others built a parallel, financial market for gas. In this market, everyone could get what they wanted: producers could sell forward production for many years, LDCs could absorb some market risk, but hedge against extreme events, and speculators could get in the middle of it all. Unlike Oscar Wyatt in the 1970s, however, a bad speculative bet by a financial gas trader would cost their firm money, but not leave people freezing through the winter.

Gas prices shot up during hurricanes, which caused production facilities along the Gulf Coast to shut down, and freak cold snaps, when demand was highest. They likewise plummeted during warm winters. In the 1990s, it was possible for a savvy trader to make tens of millions of dollars betting on the future price of natural gas. By the 2000s, that number had risen into the billions.

Prices never stayed high for long, though. A situation of oversupply—the gas bubble—stretched from the early 1980s all the way through the new millennium. This helped the transition from a regulated market to an unregulated one tremendously, as low prices meant that there was precious little pushback from consumer advocates. In fact, that the transition did not result in total market dysfunction was only possible because of the magnitude of the glut. Excess capacity existed at the wellhead, in the pipeline network, and at storage fields. FERC was thus able to brutalize pipelines for a decade without encountering the adverse effects of underinvestment, prices stayed low outside of short-lived weather events, and interruptible spot sales were rarely interrupted. Fortunately for consumers, the car had broken down right next to a service station.

## THE EARLY SPOT MARKET

At the highest level, the 1980s saw a handful of traders break up inefficiencies that were created over decades of pipelines signing long-term contracts at regulated prices. It is something of a dark period in the history, since the deals were unregulated, reporting requirements were virtually nonexistent, and the transactions were purely physical—no financial natural gas products existed until the 1990s. Today, there are more than one hundred trading points where gas is bought and sold on a daily basis, mostly electronically, and the price is known to all market participants. In the early 1980s the price was whatever had been agreed upon, and was known only to the parties in the deal.

Probably the largest inefficiency of the pipeline-as-merchant model was that producers and consumers had no idea who each other were. It is easy, in the age of Google, to forget the value of the rolodex, but as deregulation began, it was still only pipeline companies that knew who to telephone and what they were paying or receiving for their gas. Because of this, at the very beginning of the spot market, it was the pipeline companies themselves who brokered direct transactions between producers and consumers. They may not have liked it, but massive take-or-pay obligations and the cascade of FERC orders, starting with Order 380, which voided minimum bills, forced their hand. Special marketing programs allowed the pipelines to make some additional revenues transporting gas that otherwise would have stayed in the ground, and these volumes also reduced their take-or-pay liabilities to producers.

For the first time in a long time, there was a true price discovery mechanism: direct sales yielded a market-clearing price, where the producer's willingness to sell intersected with the consumer's willingness to buy. But even as the volume of spot sales increased, a second inefficiency kept market highly balkanized.

Producers and consumers got to know one another, but most transactions were still tied to a single pipeline. Transco, which runs from the Gulf Coast to New York City, might be able to broker a deal between a producer in Louisiana and a power plant in New Jersey, but it could not show the Louisiana producer a bid for gas in Chicago, nor could it show the New Jersey power plant an offer from a Colorado producer. The powerful rolodexes in pipeline company marketing offices were filled only with the names of companies who had access to their pipelines. Supposing even that a seller could get in touch with off-system customers, they would have to make arrangements for transportation on multiple pipelines, often on short, nontraditional paths. Initially, pipeline companies were hesitant to segment their systems and chop them up into ever smaller paths, and it was not clear what rates they could charge.

All this meant that while early spot sales did represent market prices, there could be myriad prices within markets characterized by nearly identical supply-and-demand dynamics, based solely on what pipeline you were connected to. To build a true commodity market, pipeline systems needed to function as an interconnected network, not as silos.

John Barr was an unlikely breaker of these silos. He was a banker at Morgan Stanley, which had, for decades, purposefully neglected pipeline companies, leaving this business to its "lesser competitors" while it did big-dollar business with major producers and utilities. Barr sought to change this, but struggled to find an in, especially because the pipeline companies were well aware, and resentful, of Morgan Stanley's history of purposeful neglect.

The in came when a prominent law firm, Akin Gump, approached Barr about starting a partnership. The lawyers felt that there was a need for independent gas marketers, whose sole motivation it would be to find the best arbitrage opportunities by optimizing transportation paths across pipeline systems. They further concluded that the structure that FERC put in place made this achievable. But they were lawyers, not traders—they needed a Wall Street partner. "The irony," Barr wrote, "is that I took Morgan Stanley into this venture not because I thought it would ever amount to anything as a business, but because organizing the venture would give me an excuse to call on pipeline CEOs. . . . If ever there was a case of doing the right thing for the wrong reason, history proved that this was it."[1]

Natural Gas Clearinghouse, or NGC, was created in June 1984 as a joint venture between Morgan Stanley, Akin Gump, and Transco Energy Company. The venture was soon joined by five other pipeline companies: El

Paso, Columbia, Colorado Interstate, United, and Houston Natural Gas.[2] The idea was for the partnership to leverage the pipeline companies' customer base and operating expertise, the law firm's knowledge of the new FERC regulations, and Morgan Stanley's market mindset to create transportation paths that jumped across the six systems, allowing gas companies to seek out the most valuable market available. The partnership operated out of Transco's offices in Houston, but each of its member pipelines seconded an executive to NGC, both to gain experience and to make sure that the deals were as advantageous as possible to their employers, who still signed their paychecks.

The idea seemed to work well, as all the incentives were properly aligned. NGC brokered its first successful transaction in October 1984, just a few months after it was formed, when it closed a deal for two hundred million cubic feet per day from one of Houston Natural Gas's storage fields in Texas to Brooklyn Union Gas, served by Transco. But despite the clear advantages of being able to jump systems and segment transport, cracks in the partnership's foundation emerged quickly.

For one thing, this was all happening before FERC issued Order 436. The pipeline companies were hesitant to cede control in the mid-1980s because some thought the spot market was a temporary phenomenon, which would exist only as long as take-or-pay issues remained unresolved. It was tempting to write off these "old codgers" as greedy, uninspired, and out-of-touch, but even unbiased and well-informed observers conceded, as late as 1988, that two, parallel markets could exist: one for sales gas and another for spot gas. Analysts from the Department of Energy wrote that year that "the market may be approaching an equilibrium between spot and system supply."[3]

NGC was also sailing against the current, as pipeline systems were actually becoming more siloed in the 1980s. They were trying to satisfy the highly restrictive take terms that they had signed with producers in an environment of rapidly falling demand, and wanted to maximize the use of their own assets in order to maintain a competitive position. In short, it was every pipeline for itself. In 1982, 40 percent of gas shipped through the interstate system traveled through more than one pipeline. By 1987 that share had declined to just 19 percent.[4] This was clearly detrimental to a broker whose key advantage was its ability to arrange for gas to jump across many systems.

Even more fundamentally, NGC's initial business model struggled because the pipeline companies wanted to keep whatever markets they could find for themselves. Supply was to be found in excess nearly everywhere, while every molecule of demand was fought over—there was no benefit to working together; no win-win situations where a company found a new market that it could not serve on its own. Barr admitted that "for pipelines to market their gas through a joint venture made about as much sense as GM and Ford merging their sales forces."[5] Indeed, this was why some of the largest systems—including Tenneco, Texas Eastern, and InterNorth—had not joined the consortium from the get-go,[6] and why NGC's member pipeline companies mostly did not second their best and brightest.[7]

The original conception of NGC—as a broker—was doomed from its birth. It asked competitors to share business-sensitive information in order to stave off liabilities, without making any actual profits. After all, spot gas was selling for far less than sales gas, and in these early days of the transition, most companies' modus operandi was still to maximize throughput of that sales gas. But already to the more prescient players, it was clear that marketing spot gas was the future. NGC would go on to prosper, just under a different business model.

From Barr's point of view, NGC was not a failure. It had not cost Morgan Stanley a dime, and he had added a few pipeline companies to his client roster, which was the reason he got into the venture in the first place. He debated whether to close up shop and go home, leaving the pipelines to figure out how best to market their gas. Instead, in 1985 Barr redoubled his efforts and relaunched NGC as a truly independent gas marketer, which took title to the gas that it bought from producers and sold to consumers. It would fulfill the merchant role that pipeline companies had held, but without owning the transportation or storage assets—no longer a broker, NGC became one of the very first pure-play gas traders.

Barr's was the art of the deal, and he was never involved with the operations side of the business. This was just as well during the first phase of NGC, because the pipeline companies were perfectly capable of operating their systems. After the companies dropped out of the partnership, however, Barr needed an operations team who knew the networks and how to move physical gas.

Chuck Watson was a great golfer, which was fortunate, because he was sure to be spending plenty of time on the course after he was let go from his job at Conoco, a large producer. He was playing in a tournament in Colorado when Barr, who had flown in to meet him, introduced himself. Barr told Watson the plan over a beer, and Watson seemed game.

Barr was chairman of the new NGC, which was now 100 percent owned by Morgan Stanley, and Watson was CEO. Watson immediately began building the company, renting office space in Houston and hiring the core team. Among the very first calls that he made was to a colleague who had experience brokering physical gas transactions.

Watson knew Stephen Bergstrom from his Conoco days. Bergstrom was the young and self-admittedly cocky executive at InterNorth, who had

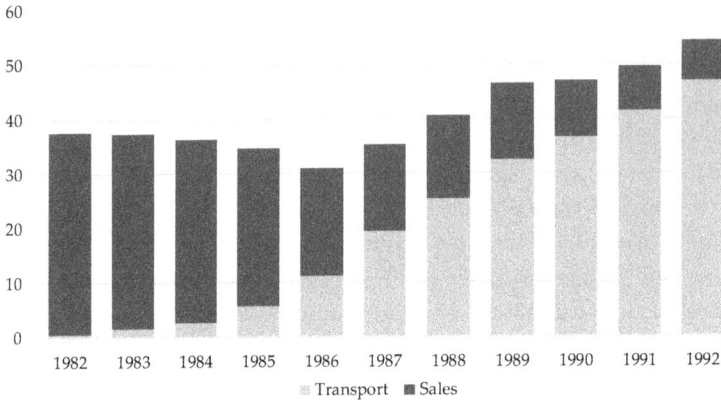

**FIGURE 11.1:** Sales versus transportation volumes on interstate pipelines, 1982–1992

matched up a Montana producer with two Iowa fertilizer plants that had been idled because it was cheaper to import from Peru. In fact, that was just the first tranche of the fertilizer deal—the second slug of production that would feed the plants in Iowa came from Conoco's acreage in Oklahoma. Conoco had a contract to sell to Oklahoma Natural Gas for $9, but that gas was not being taken. Bergstrom had come to Watson with an alternative, and Watson negotiated his way out of the $9 contract so that he could sell to the fertilizer plants for just $2.30. Better to sell at a low price than not to sell at all.

Initially, Bergstrom turned down Watson's offer to come and work for NGC. For one, the first incarnation of NGC was acknowledged as a failure, at least among gas industry professionals. For another, Bergstrom's company, InterNorth, which owned the Northern Natural Gas and Northern Border pipeline systems, had just merged with Houston Natural Gas, which owned several large inter- and intrastate systems. Bergstrom was promised the top marketing job at this new behemoth, which had only just been rechristened as Enron.

This was a promise left unfulfilled. When in 1986 it became clear that Enron had been stringing him along just so that he would not quit, Bergstrom joined Watson at NGC.[8] It was Barr who created NGC, but Watson and Bergstrom who built it into a powerhouse, and the first major independent gas marketer.

As NGC relaunched with Watson at the helm, the market for spot gas sales was on the verge of exploding in size. In October 1985 FERC issued Order 436, which laid to rest any lingering hopes of a return to the old status quo, and in 1986 take-or-pay obligations were at their peak, which meant that pipeline companies were extremely eager to extinguish these liabilities by allowing producers and consumers to interact directly. The share of volumes transported on behalf of others, a proxy for spot gas, versus volumes of pipeline sales gas, grew from 17 percent in 1985 to 37 percent the following year, and continued to increase each year thereafter. After a few years of experimentation, the latter half of the 1980s saw the market mature rapidly.

Marketing, trading, and sales all have somewhat different meanings, but in the 1980s they were mostly synonymous in the gas world. There were no financial contracts to trade, which meant that all deals were negotiated, meaning everything was done on the phone or in person. Not only were there no buttons to click to buy or sell gas, the old-school, paper contracts differed from pipeline to pipeline. You could never be sure if your supplier had agreed to the same terms as what you had promised your consumer. One trader active at the time stressed that doing business consisted of having "conference call after conference call to go over the details, because without standardized contracts, you had to trust people."[9] It was not until the mid-1990s that all the relevant stakeholders in the gas industry came together to agree on standard contract language.

In the 1980s, gas could not be properly called a commodity. Still, some standardization began to emerge. Most spot sales were for a term of one month, and the physical movement of the gas took place under interruptible contracts with pipeline companies, meaning that receipts or deliveries could be cut if the company needed gas for its firm sales customers. Since demand had dropped by so much over the prior years, there was generally enough excess capacity in the systems where such cuts were rare, but it was still the case that the quality of spot service was inferior to the quality of firm pipeline sales.

Spot-market transactions tended to coalesce at the tail end of the month, when producers and consumers were best able to evaluate how much gas they would need to sell or buy. The last week of the month came to be known as bid week, and it was during bid week when marketers like Bergstrom wore their trader hats. They were in the office, on the phone, doing deals with producers and consumers trying to maximize arbitrage and eventually arrive at a flat position, where they would be on the hook to buy the same amount as they were obliged to sell. In these days, there was relatively little intra-month swing activity, so after bid week was over and the telephones stopped ringing on the hectic trade floor, the marketers put on their elbow-patched jackets and became salesmen, often spending the next two or three weeks on the road, visiting producers and consumers, trying to drum up new business.

The first few years were highly profitable for NGC and the handful of other independent gas marketers, but before long, competition drove down margins. This competition at first came from pipeline marketing affiliates—like Transco, Tenneco, and, of course, Enron—the latter of which was, in most respects, in the 1980s, just another pipeline company.

Producers were the next group to enter the fray. Happy to sit on the sidelines for the first few years, acting only passively to market volumes by themselves, in the late 1980s producers charged into the business. They built their marketing divisions extremely quickly by offering lower prices than both independent marketers and pipeline affiliates—it was, after all, their gas, to price however they wanted. Like so many new ventures, market share was initially valued ahead of profits, which meant heavy discounts, and flat-out refusal to deal with independent marketers, such as NGC.

Margins shrank as the market became more crowded. Where before traders were able to clip dollars of profit per MMBtu, they now had to content themselves with just a few cents. Moreover, as spot sales grew to overtake firm sales by 1987, service quality emerged as an important issue. Utilities began to question what they were actually buying—could a trading shop that owned no physical assets really be relied on to deliver gas? Once again, NGC needed to adapt to a new reality.

At this point, John Barr's relationship with NGC had grown strained. He had never stopped working for Morgan Stanley, which still owned the company, and his bosses began asking pointed questions as the marketing business grew. Largely because of Barr's own efforts, Morgan Stanley was now doing big-dollar banking business with pipeline companies, and these blue-chip clients wanted to know what their bank was doing competing with them on their own turf. Meanwhile, the entrance of producers into the marketing business meant that NGC had to bulk up to survive.

Aside from lower prices, producers could boast to their clients of having guaranteed, flowing supply. Even if this was mostly optics, given the immensity of the supply glut, NGC felt that it needed to control physical gas supply in order to weather the storm. It also needed to control physical demand, as many producers would flat-out not deal with independent marketers who did not own gas-consuming assets. Morgan Stanley probably could not offer much help if it wanted to, and anyways didn't want to, so Watson and Bergstrom began persuading Barr to sell NGC to a better-suited partner. In 1989 Morgan Stanley sold NGC to two producers: Apache Petroleum and Noble Corporation. So ended Barr's relationship with NGC, the company that he had created.

This could be called the third incarnation of NGC. The first lasted for about a year, and saw it act as a broker that leveraged a working relationship with several pipelines. The second lasted for around four years, and saw it act as a trader that leveraged its employees' knowledge of various pipeline

**TABLE 11.1: TOP TEN GAS MARKETERS, 1991 (BILLION CUBIC FEET PER DAY)**

| All marketers | | | Independent marketers only | |
|---|---|---|---|---|
| Company | Sector | Volume | Company | Volume |
| Chevron | Producer | 3.0 | Natural Gas Clearinghouse | 2.8 |
| Texaco | Producer | 2.9 | Aquila | 1.1 |
| Natural Gas Clearinghouse | Independent | 2.8 | Associated Natural Gas | 1.0 |
| Enron | Pipeline | 2.6 | Tejas Gas | 0.9 |
| Coastal | Pipeline | 2.5 | Hadson Gas | 0.7 |
| Amoco | Producer | 2.4 | CanWest | 0.7 |
| Transco | Pipeline | 2.1 | Tejas Power | 0.6 |
| Exxon | Producer | 2.0 | Vesta Energy | 0.5 |
| Mobil | Producer | 1.7 | Entrade | 0.5 |
| Tenneco | Pipeline | 1.7 | Sunrise Energy | 0.4 |
| *TOTAL* | | 23.7 | TOTAL | 9.2 |

Source: Interstate Natural Gas Association of America (INGAA), Carriage through 1992, Washington DC, July 1993.

systems to buy gas cheaply and sell it at the highest price possible. In the third, Watson and Bergstrom presided over an asset-based gas trader. In the coming years, NGC would transform itself yet again, and adopt a new name: Dynegy.

By 1990 the gas market had become much more efficient. Consumers were able to secure supplies at a price corresponding to real-world, market-based dynamics, and there was enough competition among marketers to ensure that the prices they were getting was fair. But as transactions moved onto monthly spot markets and away from long-term contracts, producers and consumers were becoming more and more exposed to volatile, short-term dynamics. In fact, they were adding to the volatility.

To understand why Enron became so successful in the 1990s, it is necessary to understand the inherent volatility of a spot-only market. It is a common misconception that, in a deregulated market, the spot price is the only price that consumers pay and producers receive. In fact, in a well-functioning market, both consumers and producers prefer a portfolio of contracts with different terms and of different tenors, with only a small amount of their needs left to be transacted on the spot.

Contracts make both buyers and sellers price takers, rather than price makers—they do not influence the spot price, as they do not actively bid or offer. Importantly, this is true even if the contract is tied to the spot price itself. Suppose I sign a contract agreeing to buy ten million cubic feet of gas per day at the prevailing spot price. Isn't that equivalent to buying ten million cubic feet on the spot market on the day that I need it?

The answer is a resounding no. Regardless of the pricing terms that the contract specifies, whatever the prevailing dynamics on the day happen to be, bullish or bearish, neither party to the contract adds to them. If it is extremely cold, LDCs buying contracted volumes are not out on the spot market lifting offers, like their less-fortunate compatriots. Similarly, if it is very mild, producers selling contracted volumes are not hitting every low-ball bid. There is more stability, and less opportunity for price-gouging in either direction. Forcing all volumes to be transacted on the spot market—making every buyer and seller into a price-setter—introduces an incredible potential for volatility.

The ongoing gas bubble tamped down volatility to a great degree. It was extremely fortuitous that the market remained so loose for so long because had it tightened substantially in the 1980s, prices would likely have shot through the roof. Severe shocks might have imperiled the deregulation drive, especially if they left consumers paying high prices. Even with the glut in full effect, prices still swung around by fifty cents in either direction, which represented a nontrivial 30 percent of the total, but they always reverted quickly back to their mean.

Loose market conditions notwithstanding, utilities and industrial users did not like that they had lost the insurance embedded in long-term contracts. Likewise, pipelines resented assuming the unwanted role of suppliers of last resort—they viewed the new cadre of energy merchants as opportunists, skimming easy money off the top. There was a desperate need for risk-management tools that would allow pipelines companies' long-term firm sales contracts to expire without leaving all parties wholly dependent on spot-market dynamics.

## ENRON ASCENDANT

Today Kenneth Lay lives mostly in infamy for his role as the chairman of Enron, which came to be known as the poster child for corporate greed after the company declared bankruptcy amidst revelations of massive accounting fraud in 2001. Such a description, however, of both man and company, is akin to describing World War II as a fight between Americans and Japanese—in focusing only on how things ended, one misses some fairly important details.

Lay had been in the gas industry for three decades by the time Enron collapsed. After earning his PhD in economics from the University of Houston, he began working at the Federal Power Commission in 1971, and was promoted to deputy undersecretary at the Department of Energy in late 1972. This firsthand policymaking experience was a major part of Lay and Enron's eventual success, but it wasn't long before Lay went out into the world of private business. In 1974 he wrote to Jack Bowen, the CEO of Florida Gas, asking for a job, and got it.

Lay rose quickly, becoming president of Florida Gas just two and a half years later. He was acknowledged as a "bright young star"[10]—an independent-minded PhD in an industry long characterized by engineers who toed the company line. Indeed, Lay was one of the very first of a new breed of more creative and financially savvy pipeline company executives. He was not so transformative a force as to reshape the industry by himself, however, and he exhibited some of the same failings as other pipeline company executives, arguing "many times" that Florida Gas should begin signing take-or-pay contracts with producers in the run-up years of 1979 and 1980.[11] (It did not, to its great long-term benefit.) Lay was a big-picture guy, a not a detail-oriented problem solver, but he did seem to grasp where the industry was headed.

In 1979 Florida Gas was bought by a giant conglomerate called Continental Group. Even though this had resulted in another promotion, Lay quickly grew frustrated with his distance from Continental's power center in Connecticut. His skills as an energy statesman were being lost in a company that was selling more than one thousand different products, where the energy group made up only 11 percent of total assets.[12] And his marriage was falling apart. Lay needed a change, so he once again wrote Bowen, the man who had originally hired him into Florida Gas and had since moved to Houston to head Transco Energy Company. Bowen again extended an offer, and Lay again accepted. In 1981 he arrived in Houston as president and chief operating officer of Transco, one of the largest pipeline systems in the country.

Transco, unlike Florida Gas, had signed take-or-pay contracts with producers—lots of them. Within a year of his arrival, Lay had to confront the take-or-pay issue. It was here where his creativity and policy chops began to change how the market worked. He got FERC to approve an industrial sales program, which allowed producers to sell Transco discounted spot gas on the condition that Transco resell the volumes to customers who would otherwise have been using oil. Lay then convinced regulators to allow Transco to sell discounted sales gas to a broader customer base, which doubled its share of the Washington, DC, market, at the expense of rival Columbia Gas Transmission, which later declared bankruptcy.[13] Transco was at the forefront of these special marketing programs of the early 1980s, and was the first pipeline to join the first incarnation of Natural Gas Clearinghouse. Indeed, it was at Lay's invitation that NGC began operating out of offices on the fortieth floor of the brand-new Transco Tower in Houston.

Most innovation is borne of necessity, and it was no different at Transco. Bowen attributed the flurry of innovation to the fact that Transco was "first in the hospital"[14]—it needed to get fit or die. Regardless of how one chooses to apportion responsibility for the innovations, it was during the early 1980s when Lay began to endorse a more laissez faire, free market agenda. As the years went by, he embraced his new creed with almost religious fervor, doggedly advocating for deregulation, regardless of the circumstances. (Except, of course, when those circumstances meant deregulation would hurt his bottom line.)

Lay moved jobs one last time, after Oscar Wyatt—the cutthroat who had cost the people of San Antonio, Austin, and Corpus Christi so much money in the early 1970s—launched a hostile takeover bid for Houston Natural Gas Corporation (HNG), a large Texas intrastate system. This was the beginning of gas pipeline merger mania—a mania that would result in the creation of Enron just a year later.

HNG was not in favor of a takeover, and especially did not appreciate it coming from Coastal, Wyatt's ascendant pipeline empire. The board blamed the attack on HNG's lackluster sitting CEO, and sought to install a more impressive figure. HNG's chairman called Jack Bowen at Transco and politely asked whether it would be all right to extend an offer to Ken Lay. Bowen agreed, and in 1984 Lay was appointed CEO of HNG.

At HNG, Lay did as much to further the merger zeitgeist as any other pipeline CEO, acquiring both Florida Gas Transmission and Transwestern Pipeline, which served the California market, within just six months of his start date. While some of Lay's strategies proved misguided, in retrospect, he did understand one, important dynamic better than many of his counterparts. "I know people who are sitting around waiting for the markets to get real tight again so they can get back to the old way of doing business," he said after his first pipeline company acquisitions. "I don't think that's going to happen."[15] Lay correctly estimated that the gas glut was not going away anytime soon. His solution was to combine in order to create larger and more efficient networks.

With Lay at the helm, HNG successfully fended off Wyatt's hostile bid, although Coastal went on to acquire the giant American Natural Resources system a year later.[16] After acquiring Florida Gas and Transwestern, HNG was a much larger company, but still just a midsized fish in the sea of gas pipeline companies.

InterNorth was a big fish. The Omaha-based company owned the giant Northern Natural Gas system, which moved gas from the Permian Basin and the Texas/Oklahoma Panhandle into Nebraska, Iowa, Minnesota, and Wisconsin. It also owned the Northern Border Pipeline, which picked up a massive chunk of gas at the Canadian border and fed it to those same customers. InterNorth, twice the size of HNG in both assets and earnings, was a conservative company, which kept costs—and executive pay—low and produced steady cash flow. It also had very low levels of debt. While these were admirable qualities, they also made it a prime candidate for a corporate raider. When such a raider appeared, InterNorth CEO Sam Segnar panicked, and went looking for an acquisition target to make the company "sharkproof."[17] In April 1985 Segnar called Lay and floated the idea of a combination.

Lay and HNG were much more receptive to InterNorth than they were to Coastal, and thought that the companies were a strategic fit. Perhaps their lack of resistance really was because they perceived some compelling strategic fit. Perhaps HNG's board just didn't like Wyatt. Whatever the case, Lay began to salivate as discussions with InterNorth began in earnest. Segnar was in a rush, and capitulated on many key issues with very little negotiation. Most amazingly, Segnar and InterNorth agreed to have HNG's management team take over: Lay would replace Segnar as CEO of the combined company after just eighteen months. HNG would also have almost as many seats on the new board as InterNorth, and the new company would be called HNG InterNorth, with the Texans getting first billing. Who was buying whom? The acquisition was finalized in May 1985, and HNG InterNorth became the nation's largest pipeline system, by mileage, and second largest by volume, after Tenneco.

The more the InterNorth board learned about the details of the acquisition, the more irate they grew, with their ire trained squarely on Segnar. Most egregious to the old guard was the mounting pressure from Lay and the HNG contingent to move from Omaha to Houston. During negotiations, Lay had promised that the company would remain headquartered in Omaha "for the foreseeable future," but that promise seemed to be worth less each day. So contentious was the issue that a special board meeting was called in November 1985, where a group of consultants from McKinsey & Company would present their analysis of the issue. The consultants, headed by a man named Jeffrey Skilling, advised the company to move to Houston. The nation's largest pipeline company belonged in the nation's energy capital.

After the McKinsey presentation, the board met privately. When the meeting was over, Segnar tendered his resignation. In reality, the board had demanded it. Lay was now CEO, less than six months after the merger. In a few more, he was chairman of the board. It did not take long for Lay to appoint loyalists, who approved moving HNG InterNorth to Houston within a year.[18] InterNorth had essentially paid $2.3 billion to sell itself to HNG.

Lay wanted a slick new name for the company, to send the message that it was a new business for a new era, and not just a recombination of stodgy, old pipes. He originally chose the name Enteron, but Wall Street analysts quickly discovered that this was a medical term for the digestive tract. After much embarrassment, and several thousand dollars wasted on stationery and business cards with *Enteron* printed on them, HNG InterNorth was renamed Enron.

Things did not get off to a smooth start. The old InterNorth, under Segnar, had embarked on an ill-fated diversification push, which included buying oil and gas producer Belco Petroleum. Belco owned reserves in Peru, which that government nationalized in 1985, costing Enron more than $200 million. As for its core—US pipeline operations—the merger might eventually yield higher profits, but in the immediate aftermath of the combination nothing had changed, and the gas bubble was still wreaking havoc on all pipeline companies, Enron included. Despite its slick new name, in the 1980s, Enron was just a recombination of stodgy, old pipes. Lay's combination strategy was not yielding increased profits.

Enron would go on to become a true innovator. It would, in many ways, invent the new markets for natural gas and electricity. But anyone predicting this in the late 1980s would have been laughed out of the room. In January 1987 the company's credit was downgraded from investment grade to junk, and it nearly went bankrupt later that year after a rogue oil trader—again, inherited from Segnar's InterNorth diversification efforts—bet the house in the wrong direction. True, Enron was ahead of its peers in signing some lucrative, long-term deals, and in reducing its take-or-pay obligations, but these first few years showed that Lay, while an effective statesman, was an inept manager of a large, complex organization.

Two people turned the company around: Richard Kinder and Jeffrey Skilling. They could not have been more different.

Kinder was a lawyer by training and a bulldog by temperament. He had been college friends with Lay, and gone to work at Florida Gas while Lay was president there. He was general counsel at Enron while the company was tanking in 1986 and early 1987. In August 1987 Lay appointed Kinder chief of staff, perhaps recognizing his own inability to wade in minutiae and make the many difficult decisions required to run the company. This appointment alone, however, was not enough to right the listing ship—Kinder did not have the authority he needed to implement the major changes that were required. For example, while it was agreed at the highest levels that Enron's constituent pipelines—Northern Natural, HNG, Florida Gas, and Transwestern—should hand over their marketing functions to another division within the company and instead focus simply on transporting gas, the executives of the pipelines resisted, not wanting to relinquish their traditional responsibilities. Lay, ever the statesman, argued politely that they ought to.

Things languished until Kinder lost his patience with the intransigent, old-school pipeliners during a terse meeting in 1988. In Enron history, this became known as the "Come to Jesus" meeting, and marked the start of Kinder's reign.

Fresh off the oil trading scandal that almost bankrupted the company, Kinder ripped into everyone present, Lay included. Lay had appointed too many cronies who went behind his back, Kinder fumed, and did not even try to solve the company's very real problems. He compared these problems to alligators in the swamp, yelling: "We're going to get in that fucking swamp, and we're going to kick out all the fucking alligators, one by one, and we're going to kill them, one by one."[19]

Only after the "Come to Jesus" meeting did Kinder bring things under control. By all accounts he ran a tight ship, and quickly extinguished the infighting and bickering that had plagued the company since it was formed, one merger after the other. Kinder was sharp and supremely competent—a great steward—but Enron was still mostly stagnant: a non-innovative, regulated pipeline company without any source of market-beating growth. It needed more than a steward, and it found one in Jeffrey Skilling from McKinsey.

Skilling had an MBA from Harvard and had risen through the ranks at McKinsey, by far the country's most prestigious consultancy, faster than almost anyone on record. He was fantastically clever and equally arrogant. When the dean of Harvard Business School asked Skilling, point-blank, whether he was smart or not, he replied: "I'm *fucking* smart."[20]

Skilling learned about natural gas while at McKinsey in Houston in the 1980s, where he watched the industry's fall from grace from front-row seats. Before the take-or-pay crisis, interstate pipeline companies had constituted the second most profitable sector of the US economy, trailing only Big Tobacco. They had been cash cows for decades. Skilling, though, began doing consulting work with them just as their markets, and profits, began to evaporate. He met Lay while working with InterNorth, when he was tasked with analyzing whether the combined HNG/InterNorth should stay in Omaha or relocate to Houston.

Lay was impressed by Skilling's intellect and ability to break down complex issues to their simplest terms. Like Lay, Skilling was a big-picture guy, uninterested in minutiae, and probably temperamentally incapable of dealing with them. But there were also important differences between them. Lay was a diplomat—some might even say a glad-hander—whose real skill was communicating and forming bonds with people. Skilling was more introverted, and had little patience for anyone he considered stupid, which was most people.

Perhaps most importantly, Lay had become a believer in free markets, but was, more than anything else, a shrewd political capitalist. Whereas

Lay's faith in markets was opportunistic, Skilling was a true believer. He had looked with scorn upon the Soviet–like regulatory regime under which pipeline companies operated before FERC Order 436. This regime had, for decades, simply transferred wealth from consumers and producers to the companies, which had grown fat and lazy, enjoying large and steady profits without taking on any risk. Equally, and more importantly for Enron, Skilling saw that the collapse of the pipeline-as-merchant model had left open a tremendous commercial opportunity.

Skilling recognized that the ability to manage risk had been lost in the transition from long-term, firm contracts with pipelines to spot sales with marketers. Consumers were inherently risk-averse, and keen to lock up long-term prices at the prevailing low levels. The country's fast-growing fleet of independently owned power plants represented a huge market with potentially large margins, as they were paid a fixed price by utilities for all their future electricity sales, meaning that by far their largest source of income uncertainty was the price of fuel. Many of the plants found themselves torn between using coal or gas. Gas plants were cheaper to build, but prices were volatile and could not be hedged, whereas coal miners were happy to forward-sell fixed-price contracts. Indeed, while gas consumers across the country were being courted by multiple sellers who were eager to place volumes in the glutted, month-ahead market, none were able to find longer-dated offers. The fate of the market seemed to lie in the balance—gas might have been a superior fuel on both economics and the environment, but if consumers could not count on prices remaining low over the long term, they would opt to use coal.

The only group with the ability to forward-sell gas supplies at the time was producers, and they wanted no part of it. Ever the optimistic bunch, producers all "knew" that the oversupply was a temporary phenomenon, even if it had lasted for longer than they had predicted. They had annual price growth of 5 percent penciled into their internal projections, despite no evidence of the market getting tighter. Whenever they were approached about a long-term supply deal, the near-universal response was: Why should I commit for ten or twenty years now, when prices are at their lows?

In sum, there was an enormous mismatch between the risk appetites of producers and consumers. Bridging this gap was Skilling's strategy—one that would transform Enron from "just another pipeline company" to innovator and industry leader. It was called the Gas Bank.

In truth, the Gas Bank had begun even before Skilling quit McKinsey for Enron. Already in 1987, Enron had begun to guarantee consumers fixed prices for periods of up to ten years.[21] The problem was that this was not a scalable business—Enron was just taking a short position. Eventually, it would need to go and buy the gas that it had promised on the spot market, which would just have transferred risk from consumers to Enron. Said another way, it would have been speculation, no different from what Wyatt had done in the early 1970s. In order to build a scalable business selling long-term gas contracts and managing risk for multiple counterparties across the country, Enron needed long-term supply deals with multiple producers. In short, it needed to replicate the old pipeline merchant system, where supply commitments closely matched demand obligations—where assets matched liabilities.

Enron was able to do its first long-term deals without taking on undue risk because it owned a production company—Enron Oil & Gas, or EOG—which acted as a backstop (essentially, the early Gas Bank was just marketing EOG volumes). However, EOG only produced around 350 million cubic feet per day in the late 1980s. This was a good-sized chunk, to be sure, but not enough create a stand-alone risk-management business. To scale the Gas Bank, Enron could simply have acquired producing acreage, and considered doing just this. However, it found that producers placed premium in owning land beyond what Enron was willing to pay—there was upside from any potential future discoveries or technology improvements that might increase the reserves in known formations. Enron just wanted the gas, and not to become an enormous production company.

Skilling's innovation, which won over the previously reluctant producers, was on the financial side. Despite not wanting to commit to long-term sales, low oil and gas prices meant that producers were strapped for cash. Many could not finance themselves solely from their operating profits, so they needed to borrow. Capital, however, was difficult to come by in late 1980s, given the depressed state of the upstream oil and gas industry, and the fact that the Lone Star State was the epicenter of the savings and loan (S&L) crisis, which hit its apex in 1988. Carter–era banking sector deregulation allowed small institutions to gamble massively, which they did with gusto in the early 1980s, especially in Texas, and especially in real estate. When the oil price dropped, the bottom fell out of the real estate market and the banks went bust. This left producers doubly disadvantaged—not only were commodity prices near rock bottom, capital markets had dried up, and those institutions who had money to lend were not lending it cheaply.

Enron had an edge. A bank would embed a substantial amount of premium in its loan calculations for a number or reasons, not least because it did not know what price the producer would receive in the future, and whether it would default. Enron knew the price with certainty because it would already have made the sale–it was simply filling its short and getting flat—there was no price risk. Moreover, Enron would structure its loans such that it would still have claim to the agreed-on gas volumes even in the event of the producer declaring bankruptcy, eliminating default risk. And last, there was one less middleman, as Enron would take the gas in kind—it was not loaning dollars for dollars, it was loaning dollars for gas.

The timing was perfect—not only were consumers clamoring for ways to reduce their long-term price risk, but traditional reserves-based lenders were retrenching, leaving producers on life support. Enron began hiring reservoir engineers out of the slumbering banks and put them to work partitioning out future years of supply from existing wells that Enron could sell forward to consumers. For a large sum of capital today, Enron would receive many years of future gas supply, which it could sell to consumers for a juicy margin.

Although Skilling did not conceive of the Gas Bank on his own, he quickly became its champion. After presenting the strategy to Kinder and other senior Enron management, he was hired away from McKinsey to grow the Gas Bank and to transform Enron into a financial services company. While Kinder remained in charge of Enron's core pipeline business—its hard assets—Skilling's "asset-light" division came to define Enron's culture and investment thesis over the following decade.

The Gas Bank allowed Enron to prepay for gas supplies in the very loose market. Producers saw it as a godsend. In Skilling's words, "If you offered to buy gas at a fixed price for twenty years, they would throw you out. But if you offered to hand the producer $400 million to develop reserves, he saw you as a partner."[22] Not only was capital available more cheaply from Enron than from the banks, it also did not add to producers' debts. It was cash for gas, with no interest payments. Unlike the shady off balance sheet deals in which Enron became embroiled in later years, which anyway were purely internal devices meant to manipulate earnings and keep its stock price high, the Gas Bank was perfectly legitimate.

In 1991 Enron completed around $100 million of producer financing. The following year it was $500 million, and by 1994 the total exceeded $1.5 billion, making Enron one of the largest funding sources to producers in the country.[23] Witnessing this astonishing growth, other companies created their own versions of the Gas Bank. This was an important validation of Skilling's vision, but also drove down margins, as competition always does.

Even as it was lauded as a huge moneymaker for Enron, the Gas Bank (and its look-alikes) represented a fairly small chunk of the overall market. After all, every loan was subject to due diligence and review from a team of reservoir engineers, even though all Enron was after was gas supply, and well-capitalized companies, including the majors, did not need to turn to Enron for financing. The Gas Bank was, in the end, just a way to convince improvident producers to part with their volumes without admitting that they were selling at the lows. By 1990 the gas market had come a long way, but it would still have been a stretch to call natural gas a commodity.

## TOWARD COMMODITIZATION

In April 1990, just before Enron hired Skilling, the New York Mercantile Exchange (NYMEX) introduced a natural gas futures contract. On the face of it, this was decidedly non-revolutionary, as there were actively traded futures for most commodities other than natural gas. There had never been a need for a natural gas contract in the regulated era, since the price was set administratively. Now that market participants were clamoring for a hedging tool, it would seem an easy enough thing to create. But, as ever, some gas market idiosyncrasies meant progress was slow.

A futures contract is a contract for future physical delivery, settled in kind, not in cash. This means the contract must stipulate the delivery terms, which need to be very specific. The main US crude oil futures contract, for instance, is for delivery to Cushing, Oklahoma, which is home to the largest group of oil storage tanks in the country. It is possible to speculate using futures contracts by making sure that you zero out your position before expiration, but if you do not, you become responsible for either delivering physical barrels to storage tanks in Cushing or taking them out.

As I have said numerous times, the logistics of transporting gas are far more complicated than for oil (or for almost any other commodity, for that matter). Physical limitations in the pipeline system mean that there can be wide spreads even between locations that are not very distant from one another. With oil, these arbitrage opportunities are solved with trains, trucks, and barges. With gas, delivery must be via pipeline.

Equally, the success of a futures contract depends on liquidity. If only one or two counterparties actively trade a contract, they might as well just negotiate bilaterally. There would be no point in having a different natural gas futures contracts for each receipt and delivery point on all of the country's many gas pipelines. How, then, to choose a physical delivery point for a national natural gas benchmark?

There was heated competition, as host pipelines would earn fees on every unit that passed through. The network around Katy, Texas, emerged as an early favorite. Located just west of Houston, Katy had become the most liquid trading point in the physical spot market that had developed throughout the 1980s. It boasted interconnections with several large intrastate pipelines and storage facilities. This meant that there was little to no risk of physical congestion, which is supremely important, as futures contracts are settled via delivery. But not everyone was a fan of Katy.

Intrastate pipelines are not subject to FERC rate regulation—they can charge whatever the market will bear for transportation or storage services.

In practice, this makes intrastates some of the biggest physical traders in the market. Although they trade pipeline capacity, not gas, the distinction is financially irrelevant—if you save twenty dollars bringing your own wine to a restaurant, but the restaurant charges a twenty-dollar corking fee, you are no better off. One of the intrastates that connects to Katy comes all the way in from the Permian Basin in West Texas, six hundred miles away. The market for gas in the Permian Basin shares almost nothing in common with the market for gas in Houston. The former is a supply region, where the vast majority of production comes from oil wells, and so is subject to oil market dynamics. It also experiences large-scale freeze-offs in the winter, where liquids freeze in the pipeline, blocking the flow of gas and removing up to several hundred million cubic feet of day of capacity for several days at a time. Houston, meanwhile, is a major population and demand center, home to several gigawatts of gas-fired power plants and countless industrial facilities. It rarely freezes. The price spread between the two markets can get very wide.

If spreads blow out between two markets served by interstate pipelines, the arbitrage is collected by the shippers, be they producers, consumers, or traders. If the markets are connected by intrastate pipelines, though, the spread is the pipeline company's for the taking. While one choice is not more fair than the other per se, some powerful marketers lobbied against the "Texas country club bullshit" that was sure to ensue if Katy was chosen as the benchmark. These marketers wanted a hub served mostly by rate-regulated interstate pipelines. As luck would have it, an executive from Texaco came to the rescue.

The Sabine River divides Louisiana from Texas for a bit more than half of their shared border. When the Spanish settled the area, they named the river for its many cypress trees, or *sabinas*.[24] The river, in turn, gives its name to a parish in Louisiana, and to many petroleum-related companies, including Sabine Pipe Line.

Texaco built the Sabine pipeline to move gas from its leases in the federal waters off the shores of Louisiana to its huge refinery complex in Port Arthur, which sits on the Texas side of the Sabine River. That made Sabine an interstate pipeline, albeit a small one. The pipeline received offshore gas from the Henry processing plant, so named because it is located in Henry, Louisiana. Completed in 1967, the system was highly utilized for its first decade, but the gas shortages of the late 1970s dealt it a crippling blow.

Sabine never ran out of gas—far from it. But FERC was not happy that a Gulf Coast industrial user, the Port Arthur refinery, was getting cheap gas from federal waters through an interstate pipeline, while residential consumers were shivering in more northerly latitudes. It accused Sabine and Texaco of diverting the gas from the "real" interstate market, and demanded that the pipeline begin selling large volumes of the cheap, old gas, which qualified for a price of around thirty cents under the NGPA's pricing schedule, on the open market, and making up the shortfall by purchasing more expensive, new gas, for around $2.50. Not only that, but after the worst of the crisis was over, FERC slapped a retroactive ruling on Sabine, forcing the pipeline to calculate how much gas it had "diverted" and to sell back that gas at the low price. This settlement became known as the Sabine payback, and it left the pipe almost empty by the mid-1980s. Once again, innovation came from desperation.

A young engineer named Jagjit Yadav went to work for Texaco in 1967, the same year that Sabine started operations. In fact, one of his very first tasks was to "walk the pipe," and get familiarized with Sabine's right of way and its aboveground valves and metering stations. (Even in the 1960s, the phrase *walk the pipe* was something of an anachronism, and Yadav actually drove a truck from the Henry plant to Port Arthur.) By the 1980s, Yadav had been promoted several times, and was put in charge of monitoring the Sabine payback. This was no small task, as the settlement's total value measured in the hundreds of millions of dollars—a large sum even for Texaco, which jostled with Ford Motors and IBM for a spot among America's top five largest companies during the mid-1980s, and boasted annual revenues of between $40 and $50 billion ($95–$120 billion in 2020 dollars).

By 1986 Sabine looked like a dead asset. The terms of the settlement made it uneconomical to flow gas, so it sat idle. Desperate to do something with a perfectly good pipeline, Yadav hired a team of consultants, who suggested turning the Sabine into a "market center." Here, the United States took a page from its northern neighbor, as the first natural gas market center began in Alberta, Canada, about a year and a half earlier.

I'd like to step back and review where the market was in 1986. Physical gas trading had begun, and the spot price had become responsive to supply-and-demand conditions. Still, gas cannot be said to have been "commoditized" during the first few years of trading. There was simply too much specificity required: contracts had to include exact receipt and delivery points, pricing terms, and gas quality provisions, meaning that the contracts were not interchangeable among many parties. More fundamentally, a large, interconnected network was still being economically operated as a series of nodes. The exact amount of gas that makes it through a given node—a meter, in the parlance of the gas industry—is dependent on several factors beyond human control, such as weather and reservoir conditions. Actual deliveries will never match scheduled deliveries exactly. On some days, certain pipelines will have higher pressures than others, and the network will benefit, operationally, by moving gas around from pipe to pipe. One key issue that had not been overcome, however, was who would benefit economically?

In Alberta, the first gas market center, this was not an issue. Trans-Canada Pipelines, the operator, was the only game in town, and it owned large, underground storage caverns that allowed it to inject gas into storage when actual deliveries exceeded actual receipts, and withdraw when the opposite was true. The increasingly useless Sabine pipeline was not, at the time, connected to any storage facilities, which made operating as a market center more of a challenge. The one thing Sabine had in spades—interconnectivity with other pipelines—was only an advantage if it could figure out how to make them function as a network.

Located in the heart of the "spaghetti bowl" of Gulf Coast pipelines, Sabine connected to eleven pipelines near Henry, another thirteen facilities elsewhere in Louisiana, and nine more in Texas. Yadav, seeking to rescue Sabine from desuetude, heeded the consultants' advice, and in 1988 created the first market center for natural gas in the United States. He called it the Henry Hub.

The natural way to think of a hub is the center point of a hub-and-spoke system, but this is not the way the Sabine Pipeline is actually laid out. The interconnects included in the Henry Hub are located some distance from one another, so some ingenuity on the business side was required to make the tangle of pipelines function in a unified, hub-like manner. First, Sabine got all of the interconnecting pipelines to sign operational balancing agreements, which meant that they would resolve the imbalance between actual deliveries and scheduled deliveries amongst each other, behind the scenes, instead of via charges or credits to individual shippers at the end of each month. This was of paramount importance, since trading is far riskier when all the costs are not known fully at the time of the transaction, and since Sabine did not own storage and so could not solve this problem unilaterally. It helped that Yadav knew through experience how painful imbalance payments could be, as he was in charge of Sabine while it was accumulating "horrendous" sums with its interconnecting pipelines. Nor was this experience unique to Sabine—after a cold snap in the winter of 1989–1990, pipelines began siphoning gas out of Enron's enormous, nearly national system, prompting the ever-polite Lay to lose his calm at a meeting of pipeline executives: "You fucking stole my gas. You owe me gas."[25] The operational balancing agreements that Yadav executed separated shippers' commercial transactions from pipeline companies' operational transactions, which greatly simplified the business of buying and selling gas.

Yadav's second innovation was a service he developed that allowed buyers and sellers to trade gas as many times as they wanted, within the Henry Hub, and only pay a transportation charge once the gas physically left. For a very low fee, Sabine automatically took care of all of the title transfers. This meant that gas could be traded many times over during the course of a day.

Just like the operational balancing agreement, this service avoided a lot of unnecessary paperwork, which attracted traders, which facilitated liquidity. Last, despite the fact that the title transfer fee was very low, at only two-tenths of a cent, it clearly incentivized Sabine to do everything that it could to increase that liquidity. That meant offering transportation very cheaply, at below FERC-approved rates, which other pipelines had no reason to do.

The Henry Hub was an almost overnight success, but Yadav had con-signed himself to the fact that Katy would become the delivery point for the NYMEX futures contract. He had made the decision for the Henry Hub to operate exclusively as a physical hub in late 1988, so as "not to muddy the waters" at that late stage, since NYMEX's discussion with Katy were "quite advanced."[26] It was in jest, Yadav recounts, that he approached a NYMEX vice president at a conference in San Antonio the following spring, and asked him why he was wasting his time with the Katy folks. Handing him the one-page marketing diagram that he had printed out for the conference, Yadav posed a simple question: Why not use this market center that's been operating for several months?

Yadav could not have known how frustrated NYMEX was growing with the Katy consortium. As it turned out, half of the participants were naysayers, arguing that Katy couldn't function as a delivery point for a fu-tures contract. While much had been made over Sabine not being a natural hub, in the hub-and-spoke sense, neither was Katy. The idea was to draw a circle around Exxon's Katy processing plant and say, "That's the hub." The problem was that within that circle, there were multiple operators, and multiple pipelines with different tariffs, all of whom wanted to have their cake and eat it too. They were not, like Sabine, willing to lower their tariffs to drum up liquidity, or to give away some of their operational indepen-dence. The interstates, in particular, were "very cocky, and only told you why things couldn't be done," said Yadav, who took particular pride in the fact that no one on his team had ever worked for one.[27] The only inherent advantage Katy had over Henry—the fact that the proposed hub included storage—became almost irrelevant after Yadav got all of Sabine's intercon-necting pipelines to agree to operational balancing agreements.

Katy's several owners met time and again, but could not agree on a rev-enue-sharing stream or many other operational necessities. Representatives from NYMEX were pulling their hair out, so when Yadav handed them his one-page marketing sheet in San Antonio, he almost immediately won them over. The Henry Hub was chosen to be the physical delivery point for the US natural gas futures contract later in 1989, and trading began the following year.

It was far from clear that the Henry Hub contract would be a success. In fact, NYMEX put Henry Hub traders in the same pit in Lower Manhattan where an already-successful energy futures contract, for unleaded gasoline, was already being traded, with the thought that this would increase interest. As it turned out, the worries were unwarranted, and liquidity grew rapidly. Within just a few months of its introduction, the Henry Hub futures contract solved the main problem that the gas market had encountered since deregulation: hedging future price risk. And, unlike Enron's Gas Bank, there was no need to go out and hunt for cash-strapped producers, and then have reservoir engineers vet their acreage.

For a few years, it seemed like the market had all the tools it needed— that it had reached a steady state. As the gas bubble extended, turning into what some people called the gas sausage, prices remained low, and demand began ticking up once again. The Fuel Use Act had been repealed by Congress in 1987, leading to new power plant build, and gas was back in vogue as a residential heating fuel. This was good for all parties—producers and LDCs were happy to sell more, and pipelines were happy to transport. Demand growth, however, exposed the limitations of the gas market as it existed in the early and mid-1990s.

Physical trading was still happening almost exclusively on interruptible contracts. As demand increased, however, interruptible contracts were interrupted more regularly. This led to the emergence of wider locational price spreads, for which the NYMEX Henry Hub contract offered no protection. After 1990 the contract allowed market participants to buy or sell gas for future delivery in southern Louisiana with minimal effort, but what good was this if a Chicago utility was buying gas from a Canadian producer? Buying or selling NYMEX futures was, in many cases, a "dirty" hedge—it did not insulate fully from price risk because you were still exposed to locational price risk. The hedge would only be "clean" if you held firm transportation capacity on one of the pipelines that connected to the Henry Hub. And firm transportation capacity was getting harder and harder to come by as demand grew.

The risk of either locational price blowouts or cuts to interruptible transportation, which are two expressions of the same phenomenon—a shortage of pipeline space—was quite low during the first decade of gas trading, as the colossal drop in demand that came on the heels of the gas shortages of the late 1970s and the take-or-pay crisis had left the United States over-piped. By the mid-1990s, however, demand had recovered, and was back near the record levels set more than two decades earlier. As constraints emerged, location began to emerge as a serious risk, especially in the winter, when demand was highest.

In addition to its value as a hedging tool, the NYMEX contract sent a signal to the market. If prices for future delivery were high, the market was telling producers to bring on more supply and for consumers to use less. If

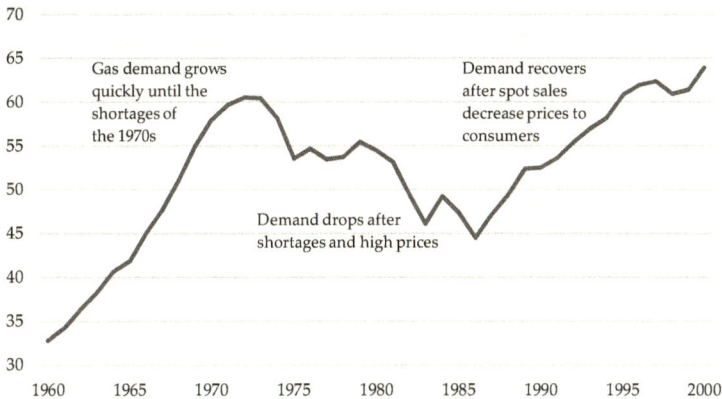

Gas demand grows
quickly until the
shortages of
the 1970s

Demand recovers
after spot sales
decrease prices to
consumers

Demand drops after
shortages and high prices

**FIGURE 11.2:** US gas demand (billion cubic feet per day)

they were low, the market was telling producers to run fewer rigs and for
consumers to use more. The gas sausage kept prices low, and consumers
were using more. New demand meant new pipelines were needed. But who
would pay for them, now that the pipeline companies were out of the merchant business? NYMEX sent broad signals to producers and consumers,
but it did not and could not send a signal to build a new pipeline from
Canada to Chicago, or from West Texas to California.

It was not immediately clear how the situation would resolve itself.
One option was to introduce new futures contracts for different regions.
While Texaco and Sabine Pipe Line were way out ahead of the pack, having created the Henry Hub market center all the way back in 1988, other
pipeline companies had followed suit, and by 1994 there were similar hubs
all over the country. This made the logistics of physical trading much
easier, and meant that the only hurdle to launching competing futures
contracts was securing enough interest among market participants to
drive liquidity. For a while, it seemed that this was the solution—that
there would be different futures contracts for different regions. Between
1989 and 1990, Enron worked to develop a set of four futures contracts—
for markets east, west, south, and north—but abandoned the idea as the
NYMEX contract took off. In 1995 the Kansas City Board of Trade tried
again, launching a western gas futures contract that specified delivery at
the Waha Hub in the Permian Basin of West Texas. Most gas from Waha,
which is the name of a large field in the vicinity, but referred to all production in the area, moved west, toward California, meaning that was
influenced by different dynamics than the Henry Hub, which sent most
of its gas north and east. The next year saw another futures contract, for
delivery in Alberta, Canada.

Ultimately, the competition faded. While there was nothing that prevented alternative futures contracts from succeeding, the network effect prevailed. One, highly liquid contract was better than many illiquid ones, and NYMEX Henry Hub futures retained prominence. A more flexible instrument was developed to deal with locational price risk that would complement, rather than supplant, NYMEX.

Part of FERC Order 636, that last order that mattered, which was promulgated in 1992, required pipeline companies to allow shippers to release their long-term, firm capacity on the open market, in order for that capacity to be used efficiently. This allowed LDCs, who had bought capacity to serve peak-day winter demand, to release the excess during the summer, when demand was lower. They could sell on the cheap, since the costs were sunk and the capacity had no value to them, which enabled marketers like NGC to offer firm service without paying the full price.

A trader and advisor to FERC who helped insert the capacity release provision into Order 636, Greg Lander, compared capacity on a gas pipeline to landing rights that airlines control at airports. He argued to FERC that the deregulation of US airlines would have been much more successful had airports been required to allow trading in landing rights. Even today, entire airlines are bought and sold solely for the landing rights they hold, which is clearly an inefficient allocation of capital. Lander urged policymakers not to make the same mistake twice, and they heeded his advice: after Order 636, capacity could be traded independently of gas.

The early days of capacity trading, however, were extremely slow-going. Pipeline companies fought the provision by requiring that shippers release their entire contract or nothing at all, which went against both the spirit and the logic of the idea. It took several years to modify the rules to allow shippers to release whatever amount of capacity suited them, and for whatever terms they pleased.

In the late 1990s, as capacity trading began in earnest, the locational market was growing in importance. Securing firm capacity provided an important tool to hedge against locational price risk. Still, these were purely physical devices, which meant they were scarce. Unlike financial products, which can be traded for hundreds of times the volume of the underlying commodity, there was limited liquidity in the capacity-release market. Plus, the market essentially went dead in the winter, as LDCs husbanded their pipe space. This segues into the second problem with the Henry Hub futures contract—namely, that it was a futures contract. All true futures contracts are settled in kind, which, in the gas market, means making or taking physical deliveries of gas in southern Louisiana.

The financial instrument that came to dominate natural gas trading, and energy trading in general, is the swap. A swap is a derivative contract, meaning that its value is derived from something else. The key difference between a swap and a futures contract is that a swap is settled via an exchange of cash, rather than an in-kind exchange of gas. You can swap commodities, currencies, point spreads on college football games, and pretty much anything else you can think of. And you can create swaps for an unlimited amount of size, since they bear no relation at all to the physical commodity. A swap is a purely financial product, no different from a bet, and there is no limit on how big the bet can be, as long as there is someone to take the other side.

The most heavily traded natural gas swap is the swap on the NYMEX Henry Hub futures contract. That may sound convoluted, because of course you could just as easily trade the futures contract, but if I want to hold a swap until settlement, I can do so without having to worry about arranging for physical molecules to show up at or leave the Henry Hub itself. More important than a swap on the Henry Hub futures, contracts, however, were the swaps that began to proliferate for regional gas prices. As capacity trading began, swaps were developed to trade out of locational price risk, often to hedge the capacity you had just bought and lock in a profit. For example, if I bought released capacity from the Henry Hub to Chicago for five cents, and could sell a Chicago locational price swap for six cents, I would have just locked in a profit of one cent.

Importantly, participants did not need to own capacity to buy or sell a swap. More often than not, as the market developed, participants bought swaps precisely because they could not get any capacity. Without physical protection, they could at least buy financial protection. As swaps began to proliferate, so did speculative locational trading.

In the gas market, the difference between the local price of gas and the price at the Henry Hub is known as *basis*. If gas on a cold day in Chicago is trading at $5, while the Henry Hub is $3, Chicago basis is $2. Meanwhile, if gas in Canada is trading at only $2.50, Canadian basis is negative 50 cents. Basis swaps allowed market participants to protect themselves fully from price swings by hedging both the outright price, at the Henry Hub, plus the basis differential. Swaps allowed any market participant to speculate on or hedge fully out of price risk, while not changing any physical operations or renegotiating any supply contracts. Moreover, because basis swaps are complementary to the NYMEX contract, they allowed everyone to benefit from a highly liquid benchmark.

Finally, by around 1998, market participants could "do it all." The proliferation of market centers in the mid-1990s had led to vibrant monthly and intra-monthly physical spot markets, there was a highly liquid futures market at the Henry Hub, an increasingly sophisticated swap market for both outright price and locational basis, and a way to optimize excess capacity on the pipeline network. Physical operations had become much simpler not only because market centers aggregated receipt and delivery points, allowing for gas to be pooled among several locations, but because in the mid-1990s a few veteran traders joined producers, consumers, and pipeline companies to form an industry group called the Gas Industry Standards Board (GISB), whose mission it was to standardize and professionalize the handshake business culture that had prevailed to that point. Lander, who was on the board, recalled that preparing and signing contracts in the early days would often take longer than the duration of the contract itself, and gas might still "go missing, two years after you sold it."[28] By 1997 the GISB had formulated a standardized contract for physical sales, which meant that consummating sales with new clients took a few days, instead of weeks. It also created a standard gas day, which meant that everyone adopted the same schedule.

The market now had all the tools it needed. Gas was a commodity. But the tools were still scattered about in different places. Henry Hub futures were traded by floor brokers, who would buy or sell gas using a series of hand movements that would dazzle even a seasoned third-base coach. Several market centers had set up electronic exchanges where physical gas and some basis swaps were sold, but these covered only certain sets of pipelines in different areas of the country. Trading gas, even in the late 1990s, felt something like visiting a souk. There was a substantial amount of opacity, which meant brokers and traders could afford to charge high fees, since the true market price was often elusive. In many ways, swaps' key strength was also their key weakness: they were custom products, which meant they did not trade as often as the benchmark. Swaps overcame both of the NYMEX contract's drawbacks—an inability to manage basis risk and physical settlement—but buying or selling one often meant picking up the phone and calling a trader, who was glad to take you for all that you were worth. Hedging using swaps thus meant losing the two things that Henry Hub futures provided: liquidity and transparency.

When Henry Hub futures began trading in commodity pits, in 1990, only three million people had access to the internet, worldwide, and the fastest personal computer available for purchase was powered by the Intel 486 microprocessor. The world had come a long way from *Tetris*, but was still a far cry from *Call of Duty*. Already by 1993, however, as gas market centers were spreading across the United States and Canada, computers had become much more powerful, and a more integral part of business. This was the first year in which gas was traded electronically.[29] By 1999 processing power and bandwidth technology had increased to an extent that made the old style of transacting begin to appear antiquated. The days of the energy souk were numbered.

By the late 1990s, Enron had become the country's leading gas trading company, on almost any metric. It was the largest physical trader and consistently boasted the most profitable financial desk. It was also a company that had wholeheartedly embraced a "we can do it better" ethos, which was born equally of dominance and of arrogance. Mildly insulted by having to transact daily on other companies' exchanges, and by paying millions in broker fees, the company's head trader suggested that Enron study building its own, internet–based energy trading platform. This made sense on some levels, but neither was it an obvious strategy. The exchange business and the trading businesses are fundamentally different, after all, with the exchange acting merely as a transactional venue, while traders act as warehouses for risk. In this analogy, the warehouses have no windows: what is inside is a closely guarded secret. What made EnronOnline different was that it moved the transactional venue inside its risk warehouse. On its exchange, Enron was on the other side of every trade.

Louise Kitchen made her name at Enron as a star European trader. In the words of her superiors, she was "just dynamite in anything she ever did," and extremely detail-oriented.[30] Kitchen's tremendous focus meant that she had learned to communicate almost as fluently with programmers in the IT department as with her fellow traders. In early 1999, Kitchen was asked to oversee the development of Enron's online trading platform. After just four months, she had a beta version up and running, and the kinks had been sufficiently ironed out that Skilling, who was by then the president of the company, announced EnronOnline to the market in October. It was launched, officially, one month later.

Skepticism about the venture came both from within and without. Skilling thought publishing price quotes was anathema to making money, as the additional transparency would erode margins. Why give away the keys to the warehouse to everyone else in the market?

The Wall Street stock analysts who assigned buy or sell ratings to Enron did not assign the project much value either because they thought the concept was fatally flawed. Unlike other exchanges, which accepted bids and offers from all participants—a many-to-many exchange—EnronOnline only showed Enron's bids and offers—a one-to-many exchange. Anyone could buy from or sell to Enron, using its live quotes, but not from any other market participant.

As it went online, then, even Kitchen did not expect anything world-changing. At best, it was hoped that the compression of margins— the narrowing spread between the bid and the offer that was certain to occur when live quotes were available at all times—would be offset by an increase in transaction volumes. Skilling had given his blessing only because he had become convinced that if Enron did not create this product, someone else would.

EnronOnline shattered these modest expectations. Six months after its launch, more than half of Enron's trades came through EnronOnline, which quickly dwarfed every other electronic energy exchange in the world. Phones still rang, and voice brokers still treated their clients to expensive dinners, but energy trading truly entered a new era in those first few months of 2000. What before had been a daily business transformed into a real-time, minute-by-minute business. Liquidity soared.

EnronOnline became a one-stop shop. There were hundreds of products available to trade (although gas and electricity dominated in terms of volume), which meant one could buy physical gas in Chicago for tomorrow or sell basis swaps in West Texas for next year, all from one screen. Kitchen's design was sleek and simple, trader-friendly, and bug-free—the iPhone of energy trading. It seemed to mark the end point for natural gas trading, and energy trading more generally. Indeed, despite almost twenty years having passed since its launch, any energy trader today would know their way around the EnronOnline platform within a few minutes, so little has the basic design changed.

The issue with the one-to-many system, however, and one of the key reasons why many thought it would fail, was that everyone that transacted on EnronOnline assumed Enron's credit risk. Of course, if there was to be a universal counterparty for energy trading, it would have to be Enron, which had won Fortune's Company of the Year award every year since 1996, and whose debt was rated solidly as investment grade by all of the three major credit ratings agencies. Suggesting that Enron would be bankrupt within two years of EnronOnline's launch would have sounded just as ridiculous as suggesting that Enron would have gone on to win Company of the Year in 1988, when Kinder began eliminating the near-bankrupt company's problems as if they were alligators in a swamp.

The other issue that EnronOnline's doubters saw with a one-to-many system is that, as the universal counterparty, Enron saw how everyone in the market was positioned. This was no different than seeing everyone else's cards at a poker table—it gave Enron's already very successful traders a breathtaking informational advantage over anyone else in the market. Skeptics were certain that traders from other firms would recognize this and take their business elsewhere. It is true that Enron's competitors set up their own online systems, but none was able to achieve the critical mass necessary to function properly. Kitchen's brilliant execution ensured that the network effect greatly outweighed the wariness of others in the market both to park so much risk with and yield so much information to a single counterparty, and one who was often a direct competitor, to boot. In 2000 profits at Enron's trading division began pouring in at a dizzying rate.

## A RECAP OF THE JOURNEY

When Fukuyama wrote "The End of History" in 1989, the Berlin Wall had not yet fallen, and the Soviet Union still existed. Fukuyama correctly presaged the end of this empire, and, just a few years later, a rapid transition to a liberal world order seemed inevitable. Western diplomats and economic advisors began preparing a one-size-fits-all checklist of reforms that would speed the process along: liberalize your markets by removing state regulations, and democratize by organizing free and fair elections. This winning formula was what made the West powerful.

Many economies, including Russia, the heart of the former Soviet Union, followed these policy prescriptions. The results were disastrous. Rather than leading to a well-functioning free market, kleptocracies emerged, where nations' wealth became concentrated in the hands of a few oligarchs. Western-style democracy as implemented by Russia's first president, Boris Yeltsin, led to an almost complete loss of government control over both the economy and civil society, and the chaos and corruption of what became known as the "Wild '90s" led directly to the rise of a more autocratic state under Vladimir Putin.

If they have done nothing else, the prior few chapters have shown that the transition from a highly regulated to a mostly deregulated market is extremely complicated, and horribly vexing, even in a country with a long history of capitalism and well-established government and regulatory institutions. By the mid-1980s the destination was clear, but it took the United States more than a decade to transition fully to a free natural gas market, where buyers and sellers could stand on their own two feet. As US-educated economists spread the gospel of capitalism in the early 1990s and asked the rest of the world to swallow the hard medicine needed to transform themselves into liberal, free-market societies, US gas pipelines continued to pass on accumulated take-or-pay obligations to consumers via regulated tariffs, and were forced to shield them from price spikes by standing ready to provide service on peak demand days.

The "free" gas market that has existed in the United States since the mid- to late 1990s is, in fact, not fully free. Interstate gas pipelines and local distribution companies are still subject to cost-based rate regulations—only the wholesale price of the commodity is fully deregulated, just as the original Natural Gas Act of 1938 intended it to be. This seemingly modest success was hard-won and required the assent of a great many players with diverging interests, which was only achieved after numerous compromises and innovations. There is some truth to the argument that other economies should have been able to leapfrog the United States in some areas, having witnessed the mistakes made here, but history has shown time and again that rapid or poorly conceived liberalization efforts do more long-term harm than good.

US policymakers were unrealistic not only in what they expected from other countries after the fall of the Soviet Union, most of which had limited experience with free markets, but in what they expected from domestic markets that remained regulated. They should have known that simply allowing producers and consumers to face each other on the spot market was only the first step of many, and a dangerous one at that. Apparently, they had not paid very close attention.

# CHAPTER 12

## THE DASH FOR GAS

The 1990s were exciting times for gas traders, as they cut their teeth in a newly deregulated market. For producers, however, these years were as dull as the decade that preceded them. The gas bubble was shrinking, but from an incredible size. Experts believed that the productive capacity of existing wells in the United States hovered around seventy billion cubic feet per day throughout the first half of the 1980s, against actual production of just fifty billion cubic feet, meaning that producers were holding back 40 percent of their potential output. By the early 1990s the surplus had halved, but still represented an overhang of around ten billion cubic feet per day, or 20 percent of total production.[1] In the 1990s, prices were stuck between $1.50 and $2.25, only jolting outside of that range on extreme—and short-lasting—weather events.

Nor were producers having a better go of it on the oil side of their operations. Aside from a spike during the seven-month-long Gulf War, oil prices held steady at around $20 per barrel, which in inflation-adjusted terms was only one-quarter of where they had peaked in 1980. The drilling bonanza that followed the 1979 crisis, along with the demand destruction, led to nearly two decades of softness. Indeed, as the new millennium approached, the up-and-coming set of business leaders, policymakers, and elected officials had never known energy shortages in their adult lives. They also shared an almost unshakeable faith in the virtues of free markets. It could be argued that the one followed from the other.

In the early 1990s, the US electricity market was still dominated by vertically integrated, fully regulated utilities. They owned the power plants that produced electricity, the high-voltage transmission systems that transported it tens or hundreds of miles, and the low-voltage distribution systems that delivered power to homes and businesses. Encouraged by the "successful" deregulation of other markets—gas, yes, but also banks, telecoms, trucking, and airlines—and a strong belief that capitalism held the cure to all the world's ills, federal and state decision-makers ploughed forward with domestic electricity market reform in the late 1990s.[2]

Hijinks ensued. In California the experiment careened off the rails, and consumers were saddled with tens of billions of dollars of unnecessary electricity payments for years to come. Other regions took a more gradual and sensible approach to deregulation, and were able to avoid California's embarrassments. Still others ignored the guidance from Washington, and enacted relatively few changes to their old systems.

States had the ultimate say over whether they would embark on electricity market reforms, and their strategies diverged greatly. Almost all, however, agreed on one thing: cheap, abundant, clean-burning, domestically produced natural gas was the fuel of the future. Virtually all new power plants built from 1998 onward were fueled by natural gas, and by the start of the new millennium, the gas bubble finally popped.

## THE CALIFORNIA ELECTRICITY CRISIS

Like most major crises, the events that led to blackouts in California and the bankruptcy of the Golden State's largest utility unfolded over many years. A simple narrative that places the blame squarely on one party is inherently flawed, and anyway, in the case of the California electricity crisis, there is enough blame to go around.

Most tellings of the crisis begin with the bill that restructured California's electricity market, AB 1890, which was passed in 1996. In fact, the seeds for the crisis can be said to have been planted all the way back in 1978, when President Jimmy Carter passed his package of energy reforms. I have spoken at length about the Natural Gas Policy Act (NGPA), especially its provision allowing gas pipeline companies to pay whatever they wanted for high-cost gas, which filled the pockets of many Oklahoma wildcatters and directly led to the take-or-pay crisis. Carter's package also included a bill called the Public Utility Regulatory Policies Act, or PURPA. PURPA was the first step taken to break the vertical integration of electric utilities. It did this by allowing independent power producers (IPPs) to generate electricity and sell it to the utility. Just like gas markets, electricity markets can be broken into three distinct business segments: production, transmission, and distribution. And, just like gas markets, history had demonstrated that transmission and distribution were indeed natural monopolies, as they displayed increasing returns to scale. Electricity generation, however, had ceased to meet this criterion. (Gas production never had.) This was not an outright repudiation of Samuel Insull's central station model—it was not as if policymakers wanted a dynamo in every basement of every building—but rather an acknowledgment of the fact that a fifty-megawatt plant might be more economical than a one-thousand-megawatt plant, and that a regulated utility held no competitive advantage over an independent company.

Since it was passed in an era of scarcity, PURPA had a strong conservationist bent. In order to become a qualifying facility and be eligible to sell energy to a regulated utility, the independently owned power plant had to be either a fairly small cogeneration facility, which used the waste heat from generating electricity for some other purpose (either an industrial process or simply to heat homes or businesses), or use a renewable fuel source, such as biomass, or waste.

As important as its direct effect on the composition of the US power plant fleet was PURPA's indirect effect on the utility planning process. Before PURPA, the utility determined when to build new power plants, how large to build them, and what fuel they would use. The state public utility commission would have a right to review the plans, but this task was largely perfunctory. After PURPA, however, utilities were required to consider a much broader array of resources, including resources that they themselves did not propose to build, and choose the least-cost option. The introduction of competition was, in most ways, a good thing, and the results spoke for themselves: throughout the 1980s around half of new generation capacity was built by IPPs, rather than the utilities themselves.[3] But resource planning also became a highly adversarial and time-consuming process.

As PURPA was being implemented in the early 1980s, energy prices were sky-high, and expected to rise. Thus, IPPs were offered long-term contracts at extremely attractive prices that escalated based on where policymakers believed the price of oil was heading. The California Public Utility Commission (CPUC) drew up the rates that it awarded to PURPA-qualifying facilities in 1982 and awarded contracts to around ten gigawatts of new power plants before oil and gas prices tanked, and the commission rescinded its offers.[4] It was, essentially, the electric equivalent of the NGPA—the CPUC's price schedule forced utilities to pay new power plants high prices for ten or more years, just as the NGPA's price schedule forced gas pipelines to pay high prices for new production. There was a notable difference in magnitude, however—whereas even the most aggressive pipelines only bought high-cost gas contracts equal to about 15 percent of their total supply volumes, the ten gigawatts of PURPA-qualifying facilities nearly doubled California's in-state generating capacity. Already in the early 1980s, regulated deregulation had left utility companies on the hook for too-high electricity prices, which they passed on to California ratepayers over the following decades.

The liabilities that California utility companies accrued from paying such high rates to PURPA-qualifying facilities were compounded by the commissioning of the nuclear Diablo Canyon Power Plant in 1985, after almost two decades of construction and massive cost overruns. Pacific Gas & Electric, the state's largest utility company, which serves essentially all of

California north of Los Angeles, announced plans to build a nuclear reactor all the way back in 1963. Costs were initially pegged at $320 million, but after sixteen years of construction and nine budget revisions, $5.3 billion had been spent.[5] Even after accounting for inflation over the years between the initial estimate and the final reality, the facility was more than five times over budget.

PURPA and Diablo Canyon left California as one of the highest-cost electricity states in the nation. More important still, California's neighbors were among the lowest-cost electricity states in the nation—the Pacific Northwest has massive hydroelectric facilities that generate at zero cost, and the rest of the West has indigenous coal resources and did not see nearly as much PURPA–related power plant build. Neither had embarked on such ill-fated nuclear endeavors. This made the Golden State ripe for reform.

Politically, then, California was a logical early mover on electricity market reform, as a high-cost island in a sea of cheap electricity. And the timing seemed ripe in the mid-1990s, as the planning process for adding new capacity had almost totally broken down. After saddling the utility companies with high-cost electricity in the wake of PURPA, in 1993 the progressive CPUC ordered utility companies to include social and environmental factors in their calculations. The companies hated this idea, as it meant higher costs, and FERC sided with them, ruling that the CPUC had overstepped its bounds. As ever, uncertainty was anathema to new investment, and wary utilities and IPPs tabled plans to build new plants. For most of the 1990s, generation additions did not keep pace with demand growth in California.[6] A clean break with the past was needed.

Or was it? It is worth stepping back and comparing gas market deregulation to electricity market deregulation, especially in California, where the latter went so wrong. The deregulation of the gas industry happened in fits and starts. The least successful phase was the regulated deregulation that followed the Natural Gas Policy Act of 1978, where prices were set administratively and pipelines—the equivalent of electrical utilities—remained in the business of buying and selling gas, rather than merely transporting it. It was only after this post–NGPA status quo proved totally unworkable that true deregulation occurred. Wellhead prices were unfrozen, the transmission grid of pipelines was converted to an open-access system, and the pipeline companies ceased their merchant function. After this round of deregulation—epitomized by FERC Orders 380, 436, and 636, as well as the Natural Gas Wellhead Decontrol Act—the industry regained its health, without any financial assistance from the government, and in an environment of decreasing costs to the consumer. What better outcome could have been imagined?

Meanwhile, when California began taking steps to deregulate, the status quo was, in truth, working fairly well, and had for several decades. Blackouts were extremely rare, caused by technical issues rather than economic ones. Prices were high, but this was largely because California's utilities had been forced by the CPUC to purchase expensive, PURPA-qualifying electricity. The high prices should have been seen as an indictment of Carter-era regulated deregulation rather than of the traditional utility compact, but the distinction was lost on policymakers.

Even more fundamentally, natural gas deregulation took place during the first years of a two-decade supply glut. There was plenty of wiggle room if things went wrong—the car had broken down next to the service station. Electricity deregulation in California, on the other hand, took place just as the market was tightening, after years of underinvestment in capacity. The failures of the resource planning process made restructuring seem urgent, but in reality, the timing could not have been worse.

With that context, it is clear that the California bill restructuring the electricity market, AB 1890, had two primary goals: (1) to incentivize the construction of new generation capacity, and (2) to prevent the utilities from overpaying for capacity. In a very limited sense, it worked well. In any broader sense, however, it was a horrible failure.

AB 1890 passed in September 1996 without dissent from either political party. Its unanimous passage was certainly related to the fact that the bill guaranteed a 10 percent reduction in rates to consumers until 2002. In other words, while the wholesale price of electricity was left to be set by the free market, retail prices were administratively capped. Originally, there was no language in the bill that capped the retail price, but that language had been penciled in to curry favor with state legislators.

It looked like a legislative slam dunk, but California's unique ballot proposition system, which allows for a direct vote on a wide range of issues, threw a spanner into the works. Opponents of AB 1890 portrayed some sections of the law as a bailout for Pacific Gas & Electric's debacle at Diablo Canyon. Proposition 9, which sought to throw out those provisions, ultimately failed, but it held up the enactment of AB 1890 for two years, during which time virtually no new power plants were built or initiated, and several retired. During this time, California had become dangerously dependent on imports of hydroelectricity from its northern neighbors. Fortunately, the rains had been forthcoming—1995–1999 were all bumper years for Pacific Northwest hydro.

By 2000 the new market was in place, and the overall mood was optimistic. The policy certainty had encouraged IPPs to begin construction on several large, new gas power plants, collectively totaling several gigawatts of capacity. Plants were also being built in neighboring states. A large chunk

California electricity demand

California electricity imports

— 1999   —— 2000

Net bullish effect of higher demand and lower imports, year on year

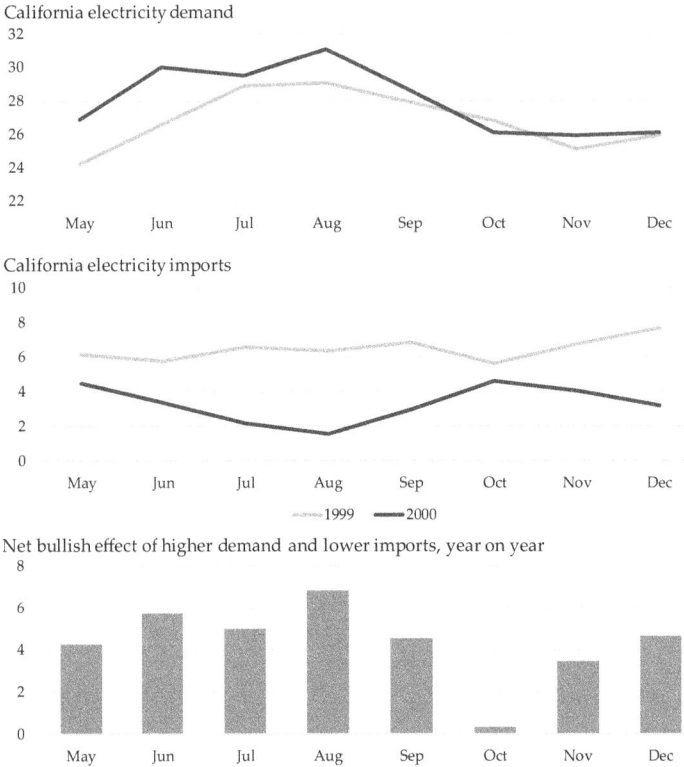

**FIGURE 12.1:** California electricity demand, imports and net effect of both, year-on-year (gigawatts)

of new capacity was due to start coming online in mid-2001, and by 2002 the tight supply/demand situation would be much more relaxed. Most importantly, though, during the first few months of 2000, electricity prices were lower than they had been before the restructuring. Everything was going to plan.

Mother Nature did not cooperate with California's electricity restructuring. Had there simply been a repeat of the prior year, weather-wise, the crisis would not have happened, or at the very least not reached such epic proportions. As the summer of 2000 began, however, there was precious little water in the vast dams of the Northwest, after a dry spell of several months, and brutal heat began descending on California. Already by May, prices were spiking to as high as $300 per megawatt hour, versus the prior few months' average of $30. By June, as the heat wave continued, prices averaged almost $150 and topped $600 on the hottest days.

Supply and demand is what it is. Even in a perfectly functioning market, electricity prices in California would have risen in response to the heat and

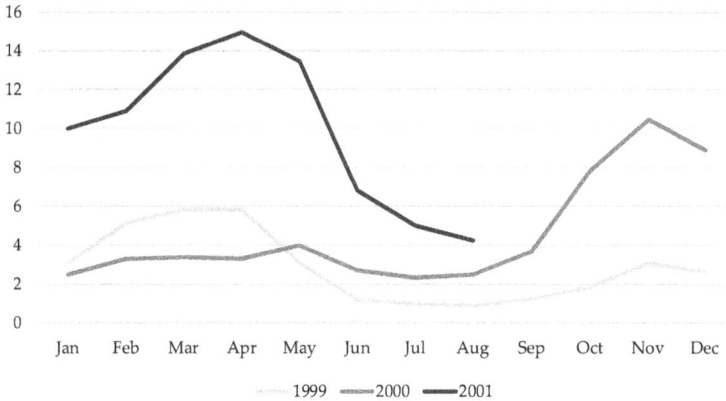

**FIGURE 12.2:** California generation capacity offline (gigawatts)

below-average hydro output, which, combined, were increasing the call on in-state generators by between four and seven gigawatts throughout most of the summer. But two flaws in market design exacerbated the situation, and increased price volatility dramatically.

First, California's rules stated that the utilities were required to purchase all electricity from the spot market. As I detailed in the previous chapter, in a well-functioning market, both consumers and producers prefer a portfolio of contracts with different terms and of different tenors, with only a small amount of their needs left to be transacted "on the spot." This preference is what made futures, swaps, and Enron's Gas Bank so successful—they all offered hedgers the ability to hedge. By forcing all business to be done on the spot market, the architects of California's electricity reform forced utilities to scramble every day to find enough supplies.

The second flaw that exacerbated the bullish situation on California's electricity market, and the one that is more talked about because of its nefarious undertones, was that there were weak and easily exploitable rules requiring power plants to stand by to provide capacity.

In combination, these two flaws were worse than they would have been in isolation. Knowing that utilities needed to source all of their supplies every day on the spot market, there was a tremendous incentive to withhold capacity, which the weak rules made it easy to do.

It remained hot through August and into September, and the average price that utilities paid for electricity remained above $100, spiking much higher on peak days. As the heat relented, however, the utility companies and the CPUC began breathing more easily. In most markets, California included, electricity demand peaks in the summer, as air conditioners run at full tilt. With the summer behind them, and oodles of new generation set to come online before next June, they felt that the bad times were over. In truth, the crisis had not yet begun.

Just as prices were bleeding down, California's power plants began going offline for maintenance at an alarming rate. By November ten gigawatts of capacity was offline, more than triple the winter prior. Despite much lower electricity demand, the California market was actually tighter in the winter of 2000–2001 than it had been in the summer. It was this massive amount of offline capacity that caused the headaches of summer 2000 to turn into a full-fledged crisis that winter.

Most of this capacity was offline for legitimate reasons. The in-state generating fleet had just had a very rough summer, where they were running old plants at very high utilization rates and throttling them up and down to coincide with daytime peak demand. This sort of behavior stresses equipment that is not purpose-built for such ramping. Indeed, among those power plant owners who had substantial amounts of capacity offline were the utility companies, who clearly had every incentive to keep their capacity online and prices low, since they were being forced to sell at capped retail rates.

Equally, however, Enron and others were manipulating the market by withholding capacity. Some of their schemes were simple, others complex. All had the effect of increasing prices.

Recordings that were not make public until 2005 revealed the blatancy of such schemes. In January 2001, during the height of the crisis, an Enron trader called a power plant near Las Vegas that supplied power to the California grid:

> ENRON TRADER. "We want you guys to get a little creative, and come up with a reason to go down. . . . Anything you want to do over there? Any cleaning, anything like that?"
>
> POWER PLANT OPERATOR. "OK, so we're just comin' down for some maintenance, like a forced outage type thing?"
>
> ENRON. "Right."
>
> POWER PLANT. "And that's cool?"
>
> ENRON: "Hopefully [laughs]."[7]

The next day, the plant was taken offline and California called a power emergency. Rolling blackouts hit a half-million consumers.[8]

Some more complicated schemes involved scheduling high power deliveries but then underdelivering, which sent the California system operator scrambling to find power at the last minute. The controls in place were so poor that Enron once scheduled 2,900 megawatts through a transmission path that only had physical capacity of 15 megawatts, which led real-time power prices to shoot up by 70 percent.[9] An official from the system operator compared the situation to a Turkish bazaar, calling it "madness."[10] Every day, it seemed, actual power supplies fell short of where they had been scheduled, and the operator needed to go out and haggle to bridge the gap.

With the passage of time and the revelation of recorded conversations, the simple question: did Enron manipulate California's electricity market? has a simple answer: yes. To the more important question of did it make a real difference, in the grand scheme of things? the answer is: not really.

The Las Vegas power plant that Enron wanted to "get a little creative" was only 50 megawatts, which represented about 0.2 percent (or 1/5,000th) of California's total load at the time. If it was the straw that broke the Californian camel's back that day, it was one among thousands. In fact, FERC, which analyzed the crisis in extraordinary detail after the fact, concluded that the most impactful manipulation that occurred was in setting the price of natural gas, not electricity.[11] The price of gas delivered to Southern California spiked to as high as $60/MMBtu in the winter. This meant that, without any manipulation on the electric side, the cost of gas generation would have been at least $700 per megawatt hour on a peak day, given the efficiency of the marginal power plants that existed at the time.

Again, much of this was a natural response to the supply-and-demand situation. Gas demand is always highest in the winter, and so some price increase is normal. It was greatly exacerbated by the fact that utilities had withdrawn much more gas from storage over the summer than usual, which left them with very low inventories heading into winter. Over the sweltering summer of 2000, California power plants burned 44 percent more gas than the prior year, and the corresponding number for the West as a whole was an even higher 46 percent.[12] Gas generators had been the only things keeping the lights on.

Making matters worse, there had been an explosion on the El Paso Natural Gas pipeline in August. The pipeline was repaired within two weeks, but capacity remained lower for the duration of the crisis due to an order from the US Department of Transportation mandating that the system be operated at lower pressures. This meant that gas utilities could not replenish their storage ahead of the winter, which put Southern California, which relies on El Paso to a much greater degree than Northern California, in a particularly precarious situation.

Last, the winter of 2000–2001 was a cold one in California and across the United States, and even Henry Hub prices, which were not subject to congestion or constraints caused by pipeline issues, were spiking as high as $10. If California wanted gas, it had to bid against utilities in Chicago, where temperatures plummeted to negative nine degrees in December.

Still, five major traders, including Dynegy and El Paso's marketing arm, admitted to manipulating the gas market by submitting fabricated data to the price reporting agency that was responsible for publishing the final spot price, against which many contracts were settled.[13] Many of them argued that they were essentially forced to fabricate their data because they knew that other traders were fabricating theirs in the other direction—that the only way to arrive at "fair" prices was to engage in virtuous manipulation. Clearly, though, this cannot hold true for all parties. A large trader and power plant owner, Reliant, also virtually controlled the market at one of Southern California's largest trading points, and was incentivized to bid prices to extraordinary levels because of some complicated contracts that it had signed.[14]

At the peak of the crisis, during the winter of 2000–2001, California's utility companies were under tremendous financial strain. PG&E was blowing through $1 billion of cash per month; its neighbor to the south, Southern California Edison, a bit less, only because it served fewer people. This brings us to the last, and perhaps most foolish design flaw of California's restructured electricity market: retail price caps. These, more than any foul play by traders, were responsible for the entire debacle.

Price caps were what had made AB 1890 so popular among legislators, as they guaranteed that prices would not rise for consumers. While consumers certainly noticed the blackouts, they were not feeling the crisis in their checkbooks. The widening chasm between the market price of electricity and the retail price was being borne solely by the utility companies, and by January 2001 they were "effectively insolvent."[15] This meant that they ceased paying their suppliers, which threatened to turn the situation into an unmitigated disaster. Throughout the crisis, there had only been six blackouts, of a few hours apiece. If out-of-state power plants turned off supplies to California, blackouts might have lasted for days, weeks, or even months. US Energy Secretary Bill Richardson issued a rare emergency order on December 14, 2000, compelling out-of-state electricity suppliers to sell to California, despite fears that they would not be paid, and followed up with a similar order for natural gas suppliers on January 19, 2000, the day before George W. Bush assumed the presidency.

In his administration's first working day in office, President Bush decided that "the federal government should let California solve its own power problem,"[16] and that it was unfair to force out-of-state suppliers to sell even when it was not clear that they would be paid. Bush's new energy secretary extended Richardson's order on January 23, but only for two weeks, and vowed not to issue any further extensions.

The Bush administration's tack appeared to lack compassion, and was especially criticized because the new president counted as a donor and a close acquaintance Enron chairman Kenneth Lay. Bush even called Lay "Kenny Boy," although this was less a sign of intimacy than it was simply

Bush's way: President Vladimir Putin was "Pootie Poot," Prime Minister Silvio Berlusconi was "Shoes," Governor Chris Christie was "Big Boy," and Governor Arnold Schwarzenegger was "Conan the Republican."

California's governor during the crisis, Gray Davis, had no nickname, at least not one that Bush was willing to air in public. As bad as Enron's traders turned out to have been, Davis was worse. He waffled throughout, refusing to take any of the admittedly difficult measures necessary to stave off catastrophe. While he was quick to point his finger at market manipulators, he refused to lift the retail price caps, which would have put the utilities back on a path toward financial health and also motivated consumers to reduce their usage. President Bush's response to the crisis seemed heartless and gave the appearance of impropriety, but it was intended to force California, and Davis in particular, to solve the problem, rather than kicking the can down the road yet again.

In late January 2001, Davis was in a corner. The big utilities were at imminent risk of bankruptcy, and the state was headed back to the Dark Ages. Politically, however, Davis still considered raising retail rates or bailing out the hated utilities to be a step too far. Once again, he solved only the immediate problem, and made the underlying issue worse. In order to keep electricity flowing into California, despite the utility companies' insolvency, Davis ordered the California Department of Water Resources (DWR) to begin buying electricity on behalf of the utilities. An agency with no experience purchasing electricity was now the largest buyer in the country, and the state of California was footing the bill.

Of course, California residents would end up paying that bill, so all Davis did was to shift the obligations out into the future. He also ended up increasing them.

Unlike the utilities, the DWR was not required to purchase all of its electricity on the spot market. While this is a good steady-state condition, allowing an agency with the ability to pass through losses to create long-term liabilities during the throes of a major crisis was another horrible idea. It was the equivalent of letting gas pipeline companies sign contracts for high-cost gas at whatever price they desired, and letting them pass all the resulting costs onto consumers automatically, which was exactly what precipitated the take-or-pay crisis. Predictably, the DWR began signing long-term contracts, voraciously, and at seemingly any price. By the summer of 2001, only one-quarter or so of electricity was being bought and sold on the spot market, down from 100 percent.

This did not mean the utilities were out of the woods: they were still obliged to sell the electricity to consumers at the capped retail rate. And they were, in theory, required to refund the DWR in full—the state was simply acting as a credit sleeve to ensure that suppliers would continue to sell electricity to California. By March and April 2001, it was clear even to Davis that rate increases would be required. The governor scheduled a televised address to Californians, where he was expected to share progress on negotiations with the utilities to get the system functioning again. At long last, it seemed, a permanent solution was around the corner.

No such solution was proposed. Instead, Davis's new plan envisaged rates rising by only a trifling amount, and required major concessions from the utility companies, including a guarantee that they would sell "low-cost regulated power to the state for ten years," and give up ownership of their transmission systems.[17] The plan, in short, was to keep prices low for consumers and soak the utility companies. Pacific Gas & Electric issued a clear rebuke to Davis's plan by declaring bankruptcy the very next day.

Meanwhile, the DWR kept signing long-term contracts, usually at rates of about $70 per megawatt hour.[18] That may seem cheap, given that prices had averaged in the hundreds of dollars for the preceding several months, and spiked above a thousand. But the spot price had been trading in the $30 range under normal market conditions, from 1998 until the crisis began in May 2000. Just like gas pipelines after the passage of the NGPA, or the CPUC after PURPA, the DWR was committing to billions of dollars of future costs by extrapolating from a highly distorted and temporary situation.

Not only were the prices in the DWR's contracts too high, the volumes were too large and the timing was idiotic. By mid-2001, several gigawatts of new gas power plants were coming online, most having been under construction since 1998 or 1999. This was known to all market participants, and it was clear that the huge slug of new capacity would prevent the extremely tight situation that had prevailed throughout the crisis from happening for at least several more years. Despite the transparency of this information and the certainty of the effect that the new capacity would have, the DWR signed several contracts that did not have deliveries beginning until 2002. Certainly, something needed to be done to get California through the period from January through June 2001, but signing ten-year contracts to bridge a six-month shortage was an absolutely ludicrous strategy, especially when the contracts did not begin until after the shortage was over.

Through the crisis, the utilities burned through between $15 and $20 billion purchasing expensive electricity on the wholesale market—sometimes for over $1,000 per megawatt-hour—and selling it at much lower retail rates of around $65.[19] The DWR, however, committed around $60 billion to long-term contracts,[20] spread over the next decade. When all was said and done, it probably overpaid to the tune of $20 billion.[21] In terms of total costs, the worst part of the crisis came just as it was ending. The financial impact of the California electricity crisis ended up being felt more acutely in the decade that followed than during the crisis itself because of the knee-jerk actions Governor Davis took trying to "save" the system from collapse.

By the fall of 2001, California spot electricity prices were back to trading in the $30 range.

## THE GAS BUBBLE, POPPED

California was not the only place where electricity demand growth had outstripped investments in new capacity throughout the 1990s. PURPA and the rise of IPPs had led to more adversarial and time-consuming planning processes across the nation. Consequently, the reserve margin for the Lower 48 states—the amount of excess electric capacity projected to be available during the hottest hour of the hottest day of the year, when air conditioning units are sucking power from the grid—had fallen from 22 percent in 1990 to 14 percent in 1998. While this was still a fairly healthy level, the downward trend was concerning, and the number for the Lower 48 masked more severe situations in the Midwest and Southeast.[22] In other words, in the late 1990s, the electricity market was characterized by conditions almost totally different to the gas market: the one was tight, the other loose.

The solution was to build more power plants. But what kind would they build?

IPPs assess investments differently from regulated utilities. Because they are risking all of their capital, they prefer investments with low upfront costs and quick payback periods. They take on as little policy risk as possible, and no technology risk whatsoever. These criteria, plus the meltdown that had occurred at Three Mile Island in 1979 and the rampant cost overruns at Diablo Canyon, meant that nuclear was not an option. Environmental rules promulgated by Congress in 1990 had made coal much more expensive, as they required operators to install costly environmental equipment on all new plants to scrub the emissions of acid rain–causing sulfur. This meant that coal was out too. And while the promise of renewable energy had captivated policymakers since the Carter administration, wind and solar remained much too expensive to compete with fossil fuels.

Gas looked like good choice for a power generation fuel even independently of the handicaps that other fuels suffered. PURPA had led to the mass adoption of cogeneration technology, which in turn led to great advances in the design of gas turbines and the highly efficient setup known as combined-cycle generation. In this setup, gas is burned to fuel a turbine, much like a jet engine, in the first cycle. In the second cycle, the waste heat is used to boil water, with the resultant steam powering another turbine. Computer-aided design advances further increased the thermal efficiency,

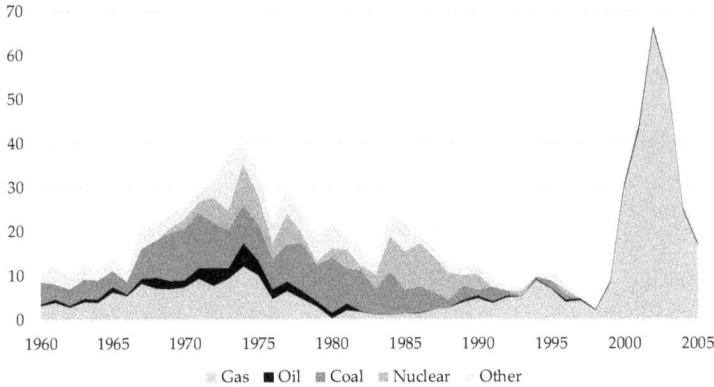

**FIGURE 12.3:** New US power plant build by fuel and year (gigawatts)

or heat rate, of combined-cycle gas plants from around 50 percent in 1990 to almost 60 percent by 2000, which made them almost twice as efficient as the single-cycle models that had been built over the prior decades.[23]

Gas had become the fuel of choice for new power plants almost immediately after Congress repealed the Fuel Use Act in 1987. In fact, calling gas the "fuel of choice" in this era is misleading—there wasn't much of a choice. Gas was so dominant that virtually no other type of plant was built, and in the five years beginning in 1999, more natural gas capacity was built in the United States than had been since the very first plant was completed in 1912. The US electricity reserve margin quickly rose back above 20 percent.

On top of all the aforementioned reasons, the dash for gas was predicated on the fuel being fairly cheap and widely available. But this was not a fact so much as a condition—equivalent to saying that traffic was better on one road than the other—the condition ceases to be true soon after it is recognized.

The rapid build-out of gas power plants was not the only indication that the gas market was on the verge of tightening back up. Demand had risen consistently since the mid-1980s, as local regulations precluding new homes and businesses from using natural gas for heating, enacted during the scarcity era, were annulled. After having dropped by more than one-quarter from its peak in 1972, or by an astounding 16 billion cubic feet per day, US gas demand finally set a new high in 1995. Production, however, had been mostly stagnant. Imports from Canada filled the gap, rising fourfold between 1987 and 2000, from 2.5 to 10 billion cubic feet per day.

Already by the end of the 1990s, the market was coming into balance, with the price of gas starting to trade in line with oil, in energy-equivalent terms, and the amount of spare productive capacity dwindling to near zero.[24] It was clear that more supply would be needed to feed the hundreds of gigawatts of new gas power plants set to come online over the next few years, but relatively few forecasters fretted over any imminent price spike. Even though Canada, which had supplied the bulk of new US demand throughout the 1990s, appeared more or less tapped out, there appeared to be gas aplenty in the ground, albeit in unconventional places.

As the millennium came to a close, natural gas was trading at around $2.50 for the next couple years in the futures market, within the historical range. Prices were higher in the winter months (around $2.70) than in the summer (around $2.35), but these spreads were not large enough to reflect any scarcity premium; they merely covered the cost of storing gas over the summer and withdrawing it in the winter. Longer-term forecasts, from both the government and consultancies, were also fairly flat, indicating that prices would rise, but slowly, and probably not for another several years.

As such, and despite all the omens, many in the market were caught unawares as prices began to rise during the summer of 2000. Demand from the electric sector was coming in higher than usual, both because newly built combined cycle plants were coming online, and because the fleet needed to make up for low hydroelectric output in the Northwest, after its first dry spell in years. By the end of June, gas prices at the Henry Hub had risen to around $4.50, and storage operators had no economic reason to inject—they would make more money by selling gas now, in the high-price environment, than buying it and waiting for winter. This of course caused storage levels to drop below normal, which would spell trouble if the United States experienced a cold winter.

That is just what happened. As November realized much colder than normal, the spot price rose above $6, pulling futures for December and January up with it. After a brief thaw, temperatures plummeted once again: December 2000 was one of the coldest on record across the United States. Spot prices and near-dated futures skyrocketed to above $10, and that was just at the Henry Hub—in the more constrained New York market, gas prices rose to almost $40, as utility companies bid against each other to secure the suddenly very limited amounts of gas on the system. In Southern California spot prices topped $60.

After winter was over, prices quickly settled back to their historical range. The gas shortage, after all, was weather-driven, and as the cold dissipated, so too did prices. Indeed, record-high prices in the winter of 2000–2001 were followed by very low prices the next, as storage was refilled to normal levels and temperatures were very mild. But the psychological impact was clear: the gas bubble was over.

If the run-up in prices in 2000 and early 2001 came as a shock, the steep decline over the summer of 2001 was a relief. Not only could consumers breathe easy—the market had shown an incredible ability to adjust and allocate scarce resources. Policymakers were pleased further as they watched the rig count tick up to levels not seen in decades: by June 2001 almost 1,100 rigs were drilling for gas in the United States, versus only 600 before the run-up.[25] Now that demand had caught up and the gas bubble was over, it was only a matter of time before production began growing once again. The market was in the midst of its first phase change—from supply-push to demand-pull—since the deregulation of the early 1980s, and it was working just as it should have been.

The production growth that was needed to balance the market was to come mainly from two frontier areas: the deep waters of the federally controlled Gulf of Mexico, and so-called unconventional resources, mostly in the Rocky Mountain region. Both of these frontiers had already proven themselves as technically and commercially viable. Deepwater production had risen from almost nothing in 1990 to three billion cubic feet per day by 2000,[26] and unconventionals from three billion to eight.[27]

All seemed to be going to plan in the frontiers, for a while, but maintaining productivity from the legacy resource base, which still accounted for more than three-quarters of total supply, proved more difficult than had been supposed. Most forecasters did not properly appreciate the extent to which North America's conventional basins had matured. Not only did new wells drilled in mature areas produce less than they had historically, but production declined at a faster rate. Producers were running faster and faster just to stand still. The price spikes of 2000–2001 had done their job—they led to a ramp-up in drilling and an increase in production—but the increase in production was not commensurate with the increase in drilling, and, by increasing prices, the market seemed to be pushing on a string.

Every year from 1998 to 2001, the Department of Energy wrote in its annual energy outlook that it "expected mid-term continuation of recent huge finds in the deep waters of the Gulf of Mexico," one of the two frontiers that would offset declines from mature areas and propel the US upstream back into growth mode. In fact, there was continued growth from deepwater wells, of around one billion cubic feet per day from 2000 to 2003, but this was more than offset by declines from shallower wells, and total offshore production had already begun to decline by mid-2001. By 2004 even deepwater production was in decline.

The Rockies performed better, but was simply not large enough to move the needle on a national level—production only grew by 2.5 billion cubic feet per day from 2000 to 2004, which was not even enough to make up for the declines from offshore. The two frontier growth areas, then, netted out to nothing, and production from the remainder of the United States remained constant, no matter how high the rig count rose.

Rockies and offshore Gulf production                                    Rest of US production

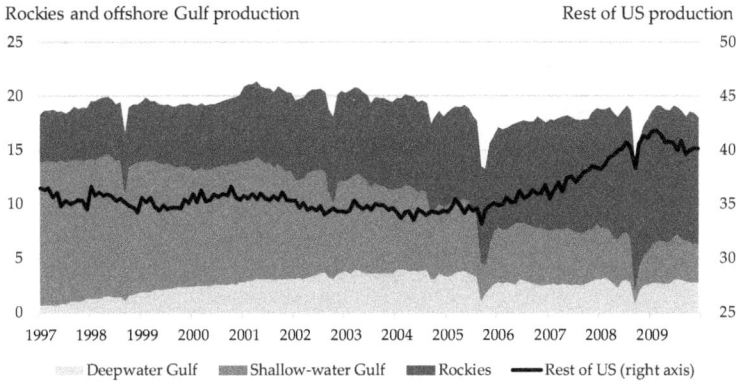

FIGURE 12.4: Production from the Gulf of Mexico and the Rockies and from the rest of the United States (billion cubic feet per day)

Prices rose again in late 2002, but this time they did not come back down.

In economics, the degree to which supply or demand responds to movements in price is known as price elasticity. It was shown in the early 1980s, after the NGPA passed, that gas supply was highly price elastic—production surged alongside prices. Unfortunately for producers, demand was highly price inelastic for the following fifteen years, meaning that, despite low prices, demand could not catch up to balance the market. This was partly because of regulations, such as the Fuel Use Act, partly because a large portion of the housing stock had just been built using electricity or oil as a heating fuel, and partly because US manufacturing never recovered after the stagflation of the late 1970s and early 1980s. The inelasticity of demand, however, was a temporary, if long-lasting, phenomenon, as the massive build-out of combined cycle gas power plants—the dash for gas—in the first years of the new millennium proved. As prices rose, forecasters expected a supply response.

They got one, but not the one that they expected, as production rose by only a couple billion cubic feet per day. By the mid-2000s it was clear that the supply side had become less price elastic than forecasters had thought. The market remained tight, and prices settled at a new normal of around $5.

## THE FALL OF ENRON

The actions of its employees during the California electricity crisis illustrate that Enron had become a ruthless trading company. Its traders did very well during the crisis: in 2000 the west power desk booked $460 million in profits, and the west gas desk another $870 million. What is even more

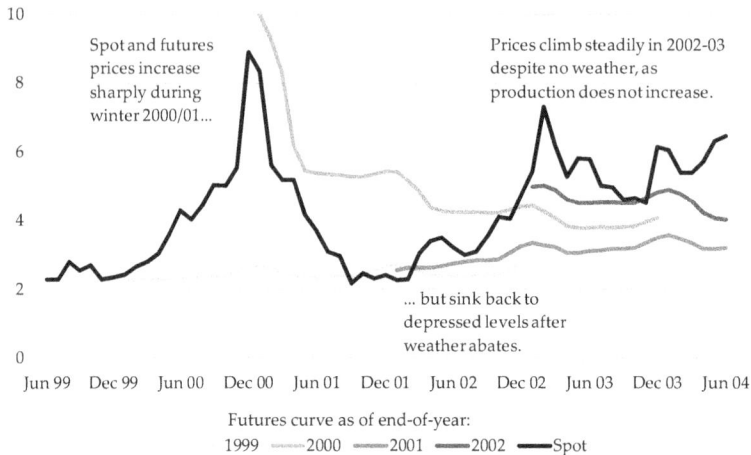

Spot and futures prices increase sharply during winter 2000/01...

Prices climb steadily in 2002-03 despite no weather, as production does not increase.

... but sink back to depressed levels after weather abates.

Jun 99   Dec 99   Jun 00   Dec 00   Jun 01   Dec 01   Jun 02   Dec 02   Jun 03   Dec 03   Jun 04

Futures curve as of end-of-year:
1999 ──────2000 ──────2001 ──────2002 ──────Spot

**FIGURE 12.5:** Henry Hub spot and futures prices

astounding is that the combined $1.3 billion that the west desks made was still only slightly more than half of the total North American trading desk's profit that year: $2.2 billion.[28]

But Enron did not collapse because it employed immoral traders, or because it profited from the California debacle. In fact, it collapsed for reasons entirely unrelated to its gas and power trading business. Boiled down to its essence, other Enron divisions had made a number of bad investments, and rather than writing down those investments after it became clear that they were impaired, the company created shambolic accounting structures in order to disguise the losses. Enron's heart—its wholesale trading business—was at the top of its game as the company went down in flames. It was its brain that had malfunctioned.

For simplicity and brevity, it is probably fair to group Enron into three business segments: gas and power trading, pipelines, and other. In the other segment were overseas investments, such as an enormous gas power plant in India, gas pipelines and a water company in Argentina, and domestic broadband and video streaming services. While gas and power had always done well, these other forays all lost massive amounts of money, for exactly the same reason: Enron never understood the businesses. While each sounded promising during McKinsey-style presentations—huge, untapped markets with tremendous growth potential and potentially fat margins—the devil was in the details, of which Enron was totally ignorant. Perhaps Enron had lost sight of what made the Gas Bank such a success—namely, that it capitalized on the idiosyncrasies of the gas market in the 1990s. Consumers desired a way to lay off price risk—to buy insurance—having lost this ability after FERC took pipelines out of the firm sales business.

Producers were unwilling to sell their production forward at the prevailing, low prices, but they were also extremely short of cash. The Gas Bank, which offered cash for gas, solved the problem, and gave Enron an edge over every bank in the country. In short, Enron filled a vacuum. Not only that, but it held a real operational and informational advantage, as it was a huge pipeline operator that could physically move the gas from the producers to the consumers with which it transacted. None of the other divisions held any competitive advantage.

Sitting atop all of its real businesses was Enron's accounting and finance division. As a publicly traded company, it was the mission of Enron's executives to keep the stock price high and rising, which meant it needed to beat earnings, quarter after quarter. In fact, it was probably this need that forced Enron into some of its other businesses, given that it had simply reached its limits within gas and power. As the anticipated profits from the new businesses failed to materialize, the accounting division started to get creative. They booked uncertain, future profits as current earnings, hoping that they would make up for it in later quarters. That did not happen, and in December 2001, Enron declared bankruptcy.

At the start of that year, this would have seemed impossible. Enron's wholesale trading division had posted several years of incredible profits, and was the envy of the energy trading world. While an outside observer could justifiably level accusations of arrogance and even depravity at some of Enron's traders, he or she would also have to acknowledge that they were smart and extremely well-informed. Jeffrey Skilling, who had virtually created the division and headed it since 1991, staffed the trading floor with MBAs from top schools across the country. These were the types of minds who could make sense of the vast amounts of data to which only Enron was privy, and turn them into profits.

The informational advantage was truly breathtaking. Through EnronOnline, which quickly became the dominant exchange for US gas and power swaps after its November 1999 launch, Enron's traders saw who was buying and selling power and gas, which meant that they knew how everyone in the market was positioned. Moreover, as Enron was still a giant pipeline company with physical assets that spanned across much of the country, they were keenly aware of both the national and regional supply/demand balances. This information asymmetry, combined with the company's appetite for risk and the extremely volatile market conditions in 2000 and 2001, virtually ensured that the wholesale trading division would post mind-boggling profits.

Despite having informational advantages that would make any hedge fund manager salivate, Enron was never predominantly a speculative trading company—it was always more involved in managing its customers' risk and providing liquidity. The differences between these three roles are subtle but important, and go a long way to elucidating how the gas markets changed after Enron's collapse.

Enron's original role as a trading company, since Skilling had come over from McKinsey and advanced the Gas Bank, was that of a risk manager, which is akin to an insurance underwriter. Enron would buy gas from producers and sell it to consumers at a higher price, executing hedges from both the bid side and the offer side. That seems like a simple enough business model, in the abstract, but in practice, Enron took on several dimensions of risk, all of which needed to be managed. For one, the tenors of the hedges were hardly ever the same—it might have bought five years of gas from a producer, and agreed to sell ten years to a power plant, with higher volumes in the summer than the winter. It might thus need storage to even out the volumetric profiles, and quickly to line up another supply deal. And of course it would buy from a producer in the field, and sell to consumers in markets, thereby taking on basis risk. Finally, in doing bilateral deals, Enron assumed the credit risk of all of its counterparties—if one of its power plant customers went bankrupt, it might need to restructure or simply tear up contracts. Having a large portfolio was essential to managing these risks, as the diversity of a large, national portfolio diluted the effect of any event that might have bankrupted a smaller competitor. The portfolio effect was a tremendous barrier to entry, both because it discouraged potential competitors from trying to get into the business, and because it gave Enron's team of originators—the salespeople/traders who structured long-term deals—more options to pitch to clients, and generally allowed them to bid higher or offer lower prices than the competitors it already had.

Indeed, its dominance in gas and electricity risk management also made Enron a large liquidity provider, or market maker. Although the roles of risk manager and market maker have a somewhat similar economic function—to warehouse risk—a market maker's profits come not from hedging on behalf of producers and consumers but rather from price volatility. Without having any directional view on where prices are heading, a market maker will always be among the highest bids and lowest offers. Therefore, while they might take on a long or short position over the short term, over the long term, their goal is to be flat, and simply be paid the difference between the bid and the offer. In providing a tight bid/offer, the market maker makes trading easier, adding essential liquidity to the market. After EnronOnline was launched in November 1999, it quickly became the largest energy trading platform in the world, vastly increasing Enron's market-making profits.

A speculative trader, unlike a risk manager or a market maker, takes a directional price view. They buy or sell contracts hoping that prices will rise or fall, or that locational or time spreads will widen or tighten. They are not performing a service for a customer, or for the market at large—they are betting on their own account. While speculative traders do add some economic value, inasmuch as they keep markets honest and efficient, by ensuring that prices do not move too far out of balance with what supply and demand conditions suggest they should, generally, speculative trading is a zero-sum game, where someone's gain is someone else's loss, and no net economic value is created.

There is a symbiotic relationship between risk managers, market makers, and speculative traders—they cannot exist without each other, and Enron was all of them, at once. So as Enron began to teeter in mid-2001, there was great worry about how the market would deal with the loss of its largest risk manager and liquidity provider. As bids and offers disappeared from traders' EnronOnline screens, one energy reporter lamented that the market seemed no longer to be able to figure out how to price anything.

Separately from the dysfunction in the front offices, where traders sat, there was anarchy in the so-called middle office, where trading shops housed their risk groups. Before Enron's bankruptcy, credit was a nonissue. Indeed, in addition to the market maker, Enron was the clearing agent, which meant that it took on the risk of default of any of its trading counterparties. But what if the clearing agent itself defaults? Suddenly, credit became a huge issue: risk groups scrutinized their trading counterparties, and imposed strict limits on how much credit traders could extend when doing deals.

Toward the end of 2001, the situation looked very scary, as it became clear that Enron was going under. As it turned out, however, the market barely skipped a beat. Just three weeks after Enron officially declared bankruptcy, the CEO of a major electric utility said: "It's like a great ship going down—there's a big splash, but the sea quickly fills in. The short-term effect is really not that substantial."[29]

Enron's demise did have some lasting effects on the market, though. First, it clearly showed the massive credit risk that came along with an Enron-style, one-to-many system. This led to a massive uptick in transactions on the more traditional, many-to-many exchanges that had struggled when EnronOnline dominated. Intercontinental Exchange, or ICE, which had been a distant second in gas and power trading, saw trading volumes double in the two weeks after the bankruptcy. So rapid was the increase, that ICE was forced to airlift in servers to accommodate the new trading traffic. As of 2020, ICE remains the dominant gas and power trading platform in the United States.

Second, Enron had always been the foremost market maker for long-term, regional contracts, with delivery dates more than eighteen months into the future and for receipt or delivery outside of the Henry Hub. The fact that it could warehouse these contracts despite their being almost completely illiquid stemmed from Enron's enormous size and the diversity of its portfolio—the company could simply absorb large chunks of risk without triggering limits or causing large price fluctuations. By controlling an incredible 40 percent of wholesale gas and power markets in the 1990s,[30] events like a major basis blowout, which might bankrupt smaller players, were essentially averaged out by offsetting positions elsewhere in Enron's portfolio. After the bankruptcy, markets for long-dated, regional gas and power dried up, making it more difficult for hedgers to lock in long-term revenues. By most accounts, the market still does not have as effective a risk manager as Enron was in the late 1990s and early 2000s.

Still, though, as the great ship crashed at the bottom of the sea, the waters at the surface were surprisingly calm.

# CHAPTER 13

## SPECULATION ON A
## GALACTIC SCALE

From the moon, the earth looks perfectly round, belying the mountains and chasms that freckle its surface. And from a point yet more distant, the planets look to orbit the sun in perfect, elliptical regularity, belying the wobbles that result from their interactions with more minor celestial bodies. The point is only that the view from a distance is cleaner than from up close. And this aphorism does as much to explain the market as the cosmos.

When people think about "the market," they tend to imagine some well-structured set of levers, pulled by highly trained technicians at just the right moment to effectuate just such an outcome. "Give me a lever long enough and a fulcrum on which to place it, and I shall move the world," said Archimedes. Buyers pull the lever in one direction, and sellers in the other. Whoever pulls more forcefully, moves the world.

While not untrue, this paints a clean picture of a messy reality.

A market—especially a market for a physical and consumable product—is not just buying or selling contracts on a computer screen or over the phone. It is not just buying released capacity on a pipeline and arbitraging a locational spread. It is not just loaning cash for gas or writing long-term hedges that guarantee a price. It is not just swapping out one set of risks for another. A free market is the amalgamation of all the above—it is a multitude of levers, set on a great many fulcrums, many of which do not seek even to move the same body. If the reader will permit a final iteration of the analogy, price in a free market is like a galaxy of stars, each one surrounded by a web of almost inconceivable messiness and complexity, with each participant yanking an unknown and ever-changing number of strands, often in different directions, and often at the same time. Participants with exactly the same information, and even with exactly the same understanding of the same information, will end up buying from and selling to each other. Like two trucks full of logs passing each other on opposite sides of the highway, the situation can be mentally infuriating. (We've already got logs!) These wobbles in the market are difficult even to perceive, and nigh impossible to predict.

I have gone to such great and whimsical lengths to illustrate the complex and overlapping nature of the market for two reasons. The first is to underscore the relative unimportance of any one set of levers as compared to the entire set. Do not get too bogged down with understanding the difference between futures, swaps, options, physical versus financial products, daily-settled products versus monthly index products, and transport arrangements. The second is to impart a sense of just how much force is required to move not just one star, or even a neighborhood thereof, but of the entire galaxy. Events of such magnitude are exceedingly rare. In the US gas market, they imply billions of dollars pushing in a single direction.

## THE RISE OF THE HEDGE FUND

Somewhat surprisingly, speculative gas trading grew large enough to move the galaxy only after the collapse of Enron, when a new type of participant entered the market: the hedge fund. Whereas even at its zenith, Enron was a bigger risk manager and market maker than pure speculator, hedge funds took massive, unidirectional bets. So big were the bets, that a single trader at a single fund could control more than half of the total open interest in the benchmark NYMEX Henry Hub contract in a given delivery month, which meant that one person's buying and selling could—and did—single-handedly determine the price.

Hedge funds traditionally pooled capital from very wealthy and financially sophisticated investors and deployed that capital much more aggressively than a traditional mutual fund. They did this first and foremost by levering themselves to put the maximum amount of capital into the market, such that cash on hand at any time was enough to meet their margin requirements, but not nearly enough to backstop the entire portfolio. They also traded in and out of stocks, bonds, or derivatives much more quickly, and at higher volumes, than mutual funds, which generally hold positions for years. Hedge funds sought not to ride out a bucking market, but to make money in both up cycles and down cycles. To do this, they complemented their long positions with short positions, which is where the *hedge* in hedge fund comes from. In addition to being able to make money in down cycles, these shorts were meant to minimize the risk of catastrophic loss. Avoiding naked risk was a key element in hedge fund strategy.

Hedge funds were also characterized by assiduous research, attention to detail, and willingness to take positions in illiquid securities. Their managers may not have been the smartest guys in the room, but they were certainly the best informed. And because they managed "sophisticated" money, they were quite lightly regulated, which gave them the ability to respond to market changes much more rapidly than other pools of capital. Sensing something was amiss, a hedge fund could immediately take action to reduce or protect its positions.

Noting that I have used the past tense in his description of hedge funds, you might be asking, "What changed, and when?" If the reader will permit another analogy, hedge funds should function in the marketplace like lions in the savannah. As a herd of wildebeest moves through, the lions pick off the slow and sickly. The tradeoff of this brutal form of natural selection is clear: the lions feast, and the wildebeest herd is kept strong.

But the ratio matters much. Too few lions, and the wildebeest herd grows weak. Too many, and there is not enough wildebeest meat to go around, and the lions starve, or turn on one another.

As hedge funds and their managers raked in billions in profits, the industry grew. Capital began to flow in from all over. And it wasn't just extremely wealthy and sophisticated investors in the capital pool—banks, insurance companies, and pension funds all wanted in on the outsized returns. Total hedge fund assets under management rose from around $100 billion in the late 1990s to $500 billion in 2002, and to over $2 trillion in 2007.[1] At this point, trading by hedge funds was estimated to account for fully one-half of total volume on the New York Stock Exchange.[2] The lions were beginning to outnumber the wildebeest.

As it went in the broader market, so it went in energy, and by 2006 the number of energy hedge funds had risen to over five hundred, from one hundred three years prior.[3]

The rise of energy hedge funds was a natural—indeed, an inevitable—outgrowth of the rise of hedge funds more generally. As the lions themselves were now acting as a herd, the central hedge fund strategy shifted from identifying "mispriced" trades to simply betting on where the herd would go next. Technical analysis, conducted by PhD mathematicians and computer scientists, replaced fundamental analysis, conducted by experts in whatever sector the companies to be invested in operated. Hedge fund returns thus became correlated to the overall market because hedge funds had become the market.

From an investor's perspective, the growth of the industry led to some soul-searching. What was the point in paying the high fees that hedge fund managers demanded if you were not getting a differentiated strategy? Then again, a growing economy—both in the United States and globally—meant that more money than ever was looking for a home. Something had to give.

Hedging, in its true sense, had been practiced in commodities trading long before the advent of the modern hedge fund. Futures markets let producers and consumers both lock in a price for a future delivery month, so that that could go about their core business without worrying about commodity price fluctuations. Once they have bought or sold enough futures contracts to account for their total expected production or consumption, they are said to be fully hedged. Agricultural futures have existed in rudimentary forms for thousands of years. In the United States, grain trading evolved throughout the 1800s such that there was a fairly functional system in place in Chicago by the 1870s that a futures trader would recognize even today. Energy futures followed about a century later, when NYMEX introduced heating oil futures in 1978, then crude oil futures in 1983. Henry Hub natural gas futures began trading on NYMEX in 1990.

Commodity futures are, in essence, an insurance market. Producers hedge against prices going down; consumers against prices going up. And it is true that investing in commodities can represent a valid strategy for individual investors, against inflation and many of the real costs that many deal with in their daily lives: filling their cars with gasoline, heating their homes with natural gas, consuming agricultural products, and paying rent to amortize buildings made of wood, concrete, and steel.

But hedge funds did not start placing billion-dollar bets in commodity markets as insurance. Rather, they did so because commodity prices, it turned out, exhibited precious little correlation to the prices of other securities. By the mid-2000s, when hedge funds had proliferated to such an extent that they were herd lions, investing in the same assets and whose returns were thus highly correlated, investors seeking portfolio diversification dubbed commodities a new asset class. It was foolish, in the contemporaneous market environment, not to have some exposure to commodities, given their counter-cyclicality. "Commodities perform best when other assets perform worst," was the simple refrain from Goldman Sachs.[4] And within the commodities space, energy reigned king. Two-thirds of the Goldman Sachs Commodity Index was energy commodities, with crude oil holding the number one spot and natural gas number two.

While it is true that investing in commodities does provide diversification, inasmuch as they are uncorrelated with moves in equity prices, this low correlation is not an absolute truth, like the laws of physics. Among the many crowded bars of New York City, there may open a new one with plenty of space, good drinks, good music, and good prices. That the bar is not overcrowded is a temporary phenomenon, not a permanent one—a condition, not a fact. In identifying commodities as an asset class, Goldman Sachs and other investment advisors were acting like critics in the *New York Times*, recommending a visit to what would soon become the hottest new bar in town.

Commodities are distinct from traditional assets in an even more fundamental way, as they pay neither dividends, like stocks, nor interest, like bonds. They are an input, and, because of this, moves in commodity price redistribute wealth, but do not create it. When prices rise, consumers "lose," producers "win," and speculators generate profits or losses depending on the position they took, but society is no wealthier. Holding a long position in a commodity pays only when you finally sell out of that position, and then only if the price has risen. In contrast, holding a long position in equity of a company that produces the commodity will pay you dividends while you hold it, such that you will make money even if the commodity price does not rise. Thus, even "dumb" investors will usually make money investing in companies, as they pay out a percentage of their earnings as dividends. To make money on commodities, though, you have to place a correct bet on direction.

In the mid-2000s there were certainly reasons to be sanguine on commodities, outside of their offer of portfolio diversification. This was the era of rapid growth among emerging economies, especially the BRICs—Brazil, Russia, India, and China. China, in particular, was industrializing at a rapid clip, and would require plenty of coal, iron, steel, oil, timber, and every other conceivable commodity, to achieve the ambitious goals laid out in the government's five-year plans. The term *peak oil* was also fresh on the lips of every talking head on Wall Street: the world was running out, and prices were heading inexorably higher.

Never mind that these dynamics had nothing to do with US natural gas. In the minds of many investors, *commodities* was no longer a catch-all term for a group of products whose at production and consumption are often highly regional and completely unrelated. They were an asset class that belonged in every portfolio.

## BRIAN HUNTER AND JOHN ARNOLD

With investors clamoring for high returns and differentiated strategies, the mid-2000s was a great time to open an energy hedge fund or add an energy desk to an existing fund. Enron's bankruptcy made this task all the easier. While it was still around, Enron dominated US gas and electricity markets, employing hundreds of traders, the best of whom were paid millions of dollars in year-end bonuses. Any would-be hedge funds who wanted to poach these star traders would thus be forced to offer multi-million-dollar guarantees just to get them in the door. The bankruptcy, however, led to a glut of energy trading talent, and eager hedge fund recruiters flew from New York to Houston to find the diamonds in the rough.

Amaranth was a hedge fund named after an antique billiards table that its owner admired, made from the deep purple wood of the Amaranth tree.[5] There was every reason to be optimistic about Amaranth's future: its owner had been a successful trader at his previous hedge fund, Paloma Partners, so much so that Paloma helped capitalize Amaranth and shared many of its support functions with what was effectively a spin-off. Moreover, from the very beginning, Amaranth embraced a diversified investment strategy, the merits of which it boasted to clients. As well it should have, as diversifying a portfolio represents the sole exception to the otherwise inviolate economic axiom "there's no such thing as a free lunch."[6]

Amaranth started by trading convertible bonds—bonds that could be converted into stock in the future—as well as the stocks of companies that were merging and of utilities. Amaranth met with considerable success over its first year and a half, and in 2002 decided to dive into energy trading.

The timing seemed perfect, as the industry was growing rapidly and there was a good deal of trading talent on the market after Enron collapsed.[7] It wouldn't cost millions just to get in the game. Amaranth's first strategic hire was Harry Arora, who had successfully traded electricity and currencies at Enron. Arora seemed to fit the Amaranth strategy well, as he favored a cautious and diversified approach—he wanted Amaranth to trade natural gas, to be sure, but also oil, electricity, and coal—and not to let one position get so large that it dominated his desk's profit and loss.

Fast forward another two years and Amaranth had entered the hedge fund big leagues, managing almost $4 billion, against the $600 million with which it began. But its performance was beginning to falter. Its investments had yielded a return of just 3 percent over the first seven months of 2004, after stellar returns of 21 percent, 15 percent, and 29 percent in the three years prior.[8] The energy desk had been making money under Arora's stewardship, however, and Amaranth thus decided to add both to the desk's bankroll and its head count, shifting more of its capital toward energy. In the spring of 2004, two years after Arora's arrival made Amaranth into an energy trader, the fund brought on Brian Hunter.

Hunter was a gas trader, pure and simple. He bet only on Henry Hub contracts, not taking any basis positions, and favored time spreads—buying one month, one season, or one year, and selling another. He was young, only thirty years old when he joined Amaranth, but already had several years of experience trading for size, having been one of the very few gas traders willing to take positions measured in the tens of thousands of lots.[9] A position of ten thousand lots means a profit or loss of one million dollars on every one-cent move, and the gas markets regularly moved by dimes, and sometimes by dollars. Hunter was a risk taker, which is by no means a bad thing for a trader. But his style was in diametric opposition to that of his boss, Harry Arora: Hunter's risk was highly concentrated among very few trade ideas. One of those few other gas traders who took on such epic risk was John Arnold.

Arnold was the reigning champion among US gas traders. Before the company's demise, he was Enron's star Henry Hub trader, and was responsible for most of the speculative risk in its gas portfolio. It is said that all VIP visits to Enron's trade floors included a stop by Arnold's desk, to watch the wizard at work. In fact, as Enron was imploding but before it became clear that it would totally cease operations, the company had set aside $55 million to be paid out as cash bonuses to retain its top traders. Arnold took home $8 million, while the nearly seventy other key employees took home $37 million between them.

After Enron, Arnold did not wait for an offer from a fund or a competing commodities trading house. He took his millions and went out on his own, starting a gas-only hedge fund that he named Centaurus.

At this point, it is worth mentioning that every gas trader knows the story that I am about to tell, at least to some extent. Arnold is a legend in the market, and Hunter is a cautionary tale. It is impossible, without knowing either man well, to gauge to what extent descriptions of them have been exaggerated so as to better fit the events that follow, but the general thrust is almost certain: Hunter was loud and brash, and craved recognition as the top trader, a position he could only win by taking over from the more inward, quietly confident Arnold.

After a stellar first year at Amaranth, Hunter was given the opportunity he so desired. The head of the fund took him out from under Arora, giving Hunter full discretion over his own, much larger, portfolio. Starting in 2005, Hunter began dominating financial gas trading in the United States, alongside Arnold and perhaps one or two other traders. By this time, Amaranth had ploughed about 30 percent of its total capital into energy trading, from its initial allocation of just 2 percent.[10] Convertible bond markets were not yielding any returns, and the fund needed Hunter to perform.

The year 2005 was a terrible one for the residents of southern Louisiana, but a great one for both Hunter and Arnold. Both traders had calculated that too little upside risk was being priced into near-dated gas contracts. Going long seemed to have a good risk/reward profile—they did not stand to lose much in a humdrum year, and stood to gain mightily on any bullish shock.

In late August, Hurricane Katrina delivered that shock, pummeling the coasts of Louisiana and Mississippi, killing around 1,500 people and causing more than $100 billion in property damages. A month later, Hurricane Rita struck, but thankfully made landfall farther west, near the Texas-Louisiana border, where there were fewer people. Both the death toll and property damages from Rita were only about one-tenth of Hurricane Katrina.

Most of the deaths and damages from the hurricanes were centered in and around New Orleans, three-quarters of which lay underwater after Katrina's thirty-foot storm surge overtopped the levees protecting the city. What mattered for gas markets, however, was that the surge and the winds

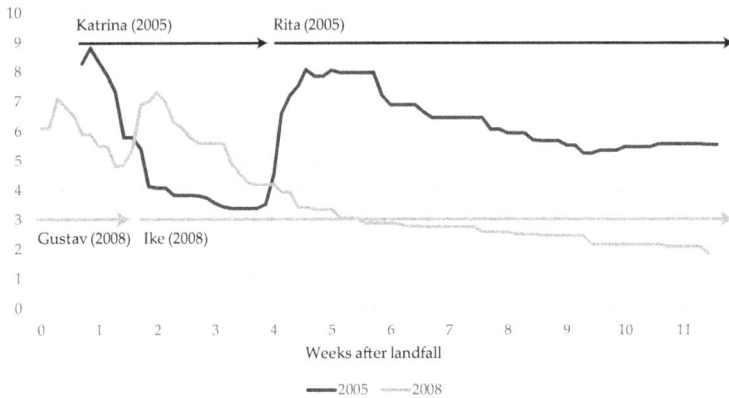

**FIGURE 13.1:** Natural gas production offline due to hurricanes (billion cubic feet per day)

also damaged or destroyed a large chunk of the Gulf of Mexico's vast oil and gas production complex.

When hurricanes approach the Gulf, producers take precautionary steps to shut in offshore platforms that they think might be in the direct path of the storm. They then send their workers back onshore on helicopters, before the winds become too high. After a weak or moderate storm, the damages are minimal, and production pops right back up to its pre–shut-in levels within a few days.

Neither Katrina nor Rita were weak storms. And they took aim straight at the heart of the Gulf Coast oil and gas production fairway. Of the 4,000 offshore platforms that were operating in the Gulf at the time, some 3,050 were in the direct path of the storms.[11] Almost no producer was left unscathed.

At its peak, Hurricane Katrina removed a whopping 8.8 billion cubic feet a day from the US market, representing 17 percent of total supplies. As importantly, even after the storm cleared, around 4 billion cubic feet a day remained offline, as the platforms underwent repairs for the quite extensive damages that the storm had wrought. Prices surged, benefiting the long positions that both Hunter and Arnold had on. A month later, and despite being less powerful, Hurricane Rita had an even longer-lasting effect on production, making Hunter and Arnold even more money.

In 2005 Hunter made around a billion dollars for Amaranth, which was sorely needed, as the fund was having a challenging time in its other markets. Hunter's earnings represented an astonishing 98 percent of the fund's total that year, making him a hero at the office.[12] He could not have been disappointed when his bonus check arrived in early 2006—it was for $113 million.

Arnold probably took home even more, as he made money not only as prices rose on the hurricanes but also collected a windfall that December, after betting that prices would fall again. In 2005 Centaurus was up 16 percent.

The staggering profits had more of an effect on Hunter, who became convinced of his own infallibility, than Arnold. While Hunter had already, unquestionably, secured a seat at the big kids' table, he wanted to be number one. No bonus figure would have been enough for him, as long as Arnold was also on the winning side of the trade. And Amaranth was happy to let its golden child give it a go.

## THE WIDOWMAKER

There are few ways to up the ante when you are already playing for billions. And there is no way to manage risk. The financial gas market simply neither was, nor is, big enough to make that sort of money without taking an almost wholly unidirectional—and practically irreversible—position. The author of a US Senate investigation into Amaranth's market power put it simply: "Amaranth was too big for the market it created."[13]

By early 2006, Hunter was the largest natural gas trader in the market. He was also the most vocal, practically advertising the positions that he took, which was in sharp contrast to almost every other trader, in any market, who generally take great pains to keep their trades anonymous, employing small armies of brokers and splitting the activity unevenly across several days or weeks. It is reasonable to think that Hunter figured that there was no way to keep his positions secret, given their size, but transcripts and interviews with colleagues also indicate that he simply liked bragging about them. Whatever the case, it is certainly true that most other gas traders during this period were simply riding Hunter's wake, as his activity was so large that it was often the primary force moving the market. Regardless of what the fundamentals said about a trade, taking the opposite side of Hunter's position could—and sometimes did—spell bankruptcy. He had been given such a long leash by Amaranth that he simply could afford to be wrong for longer than the rest of the market could afford to be right.

There were two competing trains of thought about how to trade with Hunter in late 2005 and early 2006: "Screw it, it's obvious what he has on. I'm going to be positioned in a similar fashion" or "I'm going to be at the bar until you're done."[14] For the contracts that he traded heavily, Hunter was the herd—there was an entire side business built around mimicking what Hunter was doing.

If Hunter's ego had been inflated by his 2005 trading profits of roughly $1 billion, there is scarcely an adjective to describe how he must have felt by the end of April 2006: he made $1.1 billion in that month alone, which put him up almost $1.7 billion on the year.[15] Moreover, a frustrated Arora had left Amaranth in late March, and Hunter was promoted to head of energy trading.

As things turned out, however, Hunter had precious little time to savor the moment. Even in the heady days before the 2008 financial crisis, investors were spooked by such extreme volatility, regardless of the fact that the move was in their favor. Several of Amaranth's large investors began questioning the fund's reliance on Hunter's profits, and some even withdrew money, despite incurring a penalty fee. Regardless of whether Hunter's strategy was correct, his position had simply become too large.

More concerning still was that his strategy was not correct. Beginning in May, only just after achieving the recognition he so craved as top gas trader, the rest of the herd quickly turned against Hunter, and did not let up for the next four months. By that time, Amaranth had run out of cash to post collateral, and Hunter had singlehandedly brought down one of the largest hedge funds in the world.

What was his strategy, and why was it wrong? The first part of that question is easy to answer. The second, less so.

Essentially, Hunter was betting on a cold winter—that demand would outpace supply. He expressed this trade as a November/January spread, where he was short November and long January. In other words, he would make money if the price difference between the two months widened, with January being the more expensive and November being the less expensive.

The justification for this trade was that, after the effects of Hurricanes Katrina and Rita had diminished, supplies had returned to their pre-hurricane levels, and the winter that followed had been quite warm. This left more gas in storage than was normal, which pushed prices down. In fact, being short the summer months was what had made Hunter $1.7 billion over the first four months of the year. The November short leg of the November/January trade was just an extension of that bet.

Where it got more speculative was the January long leg of the spread, especially since the spread—and the outright price—were already much higher than they had been historically. In other words, Hunter was not betting on value, as he did when he bought "cheap" gas contracts before Hurricanes Katrina and Rita. His downside in those trades was limited, while his upside was quite high. This time, his downside was large, whereas, to make money, he seemed to need a perfect confluence of cold temperatures and perhaps another bout of powerful storms.

Initially, the strategy was self-fulfilling—Hunter bought so much January and sold so much November that he himself caused the spread to widen. Every new lot that Hunter bought increased the value of his existing position. When he started putting this trade on in earnest, in February, the

November/January spread sat at about $1.25. After a furious few weeks of trading, it was closer to $1.70, and by the end of April it was $2.20. On paper, Hunter had made hundreds of millions of dollars. But by then he personally controlled 60 percent of the open interest in the November and January contracts on NYMEX.[16] It was a house of cards—had he tried to get out of the position, the spread would have collapsed. He essentially needed the contracts to expire that high, which meant keeping the market in agreement with his thesis. Luckily, he had some support from the coterie of Hunter followers, traders who mimicked Hunter's positions, who were not yet ready to abandon him. He also retained Amaranth's support, and thus plenty of cash on hand to pay the increasingly high margin requirements. Last, Hunter had successfully scared many natural buyers—utilities and the like—into buying the winter at these inflated prices, as they feared things could get even worse.

Hunter added to the position in early May, but by then the risk managers at Amaranth were starting to get queasy. After the fund sent its monthly report to its investors, which showed Hunter's eye-popping $1.1 billion monthly profits, several called to express concerns about the levels and concentration of risk; by the end of April, 38 percent of Amaranth's capital was devoted to energy.[17] These same investors had only learned of Hunter's existence the month prior, when they were abruptly told that he was being promoted to head of energy trading after Arora's resignation. They had been sold on Amaranth's risk controls and high levels of diversification. For example, one of Amaranth's larger clients, the San Diego County Employees Retirement Association—a pension fund—had invested $175 million in August 2005 despite Amaranth's above-average fees.[18] The pension fund's investment advisor had explained away these fees by saying: "A lot of the risk that we take on with an Amaranth—with any of the multistrategies— is diversified away in the portfolio. That is why we are [investing in a hedge fund]—to get diversified gains."[19]

Finally, Amaranth began pushing back on Hunter, and ordered him to reduce his positions, especially in the November/January spread. The problem, of course, was that he owned so much of the spread that the act of selling it itself would push the price down. Perhaps more importantly, it would show that even Brian Hunter had limits. Other traders would smell blood in the water.

There was another dynamic that began to move against Hunter, and that was the outright price. Speculators like Hunter mostly express views through spreads: that is the trader's main lever. Hedgers, though, care little about spreads. They push and pull the market based on the full, outright price, and natural gas for January 2007 was trading above $11. For producers, this price guaranteed them a great return on their investments. They began to increase drilling activity, and sold heavily to hedge their future output. As they did so, many of the traders who had been waiting it out at the bar began selling too.

May was an extremely painful month for Hunter, who gave back $1 billion of the $1.1 billion that he had made in April. Amaranth wrote in a letter to its investors that the experience "was a humbling experience that has led us to recalibrate how we assess risk in this business,"[20] but in reality they did little to clip Hunter's wings. Amaranth's management had looked closely at Hunter's trading and concluded that the only way to get out would involve taking a loss of another billion or so dollars.[21] That would wipe out and reverse the $700 million that Hunter was still up on the year, which would certainly spook clients. Hunter's positions had become too big to fail, and Amaranth decided to let him keep them on in hopes that the market would come around.

Hunter didn't give up on his thesis that the winter was underpriced, even if he could no longer express that by adding to the November/January spread. So, in late May, he started piling into another, even more dangerous trade—March/April—the widowmaker.

March/April is known as the widowmaker for its volatility. March is the last month of the winter gas season, and April is the first month of the summer, also known as the injection season. It is called the injection season because that is when gas starts being added (injected) into underground storage caverns, rather than withdrawn to supply peak winter demand. In April the caverns are mostly empty, which means there is almost no conceivable way to run out of places to put gas. Congestion does not exist in April, making it the cheapest month of the year.

March, on the other hand, is fraught with uncertainty. Temperatures can still get quite cold, but more importantly, as the last month of winter, March prices are affected by what has come before. After a very cold winter, there will be very little gas left in storage in March. On cold days, then, prices will need to rise quite high—perhaps above next January's levels—in order to make that gas available. There could even be physical shortages. On the other hand, after a warm winter, there will be plenty of gas in storage, and continued warmth in March might see injections begin ahead of schedule. In this case, March will trade flat, or even below, April.

Being long March/April, betting that the spread will widen, is thus a bet on a cold winter. That was Hunter's bet, despite the spread already being quite high by historical standards. Once again, he expressed his view as loudly as he could, telling other traders and taking control of 60 percent of the total market for both of the months in the spread. By August, Hunter had accumulated roughly seventy-five thousand lots of March/April.[22] Every cent that the spread moved made or lost Amaranth about $7.5 million.

Despite its lamentations of having been humbled by May's losses, the fund had actually begun adding to its positions before the month was over.

This time, Arnold was ready to pounce. Not only was it obvious to anyone paying attention that Hunter had lost money in May, word soon got out that Amaranth was actually liquidating some of its other portfolios in order to raise cash to pay for the energy desk's ever-increasing margin requirements.

Arnold and others began taking the other side of Hunter's bet, shorting the March/April spread. Initially, they miscalculated, figuring that Hunter was out of bullets. May 26 proved that he was not—on that day, Hunter bought 16,000 lots of the spread, which pushed the price up by 25 cents.[23] Everyone on the other side, Arnold included, suffered heavy losses, and began to question their assumptions: maybe Hunter could go on like this forever?

On July 31 Hunter bought another 25,000 lots of the spread, blowing the price out by an incredible 72 cents.[24] Hunter's massive trading single-handedly had taken the spread from $1.50 in mid-May to $2.50 by August. In the process, he had caused at least one large gas hedge fund that was betting against him to go out of business.[25] Arnold had also incurred major losses, but had the wherewithal to keep going, not least because his stellar, multiyear track record had earned him the trust of his investors.

The ridiculous buying spree of July 31 was the beginning of the end for Amaranth. At that point it was clear to even a casual observer that there was only one buyer of the March/April spread, which was overvalued to the point of ridiculousness. Vince Kaminski, formerly the head of research at Enron, said to Congress: "Amaranth's positions were known to the market. The market knew about it. And when I was watching the situation last year it was like watching a train wreck in slow motion. It was obvious that it would end up in a crash."[26]

Now Hunter's coterie turned against him, for the same reasons they had been with him before. They knew that selling the spread would force Hunter to stop out, forcing him to join in the selling so as to avoid further losses. They probably also knew that the position had attained such size that they would be stopping out Amaranth as a whole, by then one of the world's largest hedge funds, with nearly $10 billion under management.

After remaining buoyant throughout most of August, the March/April spread began to collapse at the end of the month, dropping from a high of $2.50 to just $1.15 by September 15. That move alone represented a loss of about $1 billion for Amaranth. There were another several hundred million of losses in summer/winter spreads for years farther into the future, which fell in sympathy with the front year (2007), and Hunter had lost an additional $600 million after getting throttled by Arnold on a bet on the

September contract. Over just a few weeks, Hunter's trades had lost more than $2 billion. The margin requirements were now higher than Amaranth could afford to pay—it had sold off all of its other liquid holdings. The fund was going under.

On Saturday, September 16, Hunter was finally forced to swallow his pride. He wrote an email to Arnold asking him whether he would buy him out of his huge March/April positions before the market opened on Monday. Arnold's response was devastating.

> The market is now loaded up on recent, bad purchases that they will probably try to be spitting out on Monday if there is a lower opening given that spread has been in free fall. In my opinion, fundamentally, that spread is still a long way from fundamental value.
>
> Over the past couple years the market has put a big risk premium into that spread yet it has paid out on expiry once in ten years. We'll be at all time high storage levels with mediocre s/d [supply and demand] and an el nino [sic]. Even though that spread has collapsed over the past 2 weeks, the only reason it's still $1 is because of your position.[27]

Arnold then gave his bid for Hunter's March/April position: 45–60 cents. Even the high end of that range, 60 cents, represented a 55-cent discount to the prior close, and would have cost Amaranth another $350 million for the front, and $300 million or so for future years.[28] If that seemed bad, Hunter's bosses had a more holistic view of the situation, as they were using the weekend to shop around Amaranth's entire natural gas book of positions. Goldman Sachs offered to take on the holdings if Amaranth paid them $1.7 billion. Merrill Lynch offered to take one-quarter of the book for $300 million, a slightly better pro-rata valuation of negative $1.2 billion. The situation worsened as word got round that Amaranth was in fire-sale mode.

Amaranth couldn't reach a deal by Monday, by which point Arnold's price projections were proven incredibly accurate. March/April was sold down to 75 cents, which, combined with other price moves, caused Hunter to take on another $600 million in losses. The spread continued to fall. The next day, Amaranth reached a deal with rival hedge fund Citadel and J. P. Morgan: it would pay $2.5 billion to get rid of its natural gas book.[29]

In a month, Hunter's trades had cost Amaranth more than $5 billion: $2 billion in losses through September 15, another $600 million on September 18, and the $2.5 billion it had to pay Citadel and J. P. Morgan to take ownership of its book.

And the March/April spread that Hunter had bought up to $2.50? It expired a few cents negative.

## AFTERMATH

There is no doubt that Hunter's trading pushed up gas prices. Eventually, however, the stars came back into alignment. The supply-and-demand situation simply did not justify his positions, and as Hunter began bleeding money, his adversaries saw blood and attacked. Speculation could not overcome fundamentals indefinitely.

But Hunter's speculative bets ended up transferring tens of billions of dollars from consumers to producers in the real economy. Even those hedgers who were not active in the futures markets were affected because the actual, spot price of gas is determined in large part by futures price levels. Suppose you operate a storage cavern, such that you have the ability to inject or withdraw gas as you see fit. If the futures price for the next month is $10 and the spot price, for tomorrow, is only $5, clearly you will buy all the gas that you can possibly inject today, and then sell it next month. And if it is December, and people want gas, but the price for January is $3 higher, you will not give it to them—not unless prices rise to make you indifferent.

Spot and futures prices as well as the shape of the futures curve—spreads between months and seasons—exert their gravity on one another. There is a natural order to them, determined by the prevailing supply-and-demand situation—the fundamentals. And while that order can be undone by large trading, the pull of gravity is unrelenting. Hunter was fighting gravity, and lost.

But he did leave a legacy. Not only as a case study in how not to trade, but as an unwitting seed capitalist for a technology that really would upend the fundamentals. Every action has an equal and opposite reaction, and in pushing up prices and transferring tens of billions of dollars from consumers to producers, Hunter did as much, if not more, than any other person to hasten the development of shale gas, which changed the face of the industry in the late 2000s.

# CHAPTER 14

## SHALE

The natural gas industry began theatrically, with an inferno outside of Pittsburgh issuing "flaming fire" that "could be seen at night at a distance of eight or ten miles" and whose "roaring sound was distinctly heard for five or six miles."[1] The Haymaker well had belched tens of millions of cubic into the air each day for months when a sightseer inadvertently set it ablaze. After it was brought under control, two armed mobs squared off to contest its ownership, and one of the Haymaker brothers was shot and killed.

By comparison, the "first" shale gas well was a dud. A company called Mitchell Energy decided to go ahead with the C. W. Slay No. 1 well in early 1981, and had finished drilling by June. But internal staffing changes meant that the well was not completed until September. When it was, it flowed a measly 120,000 cubic feet per day, such a paltry amount that the pipeline to whom it was dedicated—Natural Gas Pipeline of America—actually exercised its option not to connect it to the system. Despite the prevailing shortage, it simply wasn't worth the cost of extending the gathering lines.

In truth, though, Slay No. 1 was not the first well that produced gas from shale. While I would like nothing more than to tell the heroic story of how a group of small, independent producers from humble beginnings saw the promise of shale gas and fought doggedly against high odds to commercialize this bountiful, domestic source of energy, that story is fundamentally flawed. That is not to say that homegrown wildcatters played no role in what has been appropriately termed the *shale revolution*, but so too did the majors, and even foreign companies, drilling thousands of miles from the US mainland. Even after shale gas was commercialized, it took an army of geologists and engineers to make what began as a marginal source of new production into virtually the only source. That is not hyperbole: in the United States, from around 2009 and onward, virtually all new gas was shale gas.

The upstream oil and gas industry did not jump from conventional reservoirs directly into shale—it meandered. The main asset of Mitchell Energy, driller of that "first" shale gas well, was the Barnett Shale, which is often referred to as the granddaddy of shales. This is a fair enough moniker, but grandfathers have parents too. Before I detail who begat whom in the complicated geological genealogy of gas-bearing rock, however, it would be good to define what exactly shale is, and why it was ignored by oilmen for over one hundred years.

## BLOOD FROM A STONE

At its simplest, shale gas is gas found in a reservoir of shale rock. Most conventional reservoirs consist of sandstone or carbonate rocks, like limestone. Not to worry if you feel lost at this distinction, because, in truth, the rock type is not especially important—rather, what is important is the characteristics of the reservoir. Two, interrelated characteristics dominate in terms of their importance: porosity and permeability.[2]

Porosity measures how much oil, gas, or water the rock can hold in its open spaces—its pores—and is expressed in percent. If a reservoir has porosity of 10 percent, then ten square feet of rock has space for one square foot of fluid. (Keep in mind that gas in a reservoir is compressed, and will have expanded by the time it makes it all the way up to the wellhead.) More porosity is better, not only because it means that the reservoir can hold more petroleum, but because it also generally means the reservoir is more permeable.

Permeability measures the ability of a fluid—oil, gas, or water—to pass through a medium. In other words, it measures how well the pores are connected with one another. Porosity and permeability are obviously correlated, as more and larger pores increase the chance of the pores being connected, and the channels themselves count as pores, as they are open space.

If a good reservoir is characterized by high porosity and high permeability, then a bad one is characterized by low porosity and low permeability. I will focus on permeability because that is the characteristic that humans can actually change to extract more petroleum. Porosity only matters in the first instance: it determines how much oil and gas a reservoir can store. But since we cannot create oil or gas, which are generated over millions of years, increasing the porosity of a reservoir would not allow access to any additional petroleum.

Shale rock, by definition, exhibits lower permeability than sandstone. This is because shale is composed of silt and clay particles, which are finer than the sand particles that make up sandstone. Geologists created these naming conventions to help them classify rocks with similar characteristics, but the reality is less discrete, and more continuous. Sand is just finer gravel, silt is just finer sand, and clay is just finer silt. There is no black, nor white, only infinite shades of gray.

As these fine particles begin the process of becoming rock—lithification—they are compacted under great pressure, and build up as extremely

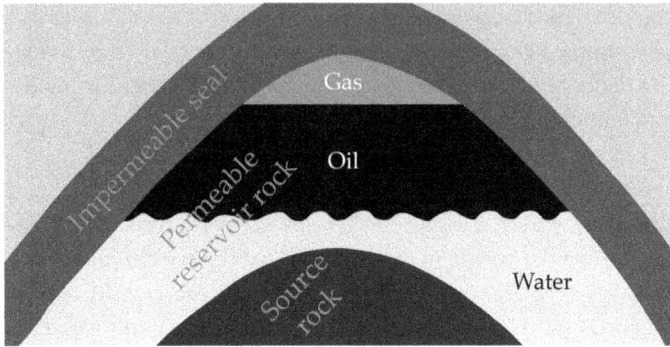

**FIGURE 14.1:** Conventional, anticlinal petroleum system

thin sheets. This is the defining visual characteristic of shale: wafer-thin layers of rock piled on each other such that there is little, if any, open space.

Jumping several million years into the future, and assuming that dead organic matter has fallen in among the particles in an oxygen-poor environment, such as at the bottom of a sea, you get petroleum-bearing rock, or source rock. To make a conventional petroleum system, this source rock needs to be covered by a more permeable reservoir rock—sandstone over shale. Over another few million years, the source rock bleeds its hydrocarbons upward into the reservoir rock.[3] The reservoir rock does not bleed off its hydrocarbons, however, as it is bounded from above by the third crucial ingredient: a low-permeability sealing rock, or caprock. Add in some tectonic-induced folding, and you have a textbook petroleum system: the classic, anticlinal layer cake.

In conventional petroleum systems, shale almost always acts as the impermeable source rock, and often also the caprock. Historically, when geologists or drillers found a source rock without a reservoir above it, they ignored it as uneconomical. In truth, however, every source rock is also a reservoir, and a seal to boot. It is just that the reservoir is of very poor quality, owing to its extremely low permeability. Again, no black, nor white, only gray.

Reservoirs of low permeability, whether they are composed of shale or sandstone or carbonates, are called tight reservoirs. Without getting into technical terms, tight reservoirs are reservoirs where natural permeability is too low for petroleum to flow in commercial quantities. Therefore, to tap tight reservoirs, we must do something unnatural.

In the popular consciousness, *shale* has come to be associated with *fracked*—shale gas is fracked gas, and fracked gas is shale gas. While the former is true, the latter is not. Gas is also fracked from sandstones and carbonates. In fact, we frack gas—and oil—from all tight reservoirs. In contrast, we do not frack conventional reservoirs because we do not need to.[4] *Conventional* does not imply any particular rock type but rather that enough natural permeability exists to allow commercial quantities of petroleum to migrate to the wellbore without intervention.

In sum, and with the knowledge that these neat classifications belie a great deal of messy variation, the following can be said:

- Conventional reservoirs can flow on their own, without additional stimulation.
- Tight reservoirs have difficulty flowing on their own, and often require stimulation.
- Shale reservoirs are super tight, and always require stimulation.

Knowing this, I can address how technology and techniques evolved to make shale gas extraction feasible.

## HORIZONTAL DRILLING AND HYDRAULIC FRACTURING

Wildcatting—searching for oil and gas in new places—is not unlike playing a game of *Battleship* on a grand scale. You know there are hydrocarbons underneath you, but not their exact location. Geologists improve the odds of success by interpreting seismic data and the bits of rock that return to the surface during drilling operations, but a great deal of luck is still involved. In a conventional, high-permeability reservoir, precision is relatively less important, as fields drain large areas. In tight reservoirs, the targets are much smaller, and coming close counts for nothing.

Since the early days of the industry, producers had drilled wells that exhibited high initial rates of production, but then declined rapidly. Not only that, when they tried to replicate their early success by drilling other wells nearby, these often produced almost nothing. Promising new discoveries bankrupted many an optimistic wildcatter, who ploughed his life savings into dry hole after dry hole, even as production at his few successful wells dropped.

What the producers did not know at the time was that they were drilling into tight reservoirs, which did not part with their oil or gas easily. Their successful wells were ones that had, quite by chance, penetrated a natural fracture: a scar in the earth's crust caused by the ceaseless movement of the tectonic plates. These natural fractures act like superhighways within a reservoir—areas of very high permeability in an otherwise impermeable space. Because of this, the wells start producing in torrents, but exhibit rapid rates of decline.

Most producers who discovered that they were operating in tight reservoirs moved on to better prospects, either in the United States or abroad, but the high oil prices of the 1970s led many to take a fresh look at these prospects, especially ones that exhibited extensive natural fracturing.

One useful thing that a geologist can tell you is how fault lines are oriented in an underground rock formation. Fractures are most prevalent along faults, so any help finding the faults will increase the odds of success. And, because tectonic forces push in the same direction for millennia, the orientation of faults is usually consistent over a small area. If you know that you are drilling into a formation with a system of north-south faults, you have learned something valuable: you are more likely to find natural fractures if you drill your wells on an east-west line. It would be like your opponent in *Battleship* telling you that all of his  or her ships are oriented vertically—clearly, your best odds for success would be to drop your bombs horizontally. Once you have located a good fault, you can then develop it along the north-south line.

Still, you are likely to waste precious time and money missing faults. Even if you are only off by a few hundred feet, the well will be dry, and it will have cost you a couple of months' wages for the rig crew, plus thousands of gallons of diesel fuel, the cost of setting up the now-useless pad site, and, finally, all the cement needed to plug the hole.

Unless, that is, you could turn the drill bit underground, and continue the well horizontally until you hit your fault. If you could do that, your odds of success would, in theory, rise close to 100 percent.

But a drilling rig is a vertically inclined creature: the engine rotates the bit via a long, hollow steel pipe called the drill string. When you imagine a driller, working with large clamps and chains on a rig in the sunbaked desert, the process you are imagining is adding another section of thirty-foot pipe to the drill string. After that work is done, the rig will commence drilling for another thirty feet, at which point another section of string is added. On it goes, for days, weeks, and sometimes months. It is this ever-growing drill string that transfers the rotational motion of the rig's drive down to the bit. Clearly, this arrangement only works if the drill string is kept straight. So how to angle your well underground?

It may come as a shock to many Americans—and to Texans, Oklahomans, and Louisianans in particular—that the first modern horizontal wells were drilled not in the US Southwest, but in France. Elf Aquitaine, which is now part of France's national oil company, Total, drilled four such wells in France and Italy starting in 1980 before anyone in the United States began to emulate them.[5]

Before I instigate a fight with any patriotic petroleum history buffs, it is worth noting that the French wells were not the very first ones to turn underground and achieve a horizontal bearing. ARCO, an American producer, began drilling horizontal laterals in the Permian Basin in 1979 to deal with an issue called gas coning, where the low pressure near the wellbore ends up sucking gas down from above and thereby decreasing oil production. Texas Eastern Transmission Company's production subsidiary also experimented with horizontal laterals that year, and noted that they increased production in "problem" reservoirs.[6] And the process for drilling these "deviated drain holes" had been around in some form or another since the 1930s.

But the French wells were fundamentally different from those in the United States, as they did not seek merely to create short laterals to deal with production issues. Rather, they extended for great lengths, allowing Elf Aquitaine to access a huge aerial extent of reservoir without drilling several wells. Since, by virtue of gravity, sediments are laid down horizontally, rather than vertically, petroleum reservoirs are shaped more or less like pancakes. The gastronomically inclined French exploited these pancake-like characteristics with their horizontal wells, rather than succumb to the limitations they imposed on vertical bores.

The key innovation that allowed them to do so, which was adopted en masse in the United States in the mid-1980s, was the use of so-called downhole mud motors. These fairly simple devices take advantage of another source of energy that the rig provides, and that the drill string transmits.

In addition to rotating, the hollow drill string pumps a steady stream of drilling mud downhole. This mud lubricates and cools the bit, and also flushes the entire well, picking up the crushed pieces of rock (cuttings, in industry jargon) and bringing them back to the surface. There, they are either discarded or analyzed by the geologist. Thus, the drill string carries two types of energy: rotational energy in its walls and translational energy through its hollow interior. The rotational energy is unusable at an angle, but the translational energy of the mud is not.

A straightforward but ingenious design funnels the mud through a chamber with a helical rotor, which rotates the bit. This design—the basis for the mud motor—is called a progressive cavity pump. It was invented by René Moineau, a Frenchman, in Paris in 1930. (The author apologizes.) In sum, Elf Aquitaine's wells were drilled downward, vertically, the usual way, with the rig rotating the drill string. Then, when it was close to its total vertical depth, the rotation stopped, and the drill string went dead calm. From this kickoff point, the mud motor took over, and the driller began to angle the bit until it was traveling horizontally, or at whatever angle he wanted it to. Mud motors had already been around for decades by the time that Elf Aquitaine used them to drill their horizontal wells, but were unnecessary and uneconomical in most conventional reservoirs. As US independents began searching for natural fractures in tight formations, this changed.

The other technology that is a sine qua non for extracting shale gas is reservoir stimulation, or fracturing. Like horizontal drilling, the technique was not new in the 1980s—drillers were fracking with nitroglycerin in Appalachia all the way back in the 1860s. But, also like horizontal drilling, the basic purpose of fracture treatments was to change fundamentally over the years, starting merely as a way to improve performance at problem reservoirs, but eventually unlocking vast amounts of new resource.

In every case, the purpose of fracking a well is to increase permeability, and thus stimulate production. What changed over time was the volume of rock that operators could stimulate. As technology advanced, so did producers' ambitions. Early fracture treatments were used to clean up the reservoir close to the wellbore, to allow oil and gas to flow in more easily. This was a fairly perfunctory task, but one that could work wonders at underperforming wells. In the 1970s, larger treatments allowed operators to improve the quality of the reservoir by establishing connections between natural fractures and the wellbore, but these were still mostly short fracks, with limited reach. It was not until the late 1990s, when engineers began experimenting with massive volumes of water, slickened with chemicals to reduce friction and maximize impact, that operators realized that they could actually create reservoir, where before there had been none.

## THE GREAT-GRANDFATHER OF SHALE

The Eagle Ford Shale, in South Texas, is one of the largest and best-known oil and gas production provinces in the United States. Many may not be as familiar with the Austin Chalk, but it was here where much of the pioneering work that would unlock shale took place. The connection between the two formations is simple: the Eagle Ford is the source rock for the Austin Chalk. But it took roughly twenty-five years to hone the tools and techniques that eventually allowed producers to cut out the middleman and go straight to the source.

The Austin Chalk was a wildcatter's graveyard that nonetheless experienced several renaissances over the course of almost six decades. Since the 1920s, it had been a minor oil and gas producer, as drillers unknowingly and haphazardly targeted some large, natural fractures in the otherwise low-permeability rock. By the mid-1930s, producers had figured out what they were targeting, and had developed a somewhat standardized fracking procedure, where they shot between 60 and 175 gallons of nitroglycerine into the well, detonated it, and then pumped in a few thousand gallons of hydrochloric acid to clean up the mess.

After mixed results led to the Austin Chalk falling out of favor, producers returned in the early 1950s. Now, nitroglycerine fracking was out, but a simple acid flush was not enough to open up the tight reservoir. Producers began to mix in sand, both to scour the formation and to prop open the resultant fractures, so they would not close up, or heal, as pressures dropped over time, blocking the oil from the wellbore. This latter purpose gave the sand its industry name: proppant.

The results were mostly unsatisfactory, which led to another exodus. But once again, producers returned, this time in the boom days of the late 1970s. Now, in addition to an initial acid wash, producers began pumping hundreds of thousands of gallons of water and hundreds of thousands of pounds of proppant down the well, under great pressure. Moreover, they were learning how to aim: running acoustic logs downhole before the fracturing began gave producers a good indication of where natural fractures were most highly concentrated, depth-wise.[7]

It was this era that saw the Austin Chalk rise in stature from a paltry lab experiment to a decent-sized development asset. Oil production had skyrocketed from just 8,000 barrels per day in 1975 to 150,000 barrels by 1981, while gas production rose from 12 million cubic feet per day to 400 million. But it was only during the next wave of interest that independent producers across the country stood up and took notice, asking themselves, *Could what is happening in the Austin Chalk apply elsewhere, and maybe everywhere?*

As evidenced by the Austin Chalk experience, fracking had already come a long way by the early 1980s, but no US producer had yet begun to drill proper horizontal wells. In fact, rather than heralding a revolution, the pioneering work in France seemed only to be a niche technology—by 1986, six years after Elf Aquitaine had completed the first, modern horizontal well, fewer than fifty such wells had been drilled worldwide.[8]

A handful of these earliest horizontals were located in the Austin Chalk, where Sun Oil Company had a legacy acreage position. Sun's story in the Austin Chalk muddies that beautiful narrative of unconventional oil and gas having been unlocked by a group of wily independents from the hardscrabble Southwest with nothing to lose.

Far from an upstart, Sun was the twentieth largest company in the United States, by revenues, in 1985, when it drilled its first horizontal well in the Austin Chalk. Nor was it young, having been founded in 1886 by Joseph Pew and Edward Octavius Emerson, those two enterprising gentlemen who had, less than a decade earlier, bought the Haymaker well and started supplying natural gas to Pittsburgh through their newly created Peoples Natural Gas Company. It was run not by a cowboy in a ten-gallon hat, but by a highfalutin Yankee: Sun Oil's president, Robert Hauptfuhrer, had attended a prestigious private high school in his hometown of Philadelphia, where he graduated as valedictorian, went on to Princeton, served two years as an officer in the navy, and then enrolled in Harvard Business School. Soon thereafter, he married Barbara Dunlop, daughter of Sun Oil's then-president Robert Dunlop. He joined Sun and, unsurprisingly, rose through the ranks quickly, and though no one questioned his intellect or work ethic, neither did Hauptfuhrer win the respect of the rank and file. One of Sun Oil's geologists captured the prevailing sentiment well when he said, "Hauptfuhrer wouldn't know oil if it dripped on him."[9]

Nevertheless, it was Sun Oil under Hauptfuhrer that had the earliest success drilling horizontal wells into tight reservoirs in the United States. Though his engineers may not have held their boss in high esteem, they were optimistic about their acreage. A small cadre within Sun Oil was convinced that they could wring healthy production volumes from the tight Austin Chalk, even though both their and their competitors' legacy wells in the field were producing next to nothing. They persuaded their managers to give them a bit of rope.

Their early work was more than a little encouraging. Production from one old vertical well, re-drilled as a horizontal, shot up by almost fifteen-fold. The results, in fact, were too good—if the news got out, other producers would quickly start amassing land in the field. And there is no doubt that it would have gotten out, as the upstream oil and gas world is filled with veritable armies of sleuths and data hounds. Stories abound of would-be spies who are given binoculars and a jug of water before sunrise and told to spend the day observing a competitor's pad site from the safety of nearby shrubs. Bluster or not, it is certainly the case that no stone goes unturned for long in the oil patch.

A more conventional way of checking up on the competition involves poring over data that enters the public record. In Texas, as in most states, production data are required to be filed with the state, for tax purposes. There is no disguising new discoveries, be they triumphs or disappointments. But these production figures are reported as monthly totals, not daily volumes. Keenly aware of the inner workings of the exploration and production world, one of Sun Oil's production managers wisely ordered the rig crew to shut in the gusher after just two days of production. When the production data was filed, this spectacular well looked like just another, humdrum Austin Chalk stripper—no one outside of Sun Oil knew that the monthly total had been arrived at in just two days. Sun Oil didn't stop there, telling crews to sleep in trailers on-site, so they could not spread the news to friends and family. The rush was on for Sun Oil to go and pick up acreage on the cheap, before anyone found out what they were on about.[10]

But just as the new, horizontal wells started coming in and posting mind-blowing results, flowing up to two thousand barrels per day apiece, when before they were luck to squeeze out a few hundred, the C-suite announced that it had a different strategic path in mind for Sun Oil.[11] These were the mid-1980s, after all—the very worst days of the oil bust. After peaking in 1980, crude prices slid steadily downward through 1985. The next year, the bottom fell out, and the price per barrel halved, to below $15 from nearly $30, after Saudi Arabia flooded the market, in part to punish intransigent OPEC members who had been ignoring their quotas. Sun, an integrated oil company, believed it was in its shareholders' best interest to spin off its exploration and production business in 1988. This left two entities: the refinery and retail company, which we know today as Sunoco, and an independent E&P named Oryx Energy. Hauptfuhrer stayed on as chief executive of Oryx, which began testing horizontal drilling in other tight US reservoirs.

The other company actively drilling horizontal wells in the Austin Chalk in the late 1980s was Union Pacific Resources. They drilled their first modern horizontal in 1987, a bit more than one year after Sun/Oryx, but quickly went on to become the most active horizontal driller in the area, and, indeed, across the United States.

Union Pacific hewed even less to the popular narrative of independent companies unlocking America's hard-to-get unconventional resources than did Sun Oil. While it was not a major, with downstream operations, it made Sun Oil look practically like a juvenile.

Union Pacific Resources was part of the Union Pacific Railroad Company, which Abraham Lincoln created in 1862 as the public company responsible for linking the great East Coast cities with the small, but rapidly growing West Coast cities.

Building transcontinental railroads was a risky business. It required mammoth sums of capital, and returns were far from guaranteed. In fact, there is a good argument that the prevailing economic situation did not justify the massive build-out that occurred. Regardless of its pure commercial value, the US government placed enormous value on the military importance of a national rail network, and was thus willing to subsidize its construction. It did not do so directly, with a cash outlay. Rather, it gave them something it had a lot of: public land. The general rule was to grant the railroad companies half of the land on the twenty miles adjacent to each side of their tracks, in plots of one square mile apiece, laid out in a checkerboard pattern so that the railroads would not own large contiguous tracts, which would have given them de facto fiefdoms.

Between 1850 and 1871, when the practice of granting public land to railroads stopped, around 125 million acres was given to the railroads,[12] or more than 5 percent of the total area of the Lower 48. Railroad companies thus became the largest landowners in the country, and it is only natural that they developed any valuable resources on those lands—be they metals, coal, oil, or gas (they did all).

It was this unlikely pair of producers, Oryx and Union Pacific, who ushered in the fourth Austin Chalk renaissance. Over a period of just a few years, their engineers worked out how to extend the reach of their horizontal laterals from less than one thousand to more than eight thousand feet—almost two miles[13]—while simultaneously cutting costs. Meanwhile, production rates continued to impress. In 1990 Hauptfuhrer remarked, "Our costs in the Chalk are now 50% more than a vertical well, but we have three to five or more times the daily production and reserves than a vertical well."[14]

Austin Chalk oil production, which had declined alongside prices since the 1981 peak, jumped quickly back up to where it had been—around 175,000 barrels per day—while gas production tripled its former record, hitting 1.5 billion cubic feet per day by the mid-1990s, representing a substantial 3 percent of the national total. By this time, Union Pacific had drilled more than 1,000 horizontal wells in fourteen different formations across the United States,[15] and the term *unconventional* had entered the lexicon of every energy market observer, from Texas, to Wall Street, to Washington DC.

But the Austin Chalk was not shale. I have said that the term *shale gas* can be a bit of a misnomer, but there was still an important distinction between the Austin Chalk and the other tight oil and gas formation being exploited in the late 1980s and 1990s and the even tighter formations whose development heralded the shale gas revolution.

## THE BARNETT SHALE: NOT ALL FRACKING IS CREATED EQUAL

Savvas Paraskevopoulous was born in a mountainous village on the Peloponnesus of Greece in 1881, the youngest of five siblings. His family were sheepherders who made a bit of extra money collecting firewood during the summers. Paraskevopoulous, having had no formal education, could neither read nor write his native Greek, and stood to inherit only a half-acre of arid land. Recognizing this lack of opportunity, he decided, when he was twenty years old, to walk seventy miles to the closest major port to make his way to America. In 1901 he secured free passage in exchange for work on a freighter bound for Ellis Island.

Straight away, while still at the docks in New York, Paraskevopoulous was offered a job laying railroad track. Any able-bodied man would do, regardless of his English abilities, and Paraskevopoulous spent three or four years working for the railroad, first in Arkansas, then in Utah. It was during this time when he made two acquaintances that would change his life. One was his foreman, a fellow Greek, who taught Paraskevopoulous to read and write, so that he could send letters to his mother back home. Another was his paymaster, an Irishman named Mike Mitchell, who had trouble pronouncing Paraskevopoulous's name. "From now on, your name is the same as mine," Mitchell told Paraskevopoulous, or so the story goes.

Savvas Paraskevopoulous, now Mike Mitchell, gradually learned some English, and opened up a shoeshine parlor in Houston, Texas, with his cousin. Then they opened another in Galveston. It was there that the now literate Mitchell was reading a Greek newspaper and saw a picture of a beautiful, young Greek woman who had just moved to Florida: Katina Eleftheriou. She had come over from Greece to live with her sister, who thought she had found Katina a husband. But a smitten Mitchell had other plans—he went to Florida and convinced Katina to move to Texas and marry him.[16] They had four children. George Phydias Mitchell, the third of them, was born in 1919.

George grew up without much, and suffered tragedy when his mother died of a stroke when he was only thirteen. But he recovered, and managed to enjoy his teenage years, partly by playing tennis, which would become a lifelong passion, and partly by spending the extra money he made selling fish he caught. After finishing high school, he enrolled at Texas A&M University, where in 1940 he graduated valedictorian of his class, with a degree in petroleum engineering. George wanted to be a geologist, but A&M did not offer a geology degree, so he satisfied himself by taking every geology elective that he could.

After college, George worked for Amoco on a rig in southern Louisiana, near Hackberry. He intensely disliked the work, and was certain to quit before long. Then the Japanese attacked Pearl Harbor, and the United States entered World War II. Eager to leave Hackberry, but equally keen to avoid combat, George enlisted with the Army Corps of Engineers, remaining in Houston.[17]

George went out on his own in the oil business after the war. In 1947 he and a partner established Oil Drilling Inc. As the name suggests, the aim was not to find natural gas. But that is just what he found, on one of his first, small leases, of just three thousand acres in North Texas, not far from Dallas. Within ninety days of the discovery, George had cobbled together another three hundred thousand acres in the area. And even though he would not begin conducting massive hydraulic fractures for several more decades, George was a bit unconventional from the very beginning, producing most of his gas from already drilled wells, which had been abandoned after they failed to produce oil.

As Oil Drilling Inc.'s sales ramped up, it began suffering from its dependence on local markets. Initially, all of its sales were made to the Lone Star Gas Company, whose purchases were highly seasonal. This challenged the economics of the whole prospect, as there were no year-round oil revenues to keep cash coming in over the summer, when Dallas residents did not need heat. Fortunately, an alternative presented itself in 1954.

Natural Gas Pipeline Company of America (NGPL), that grand dame that Samuel Insull had built fifteen years earlier, wanted to expand its system to access new supplies in Oklahoma and North Texas to feed its growing Chicago market. Oil Drilling was interested, as were two of the majors, but then the Supreme Court made its fateful ruling in the 1954 *Phillips* decision, which subjected all producers who sold gas in interstate commerce to federal price controls. NGPL was an interstate pipeline, and the two majors suddenly wanted nothing to do with it. This gave Oil Drilling a great deal of leverage, which it used to sign a favorable, twenty-year deal with NGPL in 1955, that included an upfront loan of $7 million to drill new wells.[18]

With a large chunk of new capital, a profitable producing asset, and a dependable market, Oil Drilling grew over the following decades, changing its name several times, eventually becoming Mitchell Energy & Development Corp. George Mitchell, ever the wheeler and dealer, was not content being just an oilman, and got involved in real estate in 1963, hence the *Development* in the company's name. By far his biggest deal was the purchase of fifty thousand acres north of Houston in 1964 for $6.25 million, land that ten years later would become the Woodlands. By the time of the land purchase, Mitchell Energy was producing gas from more than one thousand wells, and it became Texas's largest independent gas producer just a few years later.[19] The Woodlands grew rapidly, and today is home to over one hundred thousand people and several major oil and gas corporations.

While Mitchell Energy's main asset was always in North Texas, centered on those three hundred thousand acres just west of Dallas that George stitched together in the 1940s, it did expand into other areas, either via purchases or organically, via wildcatting. One such area of organic growth was the North Personville field, where Mitchell Energy drilled the discovery well in 1969. The initial results were promising, but development was slow.

North Personville was geologically complex and tight. Wells often came on with healthy levels of production—between one and four million cubic feet per day—but then declined rapidly, by as much as 90 percent in the first two years, a characteristic production profile for tight, naturally fractured reservoirs.[20] Mitchell Energy quickly realized that in order to get consistently good results and slow the steep rates of production decline, the wells would need to be fractured. It was here, in North Personville, a hundred or so miles south of the Barnett, that Mitchell Energy began two decades of experimentation with fracking in tight reservoirs that would culminate in the shale gas revolution.

The more advanced fracture treatments of the late 1970s used polymers to gel the water, which made it both more viscous, and thus more able to place proppant, and also prevented it from seeping into any clay in the formation. Water causes clay to expand, which decreases permeability and therefore slows production. Producers feared it like the plague. Acid was

Conventional formation: high porosity, and high permeability. No fracturing needed.

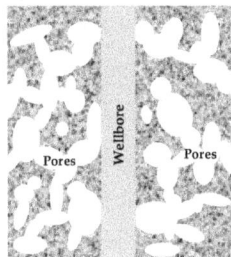

Conventional formation after reservoir damage. Short fracturing needed to re-open pores near wellbore.

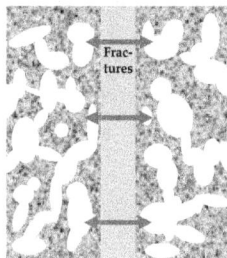

Tight formation. Massive hydraulic fracturing needed to access economic quantities of hydrocarbons.

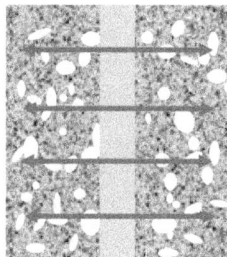

(Damage caused by fines accumulating near the wellbore, reducing flow.)

(Also shows how porosity relates to permeability.)

**FIGURE 14.2:** Short fracturing versus massive hydraulic fracturing

also used as a frack fluid, sometimes in conjunction with gelled water, and sometimes alone, and was valued for its cleaning properties, as it would dissolve the fines that accumulated over time and blocked pathways in the reservoir. Whatever the combination, these fracks shared something in common—they were expensive.

Too expensive for the tired old wells in the Hugoton field in Kansas, many of which had been producing since the 1920s. The Hugoton had entered a period of terminal decline, which ruled out large new investments in state-of-the-art fracks. There was still gas in the ground, however, so producers began experimenting with ways of fracking on the cheap.

River fracks in the Hugoton were comparatively simple. Water was trucked to the wells from the Cimarron River, where they were pumped downhole at high pressures "with little more than a few gallons of friction reducer, twenty or thirty dump trucks of river sand, and an occasional frog or turtle."[21] If the expensive, state-of-the-art treatments were akin to operations in a hospital, where surgeons used scalpels and anesthesia, river fracks were more like battlefield triage, where medics used saws and morphine. What they lost in precision, they made up for in intensity. The idea was simply to apply as much pressure to the rock as possible, to try and crack things up and stimulate a bit of additional production. To this end, the operators used friction reducer to decrease the water's viscosity, rather than using expensive gelling agents to increase it. This allowed the water to move more quickly down the wellbore and into the formation, and hence to propagate fractures more effectively. The focus was on creating a large network of small fractures, not a small network of large ones. As is usual in the energy industry, jargon proliferated quickly. These river fracks were also called *massive hydraulic fracks* because they used so much water, and sometimes *slickwater fracks*, for their use of friction reducers.

In the North Personville field, Mitchell Energy did not want to damage the still new reservoir, so it did not use friction reducer—it was still a gelled frack. But it did borrow an idea from the budget-conscious folks operating in older fields: intensity. In 1979 Mitchell Energy conducted the largest frack on record, pumping down 1 million gallons of water and 2.8 million pounds of sand. Immediately after the treatment, the well's production doubled.[22] Mitchell Energy was now in the fracking business.

The Barnett Shale was born as a play two years later, in 1981, after George read a paper that one of his own geologists published in a proceeding of the Dallas Geological Society. In it, the author suggested that the Barnett Shale was the source rock for all hydrocarbons in the North Texas region, where Mitchell Energy held a dominant acreage position. While the paper did not speculate on the commerciality of the Barnett, it was enough to plant a seed in George's mind. George himself, the CEO of the company, intervened in one of the next wells that Mitchell Energy drilled in the area, insisting that it be deepened so that it could test the Barnett Shale in addition to targeting conventional formations. This well, the C.W. Slay No. 1, was the discovery well for Newark East field, which became more commonly known simply as the Barnett Shale.

It was not as momentous as one might think. Slay No. 1 was drilled in early 1981, and was ready for completion by June, but a management restructuring put completion on hold until September. After the well was completed, it flowed a measly 120,000 cubic feet per day, so low that NGPL—the pipeline that had been purchasing the vast majority of Mitchell Energy's gas for the past twenty-five years—actually invoked a clause in its contract that gave it an option not to build a new gathering line to low-deliverability wells. The first shale gas well did not actually start production until eight months later, after Mitchell Energy built its own gathering pipeline, in June 1982.[23]

George was undaunted, partly because he and his geologists and engineers were truly optimistic about the prospects for the Barnett, and partly out of necessity. The conventional reservoirs that overlay the Barnett, from which Mitchell Energy had been producing for almost forty years, were mature, and would start to begin their terminal declines in a few more years. NGPL was well aware of this, and wanted guarantees that if it was to continue investing in new pipeline infrastructure in the area, there would be gas to fill those lines. Within a few months of Slay No. 1 beginning to flow, George directed the exploration and production group to come up with a strategy to offset declines at the company's legacy, conventional assets within the NGPL contract area.[24] In reality, this meant only one thing: figure out how to produce economically from the Barnett.

I have said that in 1979 Mitchell Energy conducted the largest frack on record in the North Personville field. But even this lacked the ambition required to exploit shale gas. The idea with both North Personville and Slay No. 1 was simply to perform a "short frack" that would punch through any damage near the wellbore and contact the natural fracture networks in the rock. This would create a small network of large, high-quality fractures. This technique was known to work in naturally fractured tight reservoirs, and Mitchell Energy's own engineers assumed all the way into the mid-1990s that "open, natural fracturing would be critical to the success of the Barnett."[25]

In short, Mitchell Energy was being conservative. Its engineers did not have the audacity to imagine that they could produce from a formation with such low permeability as the Barnett without nature's help. They considered the Barnett a tighter-than-usual tight gas play, not an entirely different animal that required entirely different techniques.

Development pressed on, both out of necessity, to satisfy Mitchell Energy's deal with NGPL, and also because the company was getting paid handsomely for the tight, Barnett gas, which qualified for the highest price schedule under the Natural Gas Policy Act of 1978. Moreover, in 1977 Mitchell Energy had renegotiated and extended the original NGPL contract for another twenty years. This meant that the company avoided the worst of the 1980s gas glut that drove so many producers out of business. (It also meant that Mitchell Energy contributed to the take-or-pay crisis, but NGPL was one of the less-affected pipelines—no harm, no foul.) It is impossible to know the counterfactual—what would have happened were it not for the NGPA's complicated price schedules or the favorable NGPL contract—but it is likely that the Barnett Shale would not have undergone continuous development without the aid of artificially high prices.[26]

There was progress over the 1980s and 1990s, but most developments were mildly disappointing. Natural fractures were present in the reservoir, but they had almost all healed, such that they were inaccessible. Techniques were found that substantially increased production, but they were too expensive to be commercial. Mitchell Energy poured hundreds of millions of dollars into Barnett wells over fourteen years without having ever earned a positive return.

In 1995 things took on greater urgency. That was the year when NGPL bought Mitchell Energy out of its favorable, high-priced contract, two years before it was due to expire. One of George's chief lieutenants recalled that "the entire Barnett program became questionable using spot market pricing at the time,"[27] which was a devastatingly low $1.50. It was only in 1995, at the pit of despair, when radical new fracking approaches began to be used.

Without a clear destination in mind, Mitchell Energy's engineers and geologists began down a path where fracture quality took a back seat to fracture quantity, and costs were cut aggressively.

Conservatism was no longer a viable strategy, and every line item underwent scrutiny. The use of expensive acid washes was dropped altogether, the amount of gel was reduced, and Mitchell Energy even began buying cheaper sand. These were incremental steps in the right direction, but it was not until 1995, when an engineer named Nick Steinsberger took over responsibilities for Barnett completions, that a more profound strategy shift began to develop.

In 1997 a petroleum engineer wrote: "The sole reasons for [using expensive gels] is to place proppant, minimize formation damage and ensure proper cleanup. In turn, the proppant has no function other than maintaining a conductive fracture during well production. What would happen though if the fracture actually retains adequate conductivity with very little or no proppant?"[28]

That engineer was not Nick Steinsberger. In fact, that engineer was not employed at Mitchell Energy at all. It was Mike Mayerhofer, of Union Pacific Resources, who had been successfully fracturing wells in a tight gas formation in East Texas in 1995, just before Steinsberger took over completions at Mitchell Energy.

Here was his radical idea: What if all of the expensive, state-of-the-art fracturing technologies were dropped completely? Would production really suffer? In essence, Union Pacific had begun to emulate an even more cut-rate version of the river fracks that Hugoton producers had begun two decades earlier, and for the same reason—to reduce costs. The title of the paper that contained Mayerhofer's prophetic quote could not have been more to-the-point: "Proppants? We Don't Need No Proppants."

Steinsberger and Mayerhofer were moving in the same direction with their frack designs, with Mayerhofer about two years ahead. In the hypercompetitive upstream oil and gas world, where binocular-wielding spies hide in the brush, willing technology transfer between unrelated companies is exceedingly rare, but Steinsberger and Mayerhofer knew each other professionally, and petitioned their superiors to allow them to open up a formal dialogue. Since the companies had very little overlapping acreage, they agreed.[29]

Union Pacific had found that it could reduce completion costs by 50 percent without any adverse effect on production by dropping the use of gels altogether, and greatly reducing, if not eliminating, proppant. Steinsberger and his predecessor had begun moving down this road, but were getting pushback from some of the more conservative engineers at Mitchell Energy. The encouraging results of Union Pacific's more mature slickwater frack program gave Steinsberger greater confidence in his approach, and the data began to convince his wary colleagues. The results confirmed Steinsberger's suspicions: Shale is tough stuff, and needs tough love. Focus on creating as large a fracture network as possible, and don't worry about creating such beautiful, large-aperture fractures close to the wellbore.

In May 1997 Mitchell Energy began to implement a large-scale, light-sand, slickwater frack program in the Barnett.[30] The new treatments not only worked as well or better than the previous techniques, they saved around $200,000 per well, which was very substantial given overall well costs that had averaged just under $1 million. There was yet another benefit to the new design: so long was the reach of these powerful new fracks, that production began to shoot up at neighboring wells.[31] Finally, using high-intensity, no-frills slickwater fracturing, engineers had found a way to coax gas from shale at a profit.

The revelations could not have come any sooner, as, by this point, Mitchell Energy had invested more than $250 million into the play and had not yet earned a positive return. It was this persistence, more than anything else, that earned George such adulation within the oil and gas industry, cementing his status as the father of shale. Despite not having been personally involved in the science, and despite having some idiosyncratic reasons for being so persistent—namely, the commitments he had made to NGPL—very few CEOs would have pressed on with a loss-making program for so long. From the Slay No. 1 well in 1981, to finally cracking the code in 1998, the Barnett had become a "seventeen-year, overnight success."[32]

By late 2000, George was eighty-one years old, and it was clear that a merger was in the company's future. With gas prices high and rising and *unconventional gas* now a buzzword on Wall Street, it was a good time to be a seller of prime acreage in the hottest play around. Earlier that year, Mitchell Energy had gone into full harvest mode, more than doubling the amount of rigs it had in the Barnett. In its last two years as an independent company, Mitchell Energy's Barnett production grew by 250 percent, to 365 million cubic feet per day,[33] and competitors were keen to get in on the action.

Devon Energy, an Oklahoma City–based independent producer, was started not by wildcatters or geologists, but by accountants. It grew through acquisitions, buying up neglected and, consequently, cheap US onshore properties in the 1970s, when the majors were focused on finding oil and gas offshore and overseas. Its founders seemed to have an uncanny knack for buying at the bottom, and the company grew steadily over time. Things at Devon picked up steam in the late 1990s and early 2000s, with a string of major acquisitions. In 1999 its revenues exceeded $1 billion for the first time. The next year, its largest yet acquisition more than doubled that number. But Devon was not finished.

Average first year gas production ('000 cubic feet per day)          New wells online (count)

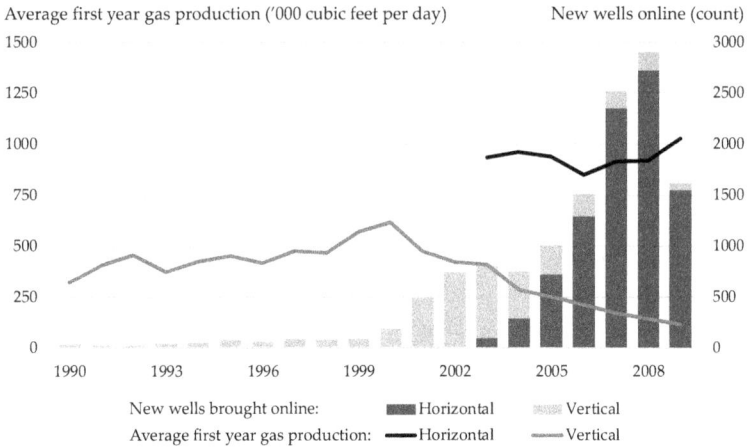

New wells brought online:   ▰▰ Horizontal   ░░ Vertical
Average first year gas production:  ━━ Horizontal   ━━ Vertical

**FIGURE 14.3:** Barnett Shale average first-year gas production and well completions

Both Mitchell Energy and Devon were in harvest mode, with Mitchell harvesting gas and Devon harvesting gas production companies. In the spring of 2001, a third-party engineering company set up a technical meeting that allowed Devon to take a deep look at Mitchell's Barnett asset, examining proprietary and theretofore secret data on geology, completion techniques, costs, and production results. While several other companies looked at Mitchell, Devon was the only one who grasped the play's potential to become something truly world-scale. The $3.5 billion acquisition was announced in August 2001, and closed in January 2002.

It was only after Devon took the reins that the Barnett realized its full potential. Mitchell Energy had begun began drilling horizontals in the Barnett in 1998, but were never able to achieve results commensurate with the high additional costs.[34] Mitchell's engineers had become master frackers, but were rather poor horizontal drillers. That all changed after the Devon acquisition. At a time when vertical wells were producing around one million cubic feet per day at their peaks, five of Devon's first seven Barnett horizontals began producing at 3.5 million.[35]

Results continued to improve over the next few years, as other operators bought positions in the Barnett and brought the entire industry quickly up the learning curve. Early horizontals, in the 2002–2004 era, produced twice as much as vertical wells, for only around 50 percent in additional costs, thanks largely to the great strides that Oryx and Union Pacific had made in the Austin Chalk in the late 1980s and 1990s. Later, new techniques made horizontals more expensive, around twice as much as vertical wells, but increased their productivity by an even greater margin, such that they began yielding four to five times as much gas as verticals. The math spoke for itself, and by 2005, vertical drilling was largely a defunct technology in the Barnett.

The frenzy that followed was both unique and characteristic for an industry defined by its boom and bust cycles. The six years after Devon acquired Mitchell Energy were the heyday of the Barnett—more than ten thousand wells were drilled, and production increased by tenfold, from five hundred million cubic feet per day at the beginning of 2002 to five billion by the end of 2008, making it the largest gas field in the United States. Devon began producing so much that it became difficult to market its growing volumes, so much so that when it heard that Tractabel, an independent power producer, was considering building a large, new, combined-cycle gas plant near its acreage, it built a dedicated pipeline lateral to serve the new facility. Devon's representatives also engaged in site selection for what would become the 750-megawatt power plant, driving an engineer from Tractabel along the farm-to-market roads of rural Wise County in a pickup truck until suitable ranchland was found, and then offering the rancher a buyout on the spot, which he accepted.

In this environment, costs rose, and people overpaid wildly for acreage. Characteristically, drilling slowed in 2009, and has not recovered since. What was unique was why drilling activity in the Barnett slowed. Just as the Stone Age did not end for a lack of stones, neither did the Barnett slip out of the limelight because it ran out of gas. Rather, producers began to apply the lessons that Mitchell Energy and Devon had learned in the Barnett to other shale plays across the United States. The results were nothing short of tremendous. The Barnett, the granddaddy of shales, had spawned powerful offspring.

## A REMARK ON THE EVOLUTION
## OF FRACTURING TECHNOLOGIES

It is natural to think of evolution as a linear process: begin with something simple, evolve to something more complex and better adapted to a changing world. Over the long term, it is difficult to argue against this clean and unwavering arc of progress. History, however, is said to be written by the victors, and often ignores those events that do not fit the victorious narrative. Even though all evolution is random mutation in the present, we reserve the word *mutant* only for variants that do not survive into the future.

Fracking took a circuitous route to end up where it is today. The "improvements" offered by complex and expensive gel, acid, and nitrogen treatments actually did more to retard progress in extracting shale gas than they did to hasten it. I did not mention Project Gasbuggy, which was a government-funded initiative that stimulated tight gas reservoirs by detonating a nuclear bomb downhole. Atomic fracking was short-lived—although successful at stimulating production, the gas was radioactive, and totally unusable.

Complex is not always superior, and the oil and gas industry has shown that evolution is not always linear, and that it is important not to discard old techniques simply because they are old.

# CHAPTER 15

## THE AFTERMATH

As Brian Hunter was busy bankrupting his hedge fund, Amaranth, in 2006, producers were happy to sell the young Canadian gas at $11 per MMBtu. It was only a few years ago that they were selling gas for $5, and only a few years before that when anything above $2 looked like great value. Wildcatters, however, have never been particularly good stewards of capital, so instead of returning the profits to their shareholders as dividends, they scrambled to find ways to spend it. As Robert Hefner and Penn Square Bank demonstrated in the early 1980s, it is very easy for a small group of Oklahomans to spend billions of dollars punching holes into the ground.

Aubrey McClendon, Tom Ward, Harold Hamm, and J. Larry Nichols were all dyed-in-the-wool Oklahomans. They and a handful of others saw how Mitchell Energy's work transformed the Barnett Shale from a tired old reservoir to the largest gas field in the United States, and realized that there were plenty of other tired old reservoirs underfoot. At $11, you didn't even have to be good, and the 2000s were characterized by one land rush after the next. After the Barnett came the Fayetteville, then the Haynesville, the Bakken, the Eagle Ford, the Eaglebine, the Brown Dense, the Tuscaloosa Marine Shale. At a certain price, every sedimentary rock in the United States seemed to be a winner. Of course, not every one was, especially when the veritable tidal wave of production sent prices crashing back down to $2.

Perspective requires the passage of time and the accumulation of consequence. In assigning relative importance to events as recent as the "shale gas miracle," I am blinded by the bias of the present. Perhaps the ample supply of gas that North America now enjoys will be remembered in years hence as a mere blip in the long history of the extractive industry. But I think not.

Sometimes numbers smother a narrative, and other times they breathe life into it. Hopefully the latter will be true here.

In the pre-shale, pre-horizontal drilling era, the average gas well in the United States produced around one hundred million cubic feet in its first year of production. In the glory days of the Barnett, its horizontal wells

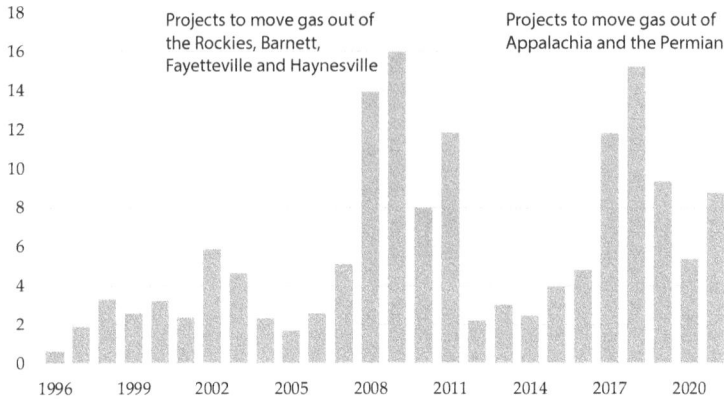

**FIGURE 15.1:** Dollars spent on new and expanded pipelines by in-service year (2019$ billion)

produced three and a half times that much. The play that one-upped the Barnett—the Haynesville Shale in northern Louisiana—did so because its wells produced close to two billion cubic feet during their first year, more than five times a Barnett well, and twenty times a pre–shale-era well. That was in 2010. Now, Haynesville wells average more than three billion cubic feet in their first year, with the largest doubling even that. In 2019 the average gas well produced twenty times more than the average gas well in 1990. How's that for resource depletion?

But the play that would truly upend the US gas market was not in Texas, Oklahoma, or Louisiana. It was back where it all started—in hilly Appalachia. Wells in the Marcellus and Utica shales of the Appalachian Basin produce just as much as their Haynesville brethren, and there are seven times more of them. The Appalachian Basin underlies twenty thousand square miles of Pennsylvania, West Virginia, and Ohio, and contains a truly mind-boggling amount of resource. Appalachia changed the energy conversation in the United States, causing developers to scrap plans for new liquefied natural gas (LNG) import terminals and instead convert existing facilities to export service.

This posed a slight problem—namely, that the network of US gas pipelines was built, beginning in the 1940s, to move gas into the Northeast, not out of it. Meanwhile, almost all of the new export capacity is located on the Gulf Coast. Any pipeline executive who fell asleep, Rip Van Winkle–like, in the mid-2000s, only to awake in 2020, would think that he had been transported to some dystopian, antipodal world where up was down and gas moved from low pressure to high. The reawakening of Appalachia as a supply region has led to the wholesale replumbing of the US gas grid.

What is perhaps most amazing is how effectively the process proceeded. Basis turned upside down, violently, which led to a wave of new pipeline build not seen since the post–World War II era. From 2008, roughly the start of the shale era, through 2019, more than $100 billion was spent on large-diameter pipelines, an average of more than $8 billion per year. In the ten years preceding the shale era, the number was $2 billion per year. Despite this tectonic shift, no changes in regulation or market design were required. FERC still oversees pipelines, which operate as rate-regulated, open-access transporters, futures are still traded at the Henry Hub, and swaps are bought and sold on a screen that is almost eerily similar to EnronOnline. After a few days of rest and recuperation, the long-sleeping pipeline company executive could go back his office and pick up more or less where he left off.

## AUBREY AND CHARIF

On the day he was buried, an Oklahoma newspaper declared Robert Samuel Kerr to have been that state's most powerful citizen. This was, almost without question, true. Kerr, raised poor in a log cabin in what was then still Indian Territory, was wildly successful in that quintessentially American way, rising from rags to riches on force of personality, never quitting, and never taking no for an answer.

Kerr's parents were tenant farmers and devout Southern Baptists who could not afford any luxuries, but instilled a strong work ethic in their preternaturally ambitious son. As a youth, Robert vowed to his father that he would someday start a family of his own, earn a million dollars, and become governor of Oklahoma. His early adult life, however, was beset by tragedy, rather than success. Commissioned as an officer in World War I, he married soon after the war ended. His wife was pregnant a few months later, but their twin daughters died at birth. The next year, the produce business that he had started burned to the ground, and three years after that, as his wife was in labor, this time with a boy, both she and the child perished. Kerr, who had recently passed the Oklahoma Bar despite running out of money before he could finish law school, dealt with his sorrow by spending long hours in his law office.

Resolute in spite of calamity, Kerr found another wife, with whom he successfully began a family, and worked ceaselessly to build his practice. As a tenacious lawyer in 1920s Oklahoma, Kerr inevitably did business with oilmen, and in that promise-rich, cash-poor industry, he earned and kept clients by accepting shares in drilling leases in lieu of coin for his legal services. It was not long before he decided that a career switch was in order. There was not much soul-searching—Kerr had become a lawyer not because of a passion for the law but because it was the best-paying job he could think of. After he saw firsthand how quickly oil could create immense wealth, he gave up his legal practice to commit fully to the oil business.[1]

In 1929 Kerr and a business partner formed a ragtag drilling company with two steam-driven rigs, and three boilers to power them. Early on, he managed to secure a meeting with Frank Phillips, founder of Phillips Petroleum, and began doing contract drilling with the already well-established producer. The work was profitable and fairly steady, but was getting Kerr no closer to his espoused goal of becoming a millionaire. Kerr then began producing his own volumes, but his timing was poor—his wells came in during the worst years of the Great Depression, and the young company teetered on the verge of insolvency. By 1937 things had gotten dire—oil prices had risen from a low of less than $10 per barrel (in 2020 dollars) when Kerr began the company to almost $20, but the company's producing assets were marginal at best. In a last-ditch effort to find the field that would breathe life into his dying concern, Kerr called Phillips Petroleum chief geologist Dean McGee, who had had remarkable success finding Phillips new fields to drill. Kerr offered McGee the title of vice president and a salary that dwarfed his own. The discrepancy didn't matter much, as McGee would either save the company or watch as it went under.

The decision paid out in spades. Soon after McGee was hired, he sniffed out a nice-sized field in Arkansas, which brought in much-needed cash and allowed Kerr to do what Oklahoma oilmen do best: use a small stream of revenues to summon great oceans of capital. The company that some years later would be renamed Kerr-McGee grew alongside wartime demand for oil to become a major producer, and later expanded into related (and some unrelated) businesses. As oil revenues began to pour into Kerr-McGee, Kerr finally found himself becoming a rich man. Having thus satisfied two of his three lifelong goals, and given that he was no more passionate about producing oil than he was practicing law, he quit the business, leaving McGee to run things, and in 1942 moved full-time into politics. In his first major political campaign, Kerr ran for and narrowly won the Oklahoma governor's seat, becoming the Sooner State's first native-born chief executive.

Aubrey Kerr McClendon was less than four years old when his great-uncle, Robert Kerr, died from a heart attack in 1963, so it cannot be said that Kerr had any direct influence on McClendon. But neither was McClendon's lineage immaterial: Aubrey would not have been Aubrey if his mother, Carole, had not been a Kerr.

Aubrey—as McClendon preferred to be addressed—grew up wealthy, but not rich, in Oklahoma City's Belle Isle neighborhood. He was just far enough away from the Kerr fortune to know what he was missing, and seem determined from childhood to pursue a lifestyle more befitting his pedigree. He started a lawn-mowing service in the late 1970s, when Oklahoma was in the midst of a drilling boom, and was able to charge a whopping $10 per hour. Every weekend, he would ride his bicycle the mile or so to Penn Square Bank to deposit that week's earnings. Smart and hardworking, he graduated as co-valedictorian, and went on to study history at Duke University.[2] What set Aubrey apart, however, was not his intelligence, but his charisma. From the get-go, he was a smooth talker and a natural leader who both oozed confidence and inspired it in others. His fate as a Kerr, of course, lay in the oil patch, but not as an oilman per se. Instead, Aubrey went looking for massive deposits of gas.

Over the first several years of the twenty-first century, it became generally accepted knowledge that US oil production was in terminal decline, and that despite the increasingly impressive volumes coming from the Barnett Shale, unconventional gas would not even offset losses from conventional fields. Traders and analysts agreed, as did the majors. Even Federal Reserve chairman Alan Greenspan was convinced of the consensus, saying in June 2003: "Rising demand for natural gas, especially as a clean-burning source of electric power, is pressing against a supply essentially restricted to North American production. . . . Today's tight natural gas markets have been a long time in coming, and futures prices suggest that we are not apt to return to earlier periods of relative abundance and low prices anytime soon."[3]

This setup may sound similar to the situation in the late 1970s, when the first generation of LNG import terminals and coal gasification facilities were built, immediately before a two-decade gas glut, which drove prices down to rock-bottom levels and made the multi-billion-dollar investments worthless. However, there were important differences. First, the gas market was not being distorted by prices set below producers' costs in the 2000s, as they were in the 1970s. Indeed, prices rarely dropped below $5 between 2003 and the onset of the financial crisis, which finally caught up to commodities in late 2008. If there was gas to be found, the market was sending a strong signal to go and find it. Second, the gas glut of the 1980s and 1990s was led more by demand falling than supply rising, as scores of new coal and nuclear power plants came online, new homes were built with electric heating, and the manufacturing sector declined as a share of the American economy. Producers were holding back as much as 40 percent of their total capacity simply because there was no place for the molecules to go. This was not the case in the 2000s, by which point producers were flowing at full volumes, while demand rose and rose. Natural gas was once again the favored home heating fuel, and coal and nuclear power plants had become much more difficult to build over the prior twenty years, for politico-environmental reasons, meaning gas was also the default fuel for new power plants.

The strong consensus on America's energy situation, and its oil and gas situation in particular, translated into bipartisan political agreement on the need to develop renewable sources of energy, with wind and solar being favored for electricity generation and corn-derived ethanol as a petroleum substitute in cars and trucks. Wind and solar tax credits were politically popular both on the East and West Coasts, where they were seen as addressing the looming threat of climate change, and also in Texas, Oklahoma, and much of the Midwest, where the wind blows the strongest and the turbines would be installed, generating jobs and taxes, in addition to electricity. Meanwhile, ethanol mandates were extremely popular in the Corn Belt, as they gave farmers a huge new market for their crop, and also among foreign policy hawks, who were uncomfortable at how dependent the United States had become on imports of foreign oil.

In the 2000s, things looked good for renewable energy, even if it meant higher energy costs for consumers, and hence a slight loss in American economic competitiveness. The acceptance of this trade-off—higher costs for cleaner, homegrown energy—was not based on shared ideology but on the generally accepted view that there was simply no alternative. As the United States and emerging economies around the world consumed and imported more oil and gas, prices would rise. This made wind, solar, and biofuels relatively less expensive. Meanwhile, after the attacks of September 11, 2001, and the souring of the relationship between the United States and Russia in the latter half of the 2000s, the energy independence argument became more important, even if it remained difficult to quantify. Wind, solar, and biofuels made the United States less dependent on oil and gas supplies from less-than-ideal trading partners in the Middle East and former Soviet Union. In sum, the fact that both the Left and the Right mostly agreed on US energy policy in the 2000s was due to the strong shared belief that domestic oil and gas production would, like it or not, account for an ever-smaller amount of total domestic energy consumption.

Aubrey had come to the conclusion that the country was running out of gas several years before Alan Greenspan's testimony gave it the façade of factualness and cemented it in the minds of investors. And Aubrey's company, Chesapeake Energy, seemed the perfect vehicle to profit from the situation.

Aubrey had formed Chesapeake in 1989 with an erstwhile competitor and fellow Oklahoman, Tom Ward. Aubrey and Ward had both been landmen, the foot soldiers of the oil and gas industry, who drive their trucks from ranch to ranch, convincing farmers to sign a lease and part with the minerals under their feet. They employed similar strategies, looking for holdout landowners in prime acreage and then using their charm and native Oklahoman bona fides to convince them to sign, when the majors had not. In 1989 they decided to join forces, and started Chesapeake Energy, named for the Chesapeake Bay, which Aubrey thought was one of the most beautiful parts of the country.[4]

Chesapeake was thus born of landmen, whom the founders placed on an equal footing with geologists and engineers. This differentiated the company from most E&Ps, where landmen were generally viewed as second-class citizens. Naturally, the best and most ambitious landmen gravitated to where they were held in the highest regard, which made assembling large acreage positions Chesapeake's key competitive advantage. It did this with some success through the 1990s, but persistently low prices and an expensive failed endeavor drilling horizontally in the Austin Chalk left the company in dire straits by the turn of the millennium. In 1999, desperate for more capital, Aubrey began visiting with potential investors. Among the meetings was one with a California–based independent power producer called Calpine, who was embarking on a massive effort to build out efficient, combined-cycle gas power plants. Calpine owned more than four gigawatts of gas-fired power plants at the time, was growing this number to more than ten over the next few years, and wanted to have twenty-five gigawatts of capacity by the mid-2000s.[5] This would make it one of the largest buyers of natural gas in the country, if not the largest. In the meeting, Calpine's CEO told Aubrey that his business model was predicated on the fact that gas would be cheap and abundant, available at $2 or less, for the foreseeable future.[6] Aubrey heard similar updates from utility companies across the country, who indicated that they expected demand to increase rapidly, after years of decline and then stagnation, as natural gas was once again the favored fuel for home heating.

Aubrey and Ward knew firsthand how difficult a time producers were having finding new supplies. Meanwhile, consumption looked to be on the cusp of a massive expansion. They became convinced that prices would rise, and embarked on a transformative, multiyear campaign to buy as much gas in the ground as they could. At first, however, they limited their efforts to conventional fields. As late as 2004, even after Chairman Greenspan's testimony, neither Aubrey nor Ward believed that the results that were coming out of the Barnett were applicable to other shales across the country.

Around the same time that Aubrey and Ward had their "Aha!" moment, when they realized that demand would outstrip supplies in just a few years, a Lebanese American businessman named Charif Souki had a similar revelation. His response, however, was exactly the opposite: whereas Chesapeake wanted to own as much domestic gas as possible, Charif wanted to import gas from abroad.

Unlike Aubrey and Ward, Charif had no oil and gas bona fides. Born into a wealthy and well-connected Lebanese family, Charif had come to the

United States for college, then worked at an investment bank, flying to and from the Middle East to raise money. He went out on his own in the late 1970s, made some millions investing in real estate and retail, and retired to Aspen, Colorado, in the mid-1980s. After a decade or so of the bon vivant lifestyle, however, which included opening up several restaurants that ultimately lost more than they made, Charif needed to work again. He got into energy precisely because the industry had been so beaten down over the prior decade, since oil and gas prices collapsed. Sensing opportunity, although without any thought-through business plan, he started an E&P focused on drilling in the Louisiana Gulf Coast. He named it Cheniere Energy, after the oak-lined ridges called *chenier* that run through south Louisiana's endless swamps.[7]

Cheniere was not successful as an E&P, but the more Charif learned about the natural gas business, the more convinced he became that demand was growing faster than supply, which meant that prices would need to rise. In 2000, with prices still low, he gave up the idea of running Cheniere as an E&P and instead began planning to build a massive LNG import terminal. From this point forward, Charif would be the anti-Aubrey—when things were good for the one, they were bad for the other. While their relationship was never adversarial, their two business models were diametrically opposed.

Recall that this would not be America's first time importing LNG from abroad. Four terminals were built during the gas shortage of the 1970s: Panhandle Eastern had built Trunkline LNG in the Gulf Coast, El Paso Natural Gas owned Cove Point and Elba Island on the East Coast, and Distrigas had the Everett terminal outside of Boston. Three were mothballed, and the other imported only very small volumes to meet peak-day winter demand. All were considered white elephants—massive wastes of time and money—that nearly bankrupted the companies that built them. And here was Charif, a man without any industry experience, looking to build more. The response from the investor community was skeptical, to say the least.

Over 2001, 2002, and 2003, Aubrey hustled to amass acreage in gas plays across the country, and Charif hustled to raise money for an import terminal. Chesapeake's holdings grew, while Cheniere stagnated, but 2004 was a watershed year for both. It was in this year when Cheniere signed LNG import contracts with international majors Chevron and Total, and when Chesapeake shifted its focus to shale.

LNG imports, which had only risen as high as seven hundred million cubic feet per day in the late 1970s, and then only for a short few months before petering out, rose close to two billion cubic feet per day in late 2003, not a month after Chairman Greenspan issued his prediction that the gas

**FIGURE 15.2:** US LNG imports (billion cubic feet per day)

market was entering a period of secular bullishness. They remained at elevated levels through the summer, as Cheniere inked its first import deals.

Just as the bull thesis was permeating the market, a paper presented at a meeting of the Association of American Petroleum Geologists in Dallas in the spring of 2004 estimated that the Barnett Shale contained more than double what had been previously thought.[8] The field was then producing around eight hundred million cubic feet per day, and that number was sure to continue rising. The paper was a wake-up call for Aubrey and Ward, who saw quite clearly that the technological advances happening in the Barnett would allow other shale plays around the country to produce large quantities of gas economically. If Chesapeake wanted to participate meaningfully in the new market, it had to own shale. This set off on an unprecedented, debt-fueled buying spree.

In 2005 alone, Chesapeake spent nearly $5 billion to buy energy properties, acquiring 1.4 million acres in the year, including 500,000 acres in the fourth quarter alone. In 2006 it spent another $4 billion and drilled over a thousand wells. This was no time for austerity: when Aubrey heard of competitors buying acreage in new plays, his stock response was: "Why aren't we in that?"[9] Wall Street loved Aubrey and was hot on shale, which allowed Chesapeake to issue more shares and truly staggering amounts of debt.

Charif was also busy expanding. Prices hit all-time highs after Hurricane Katrina in 2005, and remained elevated after the storms passed and production recovered. At prevailing Henry Hub prices, it would be highly economical to import from Qatar, Algeria, and virtually anywhere else in the world. In 2006 Cheniere began planning for the expansion of its import terminal at Sabine Pass, which was under construction on the swampy Texas-Louisiana border.

It may seem like a contradiction that the same investors gave money to both Chesapeake and Cheniere, who had opposing business models. However, in truth, the prevailing—and not unreasonable—wisdom at the time was that they could both be right. Chesapeake would benefit from the tight US gas market, as the high prices would make its acreage very profitable. Those same high prices made the spread between US and international prices wide enough to allow imports cargoes to profit handsomely. All that was required was prices staying high, which depended on domestic production growth falling short of domestic consumption growth. The market wanted Chesapeake and its competitors to grow, but not by enough to shift the bullish market conditions on which their investment thesis was predicated.

## BEYOND THE HILLS

The E&P business had always been characterized by high risks with potentially high returns, but by the early 2000s, the risk/reward ratio was becoming unattractive for independent producers. This was not due to an increase on the risk side, as the probability of drilling dry holes had dropped consistently over the twentieth century; rather, it was because the potential rewards had dropped by even more. By the twenty-first century, there were so few large but untapped conventional fields that even those drillers who stumbled upon good rock were not likely to drill more than a few profitable wells. After that, exploration had to start again. The lack of scale and repeatability meant that independent E&Ps had trouble becoming sustainable, going concerns.

Shale promised to change that. Producers drilled thousands of wells in the Barnett, and found that the rock characteristics were similar across miles and miles. Results were repeatable, which meant that shale E&Ps could, theoretically, transform themselves from high-risk, high-reward businesses to low-risk businesses more akin to manufacturers. However, during the first half of the 2000s, the Barnett was the only shale being actively drilled in the United States, and it was unclear whether it was a one-off.

The last name Zagorski is quite common in Poland and Russia. Literally, it is the combination of *za*, meaning "beyond," and *gora*, meaning "hill"—he who lives on the other side of the hill. It is also common in Pittsburgh, which is a city of hills full of Poles. Bill Zagorski grew up in the hills south of Pittsburgh. After graduating from high school, he earned two degrees in geology from the University of Pittsburgh, going on to work for one Pittsburgh driller, then another.

In 1983, still fairly fresh out of school, Zagorski began looking for promising acreage in the Appalachian Basin for his company, Mark Resources, to lease. He was successful, but the rock did not yield much gas. Profits came only because the gas was tight, and thus qualified for the highest price on the complicated NGPA schedule. Without subsidies, these formations were not worth being drilled.

Zagorski never changed jobs, basins, or cities, though. He continued to work for Mark Resources, which was then acquired by a company called Lomak Petroleum, which in 1998 changed its name to Range Resources. He continued to explore for new prospects in Appalachia, and continued to find modest volumes of gas.

This story perks up in 2004, when a colleague of Zagorski's asked him to examine a prospect in a gas basin in Alabama. The pitch, the colleague told Zagorski, was to point out the similarities between this prospect and the Barnett Shale, which by then was the hottest play in the United States.

"At that time, I hardly knew what the Barnett Shale was," Zagorski said. After doing some background research, he had what he called a "holy smokes" moment, as he realized that Range Resources had a Barnett Shale of their own, right in his native hills.[10] Fortuitously, Range had just drilled a well through the Marcellus Shale on the way to deeper sandstone. The Marcellus shared much in common with the Barnett, and geologists had long known that there was gas in its tiny, almost impenetrable pores, if only it could be extracted economically.

The well that had cored the Marcellus—Renz #1—was noncommercial. Range engineers were plugging and abandoning it, and most of the pad site—the few acres of land used to stage drilling and completion operations—had been reclaimed and reseeded. The company had already wasted a lot of money at Renz #1, and was not inclined to waste more, but the order came down from the CEO himself—Jeff Ventura wanted the Marcellus tested.

The promise of another Barnett fit perfectly with Ventura's long-term vision for his company. He believed strongly that in order to survive and thrive, Range needed to focus on so-called resource plays, where drilling locations numbered in the thousands and results were highly repeatable. He did not want to be just another wildcatter, living his life as if it were a never-ending game of roulette. He and Zagorski came up with a game plan: they decided simply to treat the Marcellus as if it were the Barnett, and not to "over-science" things. They asked a Barnett service company to perform a typical Barnett massive hydraulic fracture treatment. The results from the Renz #1, which was fracked in October 2004, were somewhere between acceptable and encouraging, but both Zagorski and Ventura seemed already to have caught the shale bug.

The next step was again to emulate in the Marcellus what was being done in the Barnett, which meant drilling horizontal wells. At around $3 million a pop, these were much more expensive than verticals, which cost around $800,000, and Range was still a relatively small company, with a market capitalization of only around $2 billion. It decided that it was willing to spend up to $200 million to prove the Marcellus as a commercially viable resource play, akin to the Barnett, and gave itself until the end of 2007 to do it.

In many ways, Ventura's vision and commitment was more remarkable than was George Mitchell's, as it was risk that he did not need to take. Whereas George had commitments to his pipeline, NGPL, and for the first phase of Barnett development was still getting highly favorable prices (until NGPL bought him out of the contract in 1995), Ventura ploughed $200 million into a single asset over three years simply because he believed it would work.

Things did not get off to a great start, as the first horizontal wells actually flowed less than Renz #1, the vertical discovery well. Zagorski noticed a trend on the first three horizontal wells, however, and, despite three observations being far short of statistically meaningful, he took a chance on the fourth well, and it came in flowing commercial volumes. The fifth, sixth, and seventh wells were also winners. After having spent more than $150 million of their $200 million, and with less than a year left, Range thought it had cracked the code.[11] In December 2007 it issued a press release that focused on the acquisition of acreage in the Barnett, but that mentioned the solid results in what had previously been a formation known only by old-school Appalachian Basin geologists: the Marcellus.[12]

In 2007 the prevailing wisdom on Wall Street was that the Barnett was a unique formation, and that hydraulically fractured horizontal wells would not work elsewhere. Late in the year, however, well results began being reported in Pennsylvania that looked eerily similar to those in Texas. If there was another, multi-billion-dollar resource play in the making, Wall Street wanted to know about it. Range, though, was not broadcasting its success. After cracking the code, it sought to amass large tracts of acreage on the quiet. The company was notoriously tight-lipped, and mostly stonewalled Wall Street analysts.

This led one of the more prominent of the Streeters to go searching for someone who would talk. Terry Engelder, a professor of geology at Penn State University, had just been named a distinguished lecturer for the American Association of Petroleum Geologists, and had spent his whole life studying the Appalachian Basin. Engelder was asked to host a webinar with more than one hundred potential investors that an investment bank had organized. One of the callers asked a simple question to which Engelder had no answer: how much gas was there in the Marcellus? After the call, Engelder examined the latest data and crunched the numbers. Penn State published them in a news release in January 2008, just one month after Range released the results of its latest, successful, horizontal Marcellus wells.[13]

What Engelder quickly realized was that the amount of resource was simply staggering. The commercially viable extent of the Barnett Shale was around three thousand square miles; the Appalachia Basin extended for fifteen or twenty thousand. What became clear only later was that the sweet spots in Appalachia, where the very best well results were concentrated, were themselves as big as the entirety of the Barnett.

In the Penn State news release, Engelder suggested that as much as 50 trillion cubic feet of gas could be recovered from the Marcellus.[14] As new well results kept pouring in, a year later that number had risen tenfold, to 489 trillion cubic feet, at a time when total US gas production was around 20 trillion cubic feet per year. The Marcellus held an enormous amount of gas, and it seemed clear that it could be produced profitably below prevailing futures prices.

Nor was the Marcellus the only new game in town. In early 2008, with gas prices around $9.50, Chesapeake Energy announced that it had accumulated two hundred thousand acres of leases in a new play called the Haynesville Shale in northern Louisiana, and that it was increasing its capital expenditure budget for 2008 and 2009 in order to develop the area.[15] Another producer, Southwestern Energy, had already began to develop a different shale play in Arkansas called the Fayetteville.

Aubrey was riding high. In early 2008, just after Chesapeake announced the Haynesville acquisitions, the company controlled an astonishing thirteen million acres and planned to drill thirty-six thousand wells over the next few years. It had grown to become the second-largest gas producer in the United States, after the venerable ExxonMobil. The land rush was over, McClendon declared, and he had emerged the victor.[16] The company was moving into harvest mode; gushers of profits were around the corner.

Aubrey lived a lifestyle that his great-uncle could only have imagined (and, as a devout Southern Baptist and lifelong teetotaler, probably never did). He bought the Seattle SuperSonics and brought them to Oklahoma City, renaming them the Thunder. He had a wine collection worth something like $100 million, and was in the process of commissioning a personal racetrack, which would be designed by a Porsche driver he had flown over from Germany, with a Tom Fazio–designed golf course in the infield.[17]

Charif, on the other hand, was dejected. LNG imports were plummeting, and nothing looked likely to bring them back. Just as the Sabine Pass terminal was nearing the final stages of construction, in early 2008, Cheniere's stock was ravaged by a wave of short-selling. Charif started selling off personal assets in order to meet margin calls, since he held so many of the company's shares. Things got dark enough that year, which also featured the bankruptcy of Lehman Brothers in September and the meltdown of broader financial and real estate markets, that Charif's friends began taking him out for long dinners just to make sure he was not alone.[18]

Aubrey and Charif had met each other at industry events, but were on different sides of the business, so had never exchanged much more than pleasantries. It was thus somewhat unexpected when in early 2009 Charif got a call from Aubrey, asking what it would take to turn Sabine Pass into an export facility.[19]

## TOO MUCH OF A GOOD THING

It was such a good story, even if it was never fully true: a motley group of homegrown entrepreneurs, through dogged perseverance, had found gas where the majors had failed, and stolen back America's birthright—cheap and abundant energy—from those who would have wielded it as a weapon against them: OPEC and Russia. Preaching the gospel most visibly was Aubrey McClendon, who sat atop a land empire that included positions in virtually every shale play in the country.

While his tenure at Chesapeake was eventually marred by several questionable financial transactions, no one can deny that Aubrey was a true believer—in himself, his company, and, most of all, in the superiority of natural gas. "I have a fossil fuel that makes other fossil fuels obsolete," he told *Rolling Stone* magazine.[20] That *Rolling Stone* would interview the CEO of an oil and gas company says even more about Aubrey than the quote. Aubrey, the shale messiah, seemed to answer many prayers at once: he promised a clean-burning, domestically abundant, and affordable fuel that could be easily plugged into existing infrastructure. While always careful not to criticize renewable energy, Aubrey pointed out that unlike electricity from wind and solar facilities—which was intermittent and so would require expensive and unproven battery storage, in addition to thousands of miles of new electric transmission lines—shale gas would move through existing, underutilized pipelines and into and out of a large and growing number of underground storage fields, out of sight and out of mind. Shale would power the dozens of gigawatts of relatively new gas-fired power plants that had been built after the Public Utility Regulatory Policies Act (PURPA) and electricity restructuring of the 1990s, and now sat partially idle as coal plants belched pollutants into the atmosphere. Aubrey touted gas's environmental credentials as compared to coal, that evil fuel of the past, and even compared to the other commodity his company produced: oil. Aubrey was a major proponent of using compressed or liquefied natural gas as a vehicle fuel, and the parking lot of Chesapeake Energy's campus in Oklahoma City had perhaps the highest concentration of cars and trucks converted to run on natural gas anywhere in the United States.

As much, and probably much more than anyone else, Aubrey was responsible for cultivating natural gas's new image as a transition, or bridge fuel, whose reliability and flexibility would enable the gradual build-out of renewable energy, not hinder it. However, as shale gas production grew, month after relentless month, and just as the fragile consensus that existed between the Left and the Right on energy policy began to crumble, the financial viability of shale drillers was being called into question.

If Aubrey's gospel was to be believed, demand for his perfect fuel was virtually limitless. Gas was set to replace coal in the domestic electricity sector, oil in the domestic transportation sector, and the United States would export what was left of its tremendous bounty to the rest of the world. This meant Aubrey could have his cake and eat it too. Despite the tremendous new volumes of supply that Chesapeake and its peers were bringing to market, insatiable demand meant that prices would remain permanently elevated. This justified the increasingly high prices that core shale acreage was demanding.

Chesapeake was on the cusp of transforming itself into a gas manufacturing company—having spent years cracking the code, it would now drill tens of thousands of virtually identical horizontal wells, enabling massive economies of scale. In 2008 it wrote in its corporate history: "Gone are the days when flamboyant 'wildcatters' drilled deep and expensive wells in the increasingly elusive search for undiscovered conventional natural gas reserves."[21] Shale had fundamentally changed the E&P business model—exploration risk was a thing of the past. Investors were urged to give producers a few years to whittle down their costs, after which time they would harvest their reserves, and become cash cows.

This was not the way things worked out.

Back in 1999, Enron sold off its profitable production company, Enron Oil & Gas, as part of CEO Jeffrey Skilling's preferred asset-light strategy, and to raise much-needed cash. EOG, as the newly independent company was called, established itself as a best-in-class E&P under Mark Papa's leadership. Like Aubrey, Papa believed that domestic gas supplies would have a tough time keeping up with demand, and increased his company's exposure to North American gas prices, though without Aubrey's reckless abandon. EOG acquired a prime position in the Barnett and several tight gas plays, and grew production steadily, from 900 million cubic feet per day in 2000 to 1.7 billion in 2007. The strategy had worked beautifully, with EOG's stock price soaring alongside natural gas prices.

That was when Papa's views diverged from Aubrey's. Seeing the proliferation of shale drilling, and especially early results from the Marcellus, the EOG chief grew concerned. Was the United States on the brink of another gas glut? And could this one be even worse than the one that preceded it,

US gas production by source

■ Appalachia  ■ Other shale  ▨ All other gas

Gas shale play production by month for first three years of development

Barnett ⋯⋯ Fayetteville ══Haynesville ══Appalachia

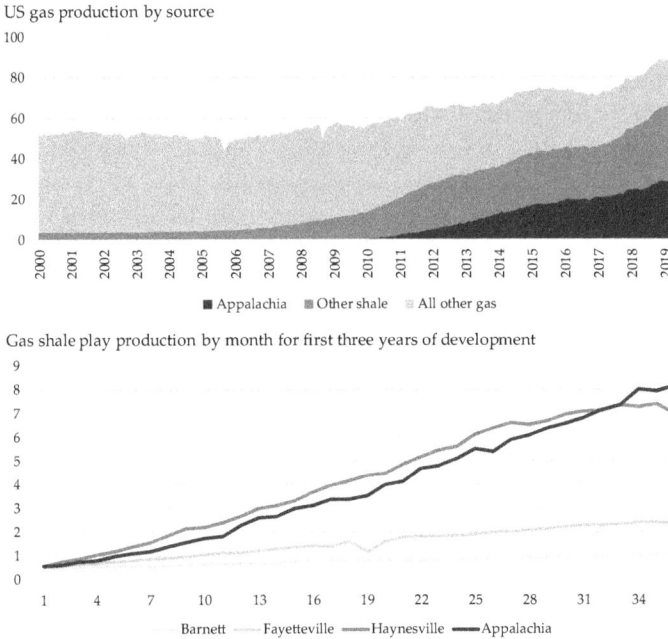

**FIGURE 15.3:** Growth of shale gas production (billion cubic feet per day)

as it was driven by a true step change in the cost of production, rather than a temporary, policy-induced drop in demand? By October 2007, Papa had made up his mind. He held a meeting at which he told his staff: "Natural gas prices are going to be weak for twenty or thirty years. We're gonna have to convert this company to an oil company or we're dead ducks. Tell your people to stop looking for gas, right away."[22] With that, EOG began the process of abandoning gas in favor of drilling for oil.

Over the next few years, every forecast of shale output was proven too low. The wells in the Marcellus and the Haynesville kept getting better and better, with initial rates of production increasing from four million cubic feet a day, to seven or eight, to ten or better, then to twenty, thirty, before the pressures became so great that operators actually had to start choking back their wells. Compare this to the Barnett, where initial rates of production remained around three or four million cubic feet per day.

Both the Marcellus and the Haynesville quickly surpassed the Barnett, and continued to grow. Total shale gas production more than doubled from five billion cubic feet per day in early 2006 to more than ten by early 2009, and then doubled again before the decade was out.

These numbers were shocking, both in terms of total volumes, and also in terms of the rate of change. Major shifts in the energy balance of a country the size of the United States generally take decades—there is an incredible amount of inertia in the system, and the numbers are too large to be moved easily. In ten short years, however, the United States was producing more shale gas than it was conventional gas. The qualifiers *conventional* and *unconventional* quickly lost their meaning as, with limited exceptions, conventional gas is no longer drilled—shale has taken over as the default resource.

In late 2008 natural gas prices began to slide, but the trend was originally written off as a side effect of the financial crisis, which brought with it an attendant sell-off of all asset classes, commodities included. The trouble was that they never recovered. By 2009 production began growing from the Haynesville and Marcellus, ensuring ample supplies for years to come. With the advent of hindsight, it was clear that the debt-fueled buying spree of the prior five was less a race to the top than one to the bottom.

Shale drillers were victims of their own success. They had dramatically overpaid for acreage during the bonanza of the mid- to late 2000s, and as they developed these acres over the next years, it quickly became evident that yesterday's tier one had become today's tier two, and tomorrow's garbage. In 2008 Aubrey seemed supernaturally to gravitate to the best acreage in each of the hot new shale plays before anyone else had even heard of them, and barren tracts of land across the United States became hallowed ground after he deigned that a well be sunk. By 2013 Aubrey was ousted from Chesapeake, the company he had founded and made into the second largest natural gas producer in the United States.

There was more to Aubrey's forced retirement than Chesapeake's consistently negative net income. He used Chesapeake as his personal piggybank, flying his wife's friends on the corporate jet, borrowing to co-invest personally in Chesapeake-owned wells while using those same wells as collateral, and essentially swapping shares for income. My goal is not to sully Aubrey's image but rather to make the broader point that while shale had changed the game, shale CEOs hadn't changed how they played it. In the end, Aubrey joined the ranks of the many spendthrift Oklahoma oilmen who preceded him, succeeded him, and were his contemporaries: he managed always to spend more than he made. Their perennial goal was to amass more acreage and grow production rather than accomplish what should be the object of every company, in any industry: to turn a profit.

While production was running away and prices were cratering, Charif, the anti-Aubrey, managed to pull off one of the best repositionings in corporate history. Aubrey had helped to plant the seed for this turnaround in early 2009, when he called Charif and asked him what it would take to remake Cheniere's idle Sabine Pass LNG import terminal into an export facility. Charif ran the sums and concluded that the idea had no merit

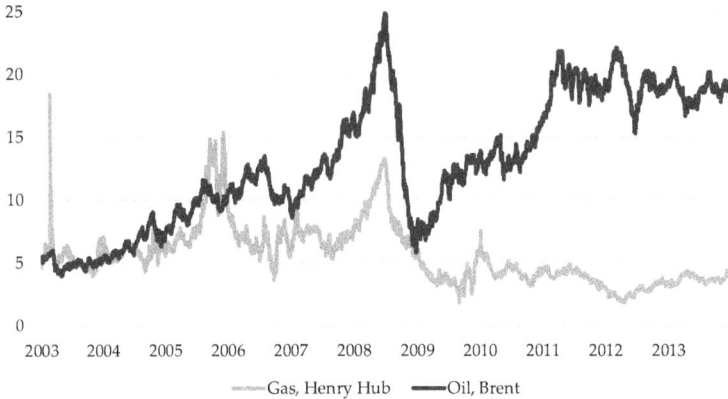

**FIGURE 15.4:** US gas versus global oil prices, 2003–2013 ($/MMBtu)

because oil prices—to which LNG contracts were tied—were too low. They too had plummeted after the 2008 financial crisis, falling from a high of over $100 per barrel in the summer of 2008 to under $50 just a few months later. Unlike gas, however, oil prices recovered fairly quickly, and by the end of 2009 were back above $75, while natural gas prices continued to slide. The widening gap between global oil and domestic gas prices, and the promise of growing supplies, made the economics of US LNG exports increasingly attractive.

The Yiddish word *chutzpah* is perhaps best defined through example, three of which follow: (1) After asking for a raise and promotion, a junior employee is given an important task by a superior. He says that he has a date tonight, but that he will get to the task tomorrow. He thanks his superior for the trust, and says that he is grateful for the raise. (2) A young man, meeting his girlfriend's parents at their house for the first time, is offered something to drink. He asks what kind of wine the parents have. Being shown the selection, the young man suggests that they open the bottle of Margaux, which was a ten-year anniversary present, saying that it is a special occasion, after all. (3) A businessman who has just raised more than a billion dollars to import a commodity tells his investors that he has lost all their money—that commodity will not be imported, after all. But he wants to share with them a great new investment opportunity: for ten billion more dollars, he can export the same commodity. Such was the chutzpah of Charif Souki.

The economics of exporting US liquefied natural gas, however, were compelling enough to overcome the irony. Gas demand was growing globally, and the United States was seen as an attractive alternative both to traditional supplies—such as Qatar, Algeria, and Malaysia, and newer entrants—such as Russia and Australia. Geopolitical and security risks were low, and Charif was ready to offer something that no other supplier would: a Henry Hub–indexed sales price. Since time immemorial, LNG had been bought and sold via long-term, oil-indexed contracts, chiefly because the fuels were substitutable; oil was a large market where prices were transparent, ensuring that neither side was getting ripped off by the other; and, later, because future oil prices could be hedged. Cheniere's sales pitch was totally different from that of a traditional LNG supplier: the company would buy gas in the United States and sell it abroad at a markup, with costs plus a small margin passed through. Buyers were thus getting not only the security and low geopolitical risk of buying from the United States, but price diversity.

In April 2010 Charif pitched to his board the idea of converting Sabine Pass to an export terminal.[23] At $8 billion for construction costs alone, the price tag was high, but it was lower than building from scratch—the tanks, berths, and pipelines were already in place, and all that was needed were four, giant liquefaction trains to supercool the gas. These first four trains would allow the facility to export around three billion cubic feet per day of natural gas. In June 2010 Cheniere went public with the idea.

The response was muted. Most of Cheniere's existing institutional investors had already accepted that they would not be getting their money back, and some on Wall Street publicly ridiculed the export idea. Even if exports did make long-term economic sense, which was far from clear, would the government allow them? Oil and gas exports had been allowed only on a case-by-case basis for the prior several decades, which did not matter much, as the United States was an importer. Plus the Obama administration was walking a tightrope on energy policy. Natural gas was good inasmuch as it was displacing coal and oil for domestic consumption—it was harder to justify exporting the fuel abroad, which would increase prices at home and make gas less competitive with coal—to President Obama's progressive, climate change–wary base. How could Cheniere Energy, a single-asset company that was slowly shedding cash, and was seen as unlikely to make a billion-dollar debt payment due in a few years, muster the political clout to obtain an export license? Better to leave such feats to the likes of ExxonMobil, Chevron, and Shell.

But Charif kept pushing, traveling the world in search of buyers, hawking the superiority of US liquefied natural gas, and as he did, oil prices kept rising and US gas prices kept falling. In October 2011 Cheniere signed its first LNG sales contract with BG Group, a major LNG player that was bought a few years later by Shell. More deals followed, including with utilities in Japan, the world's largest LNG importer, and India. As importantly, regulators at both FERC and the Department of Energy approved Sabine

Pass's application to export LNG, and allowed it to export to all foreign countries, regardless of whether they had signed a free trade agreement with the United States. Charif's chutzpah paid off, and construction on Sabine Pass's export facilities began in August 2012.

Even before then, Charif had competition. Within a year of him signing a contract with BG Group, four US liquefied natural gas import terminals had decided to convert to export operations. Ninety percent of the proposed export capacity was located on the Gulf Coast, and that number grew over time, while only small amounts were proposed for the East Coast, at Cove Point, Maryland, and Elba Island, Georgia.

## REPLUMBING

Part of Charif's sales pitch were the great redundancies that the US natural gas grid offered, especially on the Gulf Coast, which is a virtual spaghetti bowl of overlapping pipelines. Whereas an incident at a more traditional, vertically integrated LNG export project, with dedicated wells, pipelines, and processing systems, could lead to a total shutdown of operations, LNG facilities on the Gulf Coast could source gas from a plethora of pipelines, thanks to the overbuilt, interconnected grid. This was true, but it was also the case that growth in supplies and growth in demand were happening in very different parts of the country.

Jeff Ventura had gotten his wish—he had turned Range Resources into a gas manufacturing company. Range turned out well after glorious well in the Marcellus Shale, and had decades of growth ahead of them. Geological risk was a thing of the past. What Ventura had not expected was that with so little risk, so much resource, and very few barriers to entry, prices would come down to marginal costs. Said more simply, without risk (or preferential regulation), there can be no reward.

The Marcellus Shale seemed like a godsend for the United States, as high gas production gave the country a cheaper fuel compared to importing countries in Europe and Asia. The kicker was that being located in the Northeast, close to demand centers, Marcellus producers actually received higher gas prices than their brethren in the Barnett, Haynesville, and Fayetteville Shales. The Three T pipelines—Tennessee, Texas Eastern, and Transcontinental—that had been built to move gas more than a thousand miles from the Gulf Coast to the Northeast were now receiving supplies in Pennsylvania, only one or two hundred miles from markets. Marcellus producers could get their gas to consumers in New York City, Philadelphia, and Boston much more cheaply than Gulf Coast producers.

It wasn't long, however, before production from the Marcellus and a neighboring formation called the Utica Shale, also in the Appalachian Basin, began to outgrow local markets, large though they were. Demand in the Northeast was rising, mostly as a result of coal power plant retirements, but this was a drop in the bucket compared to supply growth. The Marcellus and Utica produced almost nothing in 2008, but spit out ten billion cubic feet per day by 2013, twenty by 2016, and thirty by 2019. Quite suddenly, the premium that Marcellus producers received turned into a discount: for the first time since the 1920s, the Northeast had too much gas. The Three T's began to do the unthinkable, and reverse—backhaul, in pipeline-speak—sending gas from north to south, toward Sabine Pass.

Before 2014, basis—the difference between prices at some regional hub and the benchmark price at the Henry Hub—was not systemically important. Basis was positive in the winter in congested market areas, like New York City and Boston, and negative in production areas, like Canada and the Rockies, but the best traders—like John Arnold, and, for a time, Brian Hunter, traded flat price—NYMEX Henry Hub contracts. As Appalachian production surpassed the capacity of the local pipeline network, basis emerged as an issue almost as important as the flat price. Henry Hub prices might be $3, but producers in the Marcellus were receiving $1.50, or even less. For the first time in several decades, new, long-haul pipelines needed to be built.

Technically, this did not present a challenge, as even a major new gas pipeline can be built in about a year, depending on terrain and specifications. However, building energy infrastructure in the Northeast proved to be a more daunting political and legal issue than developers anticipated, largely because natural gas had fallen out of favor with a growing and increasingly litigious environmental movement.

In 2007, when shale producers were still beloved by Wall Street, Aubrey and other Chesapeake employees began donating money to the Sierra Club, the largest and best-known environmental organization in the United States. This was a strange fellowship from the beginning, but Aubrey and the Sierra Club shared a common enemy—coal—and the donations were made specifically to the Sierra Club's Beyond Coal campaign, which aimed both to prevent any new coal plants from being built and hasten the retirement of existing plants. *The enemy of my enemy is my friend*, was the thought at the Sierra Club, and it was shared by many environmentally minded individuals across the country.

In early 2010 the Sierra Club's new executive director, Michael Brune, reexamined the relationship between his organization and the gas industry, and made the decision to stop accepting donations from Aubrey, Chesapeake, and, indeed, any oil and gas producer. "It's time to stop thinking of natural gas as a 'kinder, gentler' energy source," Brune wrote. "As we phase out coal, we need to leapfrog over gas whenever possible in favor of truly clean energy."[24] Gas was a fossil fuel, after all, and burning it released carbon dioxide into the atmosphere, although only about half as much as coal.

Not long after the Sierra Club stopped accepting Aubrey's money, I attended a meeting with representatives from the Sierra Club and the Union of Concerned Scientists, a science advocacy organization strongly in favor of regulations to stem climate change. The meeting was technical in nature—we discussed which coal plants were the most susceptible to retirement, based on what environmental equipment they had already installed, and whether they could afford to comply with upcoming emissions rules. On the elevator ride down to the lobby, after the discussions, I asked how the Union of Concerned Scientists' representatives felt about shale gas, as it was certain to represent a growing part of the US energy mix going forward. The answer was a single word: "Conflicted."

By 2012 environmentalists had turned on natural gas. It was no longer a bridge to a renewable future—it was a bridge to nowhere—and fracking had become an ugly word. In May 2012, ten years after launching the Beyond Coal campaign, which was funded in part by donations from natural gas producers, the Sierra Club began a new program: Beyond Natural Gas.

New Jersey had banned fracking in 2011, although New Jersey has no oil and gas deposits, so the move was largely symbolic. The governor of New York's decision to impose a similar ban in December 2014 was more impactful. The Empire State sat on top of hundreds of thousands of acres of Marcellus Shale, and, indeed, the formation had gotten its name from an outcrop near the town of Marcellus, in upstate New York. That New York's Marcellus acreage was not of prime quality was beside the point—this ban represented a clear win for the environmental lobby and a shot across the bow for the industry.

Nationally, however, efforts to ban fracking were going nowhere, for several reasons. First, the states regulate surface and permitting activities, and gas production taxes were filling their coffers. Many pundits argued superficially that lawmakers in "new," and sometimes politically progressive producing states, such as Pennsylvania and Ohio, would be conflicted with regards to the gas industry, and fracking in particular, and thus amenable to imposing additional regulations. The pundits seemed to forget that Pennsylvania and Ohio had been producing oil and gas for over a century, and were quite familiar and comfortable with mineral extraction, although admittedly never before on the scale of how the Marcellus and Utica were being developed in the 2010s. Even if lawmakers had been leery of producers, the US economy—and the economies of Pennsylvania and Ohio—were in the midst of a deep recession in the late 2000s and early 2010s, and energy was a rare bright spot. While moving to curtail oil and gas drilling might have been politically popular just a few years earlier, it was now certain to be viewed as overwrought, job-killing regulation. Third, and most directly, the Energy Policy Act of 2005 explicitly banned fracking from being regulated by the Environmental Protection Agency (EPA) under the Safe Drinking Water Act, unless diesel was used in the fluid, which it is generally not. This provision became known as the Halliburton loophole, as it is believed to have been inserted into the legislation by Vice President Dick Cheney, who was formerly the CEO of Halliburton.

Stymied upstream, where they could not find purchase, environmentalists took the fight downstream, reasoning that they could eliminate the problem of supply—too much fracking and fossil fuel extraction—by cutting off demand. The Sierra Club and several other organizations opposed the sole LNG export project in the Northeast, at Cove Point, Maryland, filing a motion for rehearing just weeks after FERC authorized the construction of the terminal, on the grounds that the agency had neglected its responsibility to conduct a thorough environmental review of the project. The groups argued that FERC needed to consider the environmental impact not only of the export facility itself but also of the drilling and production activities that would be required to feed it with gas.[25] FERC interpreted its regulatory responsibility narrowly, while the Sierra Club asked it to set energy policy more broadly.

Once again, the environmental lobby was disappointed, as FERC denied the request for rehearing, declaring that "future Marcellus Shale production is not an essential predicate for the Cove Point Liquefaction Project," and that "development of the Marcellus Shale region will likely continue regardless of whether Cove Point Liquefaction Project is approved."[26] While FERC's arguments were not logically ironclad—there are many legal precedents where the totality of circumstances hold sway over the narrow issue at hand, and where causal links are established between processes that are much more tangled than the production, transportation, and export of natural gas—the agency was playing it safe. Had it agreed with the environmental groups that the decision to authorize exports from Cove Point should be considered only after a full, wellhead-to-tanker environmental analysis, it would be giving the groups precedent to resist nearly any infrastructure project on similar grounds. Not only that, conducting the analysis would be a fraught task, akin to what FERC's predecessor, the Federal Power Commission, faced in trying to regulate producer prices after the 1954 *Phillips* decision. How much gas would each well produce? What percentage would escape, venting into the atmosphere as methane? How much diesel would be used by each drilling rig and completion crew? Where were the tankers coming from, and how much fuel had they burned coming to Cove Point? Where would they go afterward? FERC knew it could not assume such a broad regulatory role, and wisely abnegated responsibility.

Environmentalists were undeterred, and as new pipeline projects were proposed to transport extremely discounted Marcellus gas out of the region, they thought they had finally found a winning strategy. Unlike production, interstate pipelines were federally regulated, requiring FERC to issue a positive environmental impact statement and grant a construction license. They were also usually regulated by states, as any stream or river crossings made pipelines subject to the Clean Water Act, which gives states the right to issue or deny water permits. Last, pipelines naturally stirred up more grassroots opposition than point sources, like terminals or plants. They crossed hundreds of miles of land, most of whose owners were not being made rich by the hydrocarbons under their feet, and many of whom were angry at pipeline companies' tactics of offering lowball payments for easements while brandishing the threat of condemnation via the eminent domain rights that the Natural Gas Act granted them. At the sixty-acre Cove Point LNG site, a few neighbors might complain about the noise during construction, whereas hundreds or even thousands of landowners might protest a pipeline that crisscrossed scores of townships, counties, and states.

Challenges to new pipeline projects came from all directions. Most vocal were landowners whose land was being crossed, but as the issue of climate change became more well-known, generally accepted, and important, even landowners who were not being directly affected by pipelines began concerted opposition efforts, aided by the ease of communicating and organizing via social media. Formal requests for rehearing and lawsuits were also filed by the likes of the Sierra Club, but these were less effectual than the thousands of protests by landowners.

Pipeline companies' responses to the more-organized-than-usual opposition varied. Some were well-versed in the ways of the crowded and more environmentally sensitive Northeast, and worked earnestly with local communities to ease their concerns. Others, accustomed to building infrastructure in the Southwest, where concerns were muted and state and local governments were friendly, took a downright combative tack.

One such combative developer was Energy Transfer, whose Rover Pipeline project serves as a prime example of the unanticipated struggle to get surplus gas out of Appalachia and move it to markets where it was needed.

The company that would build Rover—Energy Transfer—was founded in 1995 by Ray Davis and Kelcy Warren as a small midstream company that owned two hundred miles of intrastate pipelines in East Texas. It grew rapidly over the mid-2000s, after Enron's bankruptcy allowed it to pick up some prize assets on the cheap, and then prospered as its pipelines carried more and more volumes out of the Barnett Shale. The company branched out into the federally regulated, interstate market when it purchased

Transwestern Pipeline—another former Enron asset—in 2006, and continued to grow both organically and via acquisitions over the next several years, until it was a heavyweight in the midstream industry. In 2007 Ray Davis stepped down as CEO and became a co-owner of the Texas Rangers professional baseball team, leaving Warren in charge.

Warren had grown up in the oilfields of East Texas, where his dad was a welder. He attended the University of Texas at Arlington, but flunked out his freshman year after having a bit too much fun. He spent the next few years in the fields, fixing pipeline leaks and helping welders like his father, attending night classes at a local college in Kilgore.[27] This early failure and the years of toil that followed helped make Warren a hypercompetitive and pugnacious, succeed-at-all-costs businessman, which served him and Energy Transfer well for many years.

Throughout 2012 and 2013, as Appalachian production continued to grow, existing pipeline systems—the Three T's and others—picked off the low-hanging fruit, doing what they could to reverse flow using steel already in the ground. At first, they let the gas flow under its own pressure from north to south, and then reconfigured piping at existing compressor stations. These were low-cost ways of providing north-to-south transport using existing infrastructure, but by 2014 this was not enough. In June of that year, Appalachian basis plummeted to negative $1.50—producers in Pennsylvania, West Virginia, and Ohio were receiving half the price for their gas as were their peers in the rest of the country. The basin was constrained, and producers were competing with each other for the same capacity.

In June 2014 Energy Transfer announced that it had secured commitments from producer-shippers to build a massive, 3.25 billion cubic foot per day pipeline out of Appalachia. Two, forty-two-inch pipelines would pick up gas from southwest Pennsylvania, northern West Virginia, and eastern Ohio, and take it to Michigan, Canada, and even the Gulf Coast, the latter possible by using some empty space on an existing system. Rover, as the new pipeline was called, was double the size of the other greenfield projects being designed to relieve producers in the Marcellus and Utica shales.

Over the four long years that it took to complete the pipe, which was delayed by numerous violations and deficiencies, Rover and two other Energy Transfer projects did more to damage the pipeline industry's reputation than the Sierra Club could possibly have wished for. Even pro-development agencies and individuals were shocked by how brazen and cavalier the company could be, and how it seemed accountable to nobody. The following pages show the downside of allowing pipelines to wield the power of eminent domain, and perhaps that their construction ought to be regulated more strictly. However, they also show that the amount of effort required to build a FERC-regulated, interstate pipeline has increased dramatically since the last time a major infrastructure build-out had occurred, decades earlier, in the postwar era.

Rover began by going through a somewhat informal pre-filing process with FERC, then submitted its formal application in February 2015. Almost immediately, thousands of public comments began pouring in. The many interventions, along with the project's large scope, meant that FERC would naturally be slow to grant a construction license. As the months ticked by, progress was nonexistent. Already by November 2015, Energy Transfer and the Rover's shippers began to grow concerned and petitioned FERC to act quickly to authorize the project by the summer of 2016 so that they would have ample time to start construction works and, importantly, have the entire winter of 2016–2017 to clear trees. The winter tree-clearing was a particularly important deadline, as Rover's path traversed forests that are the summertime nesting areas of two species of bat, one on the endangered species list, and the other listed as threatened. Tens of thousands of trees needed to be cleared between October and March: if a single one remained on April 1, blocking Rover's path, the entire project would need to wait until the next winter.

FERC issued its final environmental impact statement in July 2016, after almost a year and a half, at which point it was required to wait for a minimum of thirty days before authorizing the project. But again, the agency was flooded with hundreds of comments and protests. It might have acted sooner to authorize construction, but a disgruntled resident who had been following the project noticed something that ended up causing Rover a great deal of headaches. The company had bought a plot of land as a site for one of their compressor stations on which there was an 1843 home of some historical note. The Ohio State Historic Preservation office had earmarked the so-called Stoneman House, named after the family for whom it was built, for inclusion on the National Register of Historic Places, and was concerned that a compressor station would sully the picturesque setting. Rover discussed the house with the Preservation Office, and told both it and FERC that the company was committed to a solution that "results in no adverse effects to this resource."[28] Shortly thereafter, Rover tore down the house.

Worse even than bulldozing the Stoneman House, Rover had been caught lying: on June 15 it had written the Preservation Office to tell them that they intended to remove the house, when in fact they had done so two weeks prior—the disgruntled neighbor's photographs proved it. With egg on its face, Rover ended up committing to pay $1.5 million for preservation works across Ohio—a trivial sum for a $4 billion project. However, the incident cost Rover more time while FERC decided on an appropriate response (indeed, the delay itself was probably part of the penalty). In the end, FERC ended up authorizing the project, but only on February 2, just eighty-eight days before the end of tree-clearing season. Moreover, FERC did not grant Rover a blanket construction certificate, saying that although "the benefits that the Rover Pipeline Project . . . outweigh any adverse effects on existing shippers, other pipelines and their captive customers, and on landowners and surrounding communities . . . because of Rover's intentional demolition of a house that was identified as eligible for listing in the National Register of Historic Places . . . we deny Rover's request for a blanket certificate."[29] This meant that Rover would have to ask permission for every new construction phase of the project and wait for FERC's blessing.

Rover quickly went to work, felling trees at an incredible pace, enlisting every man in Appalachia with a chainsaw—of which there are many—as a contractor. At least one local man went to the hardware store, bought a dozen or so chainsaws, enlisted the help of a few local high schoolers and college students, bid on the labor, and won. Rover successfully finished clearing trees before the bats returned from their underground hibernacula in April.

In February 2017 Rover also began the process of drilling under rivers. The preferred method of crossing large water bodies or important roads is to drill underneath them, horizontally, in a process called horizontal directional drilling. While the drilling is happening, mud is pumped through to reduce friction and carry debris out of the hole, just as it is while drilling an oil and gas well.

In April 2017 drilling mud began seeping up from the ground into protected wetlands near the Tuscarawas River in Ohio. It turned out that the mud that Rover had been using to drill under the river was migrating underground, and two million gallons of it had found a way out. Pictures of contractors wielding giant Shop-Vacs slurping up brown mud amidst an otherwise pristine forest found their way onto local newspapers across Appalachia, and the Ohio EPA demanded that Rover stop drilling and address how such a large volume of mud had made its way into the wetlands. Rather than evoke contrition, Rover claimed that the Ohio EPA had no jurisdiction in the matter, and carried on with construction works.

FERC did not countenance this slap in the face of a fellow regulatory agency, and in May ordered the pipeline company to cease drilling operations at certain sites. While Rover tried to rejigger its construction schedule, the unalterable reality was that it had to cross several rivers and streams, some of them quite large. The horizontal drilling process can take several months to complete, and a pipeline that is 99 percent complete may as well be 0 percent complete—it is an all-or-nothing asset.

The next few months devolved into a shouting match between Rover and the Ohio EPA, with FERC stepping in every so often to either allow or disallow construction. For its part, the Ohio EPA accused Rover of "systemic and unprecedented noncompliance,"[30] with the director writing that while he understood the significance of the project and acknowledged his state's support for oil and gas development, he could "not explain how disappointed [he was] with the continued trend of Rover causing environmental damage in Ohio."[31] Again, Rover chose not to take the high road, instead accusing the Ohio EPA of "spreading false information and innuendo."[32] When the agency charged that it had detected above-acceptable levels of toxic chemicals in the Tuscarawas River, the company wrote back that the chemicals were there before Rover came, and chided the agency for ignoring "the unfortunate historical truth that the Tuscarawas was used for decades as an industrial waste site."[33]

Eventually, Rover steamrolled over its issues in Ohio, but it was not long before it ran afoul of another pro–pipeline state agency, the West Virginia Department of Environmental Protection, which in March 2018 issued a cease-and-desist order after a series of small landslides where Rover had recently laid pipe. Rover had either not installed or improperly installed so-called erosion-control devices to allow rainwater to drain cleanly, and heavy rains caused damage not only within Rover's prescribed rights-of-way but outside of them, onto private land.[34] Again, without apology, Rover rushed to fix the immediate issue. It quickly refilled and stabilized the earth about the pipeline, but did not make good on its promises to revegetate the areas it had dug up, as FERC had ordered. It began flowing gas anyway, knowing that FERC would probably not shut down operations for such a minor breach of good faith. After all, it had strong-armed its way through whatever other opposition it had encountered, with mostly successful results.

Ohio and West Virginia are energy-infrastructure-friendly states, and FERC is a development-friendly agency. Rover had made enemies of all three, tarring the reputation of the entire industry in the process. The aforementioned events prove as much, but, in Rover's defense, they also show how much more difficult building pipelines had become over the prior few decades. In the 1940s and 1950s, neither bats, nor mud, nor landslides would have warranted the ire of regulatory agencies, if they were noticed at all. When full service on Rover finally began in November 2018, more than four years after the pipeline received its final investment decision, it was a Pyrrhic victory for Energy Transfer, and a defeat for the pipeline industry in the Northeast as a whole. Future projects would face a much more stringent process, with judges and regulatory bodies who had learned through experience not to trust pipeline companies. As of 2020, there are four major new pipeline projects in the Northeast. All have suffered multiyear delays, and some may never be completed, despite the billions having been spent, and the availability of practically limitless gas supplies.

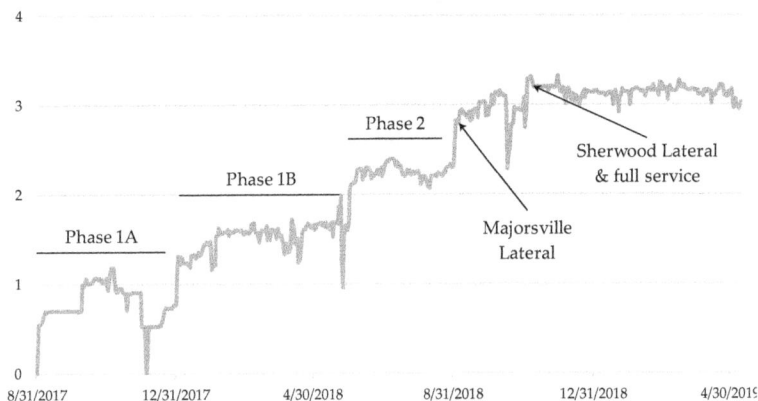

**FIGURE 15.5:** Flow of gas on the Rover Pipeline (billion cubic feet per day)

Rover's largest shipper was a company called American Energy-Utica (since renamed Ascent Resources), which Aubrey McClendon had formed after his ouster from Chesapeake. The pipeline helped the company grow into the largest privately owned gas producer in the United States, flowing over two billion cubic feet per day. Unfortunately, Aubrey did not live to see the pipeline come online or his fledgling company grow. On March 1, 2016, he was indicted by the US Department of Justice for rigging bids to purchase acreage below market prices. The following day, the largest investor in American Energy cut off all future funding. Hours later, Aubrey drove his car into a concrete wall at high speed and died.

Aubrey would at least have enjoyed seeing the first few LNG export cargoes leave Sabine Pass, which they began to just one month before his untimely death. Charif Souki, though, enjoyed it even more. In 2014 the anti-Aubrey had been the highest paid CEO in the United States, earning a staggering $142 million, and millions more in the years that followed. As time went on, another, and then another liquefaction train at Sabine Pass went into operation, sucking more and more gas out of the Northeast.

Despite the exports, gas prices remained low, both at the Henry Hub and in Appalachia, and the stocks of gas-focused E&Ps continued to be sold off. Most of this was due to their consistently negative free cash flow—E&Ps had all bet their futures on gas prices of at least $3, a level that they attained only rarely through 2020. Environmentalists, aided unknowingly by Kelcy Warren and Rover, have certainly retarded progress on many major pipeline projects, but enough squeaked through to keep the market glutted, and the new LNG export facilities well supplied. Between June 2014—when Energy Transfer secured commitments from shippers and decided to build Rover, and September 2017—when it began to flow gas, Appalachian gas producers' stock prices fell by 78 percent. They continued to fall over the next few years. Mark Papa had been right when he said that shale producers had "ruined the [gas] market for twenty years" back in 2007.[35]

The fact that oil and gas producers were losing money had never stopped investors before. So why did their stock prices take such a drubbing over the last years of the 2010s?

For one, investors had other options as the United States emerged from the recession of the early 2010s—oil and gas was no longer the only real sector of the economy where yield could be found. But perhaps more importantly, the shale miracle was too miraculous—it really did transform oil and gas producers from wildcatters into manufacturers. In the pre-shale era, one in a few hundred wildcatting E&Ps would stumble upon a truly superlative reservoir and make all its investors rich. Its founders might even get Big Rich. In the post-shale era, however, oil and gas stocks are no longer lottery tickets, forcing investors to focus more on costs than production volumes.

The death of the wildcatter is a tragedy that was decades in the making. Certainly, geologists still prospect for oil and gas, but wildcatters are not simply prospectors. A true wildcatter is, above all else, a salesperson, whose sole product is the promise of immeasurable wealth. He will convince you that a great ocean of oil and gas lies underfoot, and that the people are hungry for it. His will be a generational asset; it will make your children and grandchildren rich; roads and schools and parks will bear your name; dignitaries will call upon you. There is no room in the twenty-first century for wildcatters, though, not when the only promise anyone will believe is of "low-risk, repeatable results."

## EPILOGUE

Aubrey's death was viewed as the end of an era. In a way, it was. It was as if he spoke for the entire industry when he drove his car into that concrete wall. There was no place left for the wildcatter anymore.

In another way, it was just another inflection point in the endless cycle of the market. Mark Papa had been way out front when he ordered his company to stop looking for gas and go find oil, but within a few years he was joined by every other E&P CEO, including Aubrey, Jeff Ventura, and other erstwhile true believers in the superiority of natural gas. Shale drillers went right back to doing what they did best—spending more money than they took in—only now they spent it on oil. By the mid-2010s, the United States began exporting crude oil for the first time since the 1980s, thanks to rapidly growing production from the Permian Basin of West Texas, the Eagle Ford of South Texas, and the Bakken of North Dakota. Whether the new strategy will create capital or destroy it, as the US upstream oil and gas

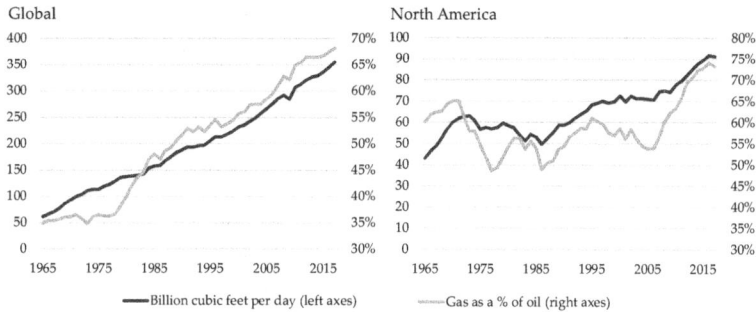

Global | North America

—Billion cubic feet per day (left axes)    -----Gas as a % of oil (right axes)

**FIGURE 15.6:** Global and North American gas consumption

industry has done so spectacularly for the past fifty years, is not my place to say. What is certain is that the technical success of shale drillers, combined with their financial missteps, have been of massive benefit to consumers of energy the world over. Energy is fairly cheap in historical terms, and cheaper still relative to what would have been the case were it not for the combination of horizontal drilling and hydraulic fracturing, which is what we really mean when we say *shale*.

Natural gas is no longer a boring utility fuel. The industry is, at long last, a strategically important sector of the US and global economy, on a par with oil, and growing in importance more rapidly than its venerable older cousin. In North America gas consumption is now equivalent to almost 75 percent of oil consumption, in energy-equivalent terms, up from just 50 percent during the doldrums of the 1980s. Globally, the growth has been even more impressive, with gas consumption more than tripling since the 1970s, while oil consumption has risen only 50 percent or so.

Whether this has been a good thing for the environment is debatable. The glut of natural gas throughout the 2010s brought electricity prices crashing down, which served as a coup de grace for many older coal plants, which found themselves unable to meet ever tighter environmental standards and compete with efficient, combined-cycle gas plants. The share of electricity generated from coal in the United States sank from 45 percent in 2010 to just 27 percent in 2018—a truly astounding drop, given the size of the numbers involved. This was a clear win for the environment, especially considering that only a bit more than half of the drop in coal generation was backfilled by natural gas, with the rest going to wind and solar. Rather than wither away and die, renewable generation grow strongly in spite of dirt-cheap gas, partially because of continued government subsidies,[36] partially because competition among manufacturers and economies of scale caused the cost of wind turbines and solar panels to drop, and partially because small-scale solar did not compete with gas on the wholesale market but rather with higher retail prices, given that much of it is installed on rooftops. Perhaps as important a reason as all of the above, however, is that some corporations and individuals simply chose to pay more for clean energy, causing installations above and beyond what policy required utilities to build.

Assessing the counterfactual—what would have happened absent the commercialization of shale gas—is difficult, but it is safe to say that while the amount of additional wind and solar installations would have been higher, less coal would have retired. At a very high level, shale gas has probably been a wash, in terms of carbon dioxide emissions. Supporters will say, rightly, that it has led to much lower energy prices. Detractors will note that methane is a much more potent climate-warming gas than carbon dioxide, and that much more gas leaks or vents than the industry is either aware of or admits, which is also true, although the statistics the detractors cite are generally dubious. They will also point out that new gas infrastructure has locked us in to a future with more fossil fuels, which is debatable. If no new pipelines had been built out of Appalachia, and basis remained deeply depressed for many years, new sources of demand would pop up, just as the first gas-fired power plants were built in the Southwest during the 1920s and 1930s, to take advantage of the long-lasting glut. There is no way to keep open a wide arbitrage without pervasive regulation.

The energy executive who fell asleep, Rip Van Winkle–like, in 2000, back when Enron dominated gas trading and shale was still considered a lab experiment, would be shocked by the gas market of 2020, to be sure, but it would not take him long to adjust to the new realities. The market structure is remarkably similar—all that has changed are the names and numbers.

What he would have a much harder time wrapping his head around are the new perception of natural gas—savior to some, pariah to others, but familiar to all. Natural gas, and energy more generally, was of marginal concern to anyone outside the industry until the confluence of peak oil, climate change, and fracking. *Rolling Stone* magazine was not interviewing energy CEOs in the 1990s. Now everyone has an opinion, and social media quickly gives people whatever facts their arguments require. Energy executives are now required to be arbiters and statesmen, in addition to engineers and businessmen, roles for which many are ill-equipped, and which most assume uneasily. Aubrey was the exception in this regard, and not the rule.

In the words of Yogi Berra: "It's tough to make predictions, especially about the future." Whether natural gas is the fuel of the future, or will see its market share quickly eviscerated by an ever-growing share of renewables, or perhaps a nuclear renaissance, is hard to say. It probably will serve as a bridge to a future where renewables play a more dominant role, but the bridge is likely to be a very long one, and we are very likely still driving upward, to the crest.

It is almost as difficult to place the events of the present in an historical context as it is to predict the future. North America is now awash in natural gas and oil, which has turned the energy conversation on its head. Rather than trying to minimize the costs of scarcity, we are trying to maximize the profits of abundance, and arguing along the way what steps should be taken to protect the earth. Everyone still agrees that oil and gas are finite, but so is sand, and we don't talk much about peak sand. Shale gas and fracking has led to a new consensus: that there is an essentially limitless supply of hydrocarbons in the world, as far as anyone currently alive is concerned, and that the ultimate constraint will probably not be driven by limited supply but falling demand.

As this book goes to press in 2020, COVID-19 is showing us what falling demand looks like in extremis. Every sailor and cyclist knows that it is harder to feel the wind when it is at your back than when it is at your face. It would have sounded unthinkable to tell oil and gas producers that they had the wind at their backs in 2019, but none would argue it now.

Shale producers have been astonishingly successful at producing oil and gas. They have been equally unsuccessful at producing profits. For several years, though, capital markets remained sympathetic to their temporary plight. The bull market in natural gas was always just around the corner: more coal plants would retire, more LNG would be exported, and declines from conventional resources would kick in, giving shale producers a larger addressable market. Meanwhile, producers' giant shift away from all but the most economical dry gas plays and into oil-prone formations had been completed by 2019. While gas was still in the doldrums, oil prices remained relatively high. Dollars and drillers headed as quickly as they could to the mighty Permian Basin in west Texas and southwest New Mexico, where there was not a silent patch of hardscrabble to be found amid the constant buzz of rigs and pumpjacks. Any producer would have admitted that it had taken them longer to turn profits than they expected, but boasted that the industry was in a stronger position in 2019 than it had been for many years.

Everything changed in early 2020. A novel strain of coronavirus, a severe and fast-spreading respiratory syndrome, quickly spread from China across the world in late winter and early spring. Economic activity, and especially travel, so important to energy demand, plummeted, as official and unofficial quarantines kept people around the world in their homes. Oil consumption quickly dropped from more than one hundred million barrels per day into the low eighties, levels not seen since the 1990s. Oil futures tanked, at one point trading well below zero. As if a once-in-a-century global pandemic were not bad enough, the preceding winter had been extremely warm across the entire Northern Hemisphere, which meant that US gas producers not only had to contend with the glut at home but saw exports drop as well, now that the United States had become a major exporter of LNG. The brand-new export facilities that were built to move $3 US gas to foreign markets that would pay $6 or more now find that there is almost no premium to be had in the rest of the world. The futures market in mid-2020 allows consumers to buy US gas for less than $3 essentially forever. The bull story is dead, and every producer can feel the wind blowing straight in his face.

In the long-term, consensus has always proven to be wrong, and there is no reason to think differently this time around. The symbiotic miracle of horizontal drilling and hydraulic fracturing is not enough to halt the commodity cycle—a shortage will return, one day—but the effect has been more than just impressive. For an energy wonk, it seems to have been almost supernatural.

All the way back in 1944, Supreme Court Justice Robert Jackson wrote that prices should be set at prevailing market conditions, rather than at original costs, because "given sufficient money, we can produce any desired amount of railroad, bus, or steamship . . . But the wealth of Midas and the wit of man cannot produce or reproduce a natural gas field."[37]

For now at least, the wit of man seems to have carried the day.

# NOTES

## CHAPTER 1: THE SMOKY CITY

1. US Department of the Treasury, Bureau of Statistics, *Report on the Internal Commerce of the United States for the Fiscal Year 1881–82: Commerce and Navigation* (Washington, DC: Government Printing Office, 1884), 152.

2. Excepting one small factory built in 1792 that was no longer operating.

3. US Department of the Treasury, *Internal Commerce*, 144.

4. Reginald Arthur Mott, *The Triumphs of Coke: To Commemorate the Jubilee of the Coke Oven Managers' Association, 1915–1965* (n.p.: Coke Oven Managers' Association, 1965), 253.

5. Western Pennsylvania Conservancy and Mount Washington Community Development Corporation, *Master Implementation Plan for the Mount Washington "Emerald Link,"* November 2005, http://waterlandlife.org/assets/Section_II_word.pdf.

6. Cliff I. Davidson, "Air Pollution in Pittsburgh: A Historical Perspective," *Journal of the Air Pollution Control Association* 29, no. 12 (2012): 1037.

7. Hax McCullough and Mary Brignano, *The Vision and Will to Succeed: A Centennial History of The Peoples Natural Gas Company* (n.p.: Peoples Natural Gas, 1985), 8–9.

8. John Newton Boucher, *History of Westmoreland County, Pennsylvania*, vol. 1 (New York: Lewis, 1906), 550.

9. McCullough and Brignano, *Vision and Will*, 10–11.

10. "Gas Replacing Coal: The Great Change Which is to Take Place in Pittsburgh," *New York Times*, June 23, 1885, 1.

11. A more technically correct name for this type of manufactured gas is carburreted water gas.

12. Robert E. Schofield, "The Science Education of an Enlightened Entrepreneur: Charles Willson Peale and His Philadelphia Museum, 1784–1827," *American Studies* 30, no. 2 (Fall 1989): 21–40.

13. Christopher J. Castaneda, *Invisible Fuel: Manufactured and Natural Gas in America, 1800–2000* (New York: Twayne, 1999), 21–24.

14. Louis Stotz and Alexander Jamison, *History of the Gas Industry* (New York: Stettiner Bros., 1938), 16.

15. Stotz and Jamison, *History of the Gas Industry*, 9–10.

16. McCullough and Brignano, *Vision and Will*, 15.

### CHAPTER 2: GOING BACK TO SMOKE

1. John J. O'Connor Jr., *The Economic Cost of the Smoke Nuisance to Pittsburgh*, (Pittsburgh: University of Pittsburgh, 1913), 13.

2. Andrew Carnegie, "The Natural Oil and Gas Wells of Western Pennsylvania," *MacMillan's Magazine* 51 (1885), 212.

3. Davidson, "Air Pollution in Pittsburgh," 1038.

4. Davidson, 1038.

5. David A. Waples, *The Natural Gas Industry in Appalachia* (Jefferson, NC: McFarland, 2012), 50.

6. Castaneda, *Invisible Fuel*, 47–48.

7. McCullough and Brignano, *Vision and Will*, 10.

8. James A. Glass, "The Gas Boom in East Central Indiana," *Indiana Magazine of History*, December 2000, 314.

9. Waples, *Natural Gas Industry in Appalachia*, 157.

10. Chuck Knox of Knox Geological LLC, emails to author, August 22–24, 2016.

11. Waples, *Natural Gas Industry in Appalachia*, 187.

12. G. J. Tankersley, "The Story of Consolidated Natural Gas Company: Innovation, Ingenuity and Accomplishment," in *Newcomen Society* (Princeton, NJ: Princeton University Press, 1980), 8.

13. Tankersley, "Consolidated Natural Gas Company," 11.

14. "Akron, Ohio, Takes Its First Step Towards Manufactured Gas," *American Gas Engineering Journal* 111 (1919): 123.

15. Dan Steward, *The Barnett Shale Play: Phoenix for the Fort Worth Basin* (Fort Worth: Fort Worth Geological Society and the North Texas Geological Society, 2007), 14.

16. American Oil and Gas Historical Society, "First Texas Oil Boom," December 19, 2019, http://aoghs.org/petroleum-pioneers/texas-oil-boom/.

17. San Joaquin Valley Geology, "Famous Gushers of the World," October 12, 2015, http://www.sjvgeology.org/history/gushers_world.html.

18. Stotz and Jamison, *History of the Gas Industry*, chapter 32.

19. Harold F. Williamson, Ralph L. Andreano, and Carmen Menezes, "The American Petroleum Industry," in *Output, Employment, and Productivity in the United States after 1800*, ed. Dorothy Brady (Cambridge: National Bureau of Economic Research, 1966), 393–94.

20. Travis Mayo, "Monroe Was World's Natural Gas Capital," *News-Star*, April 8, 2016, http://www.thenewsstar.com/story/news/local/2016/04/08/monroe-worlds-natural-gas-capital/82700550/.

21. Waples, *Natural Gas Industry in Appalachia*, 116 (for West Virginia), 121 (for Appalachia).

22. Waples, 101.

23. Hope Natural Gas, *From Gas Lights to New Energy Heights* (n.p: Consolidated Natural Gas, 1998), 8–10.

24. Franklin K. Lane speech to Natural Gas Conservation Conference on January 15, 1919, published in "Plan to Save Natural Gas," *Gas Record* 17, no. 2 (1920): 63–64.

25. "Plan to Save Natural Gas," 64.

26. Steward, *Barnett Shale Play*, 21.

27. N. D. Bartlett, "Discovery of the Panhandle Oil and Gas Field," in *Panhandle Petroleum*, ed. Bobby D. Weaver (Canyon, TX: Panhandle-Plains Historical Society, 1982), http://www.pan-tex.net/usr/p/pampa-hist/dis.htm.

28. Mayo, "Monroe Was World's Natural Gas Capital."

29. Kansas Geological Survey, "Public Information Circular," December 1996, http://www.kgs.ku.edu/Publications/pic5/pic5_3.html.

30. Federal Trade Commission, *Report to the Senate on Public Utility Corporations*, Senate Document no. 92, 70th Cong., 1st sess., part 84-A (Washington, DC.: Government Printing Office, 1935), 93–95. Taken from Christopher J. Castaneda, *Regulated Enterprise: Natural Gas Pipelines and Northeast Markets, 1938–1954* (Columbus: Ohio State University Press, 1992), 25.

## CHAPTER 3: THE RISE OF THE POWER TRUST

1. Richard Moran, *Executioner's Current: Thomas Edison, George Westinghouse, and the Invention of the Electric Chair* (New York: Knopf Doubleday, 2007), xxi–xxii.

2. Robert L. Bradley Jr., *Edison to Enron: Energy Markets and Political Strategies* (Hoboken, NJ: John Wiley & Sons, 2011), 38.

3. Bradley, *Edison to Enron*, 59.

4. Bradley, 62.

5. Bradley, 73.

6. Bradley, 87.

7. Bradley, 122.

8. Werner Troesken, "The Institutional Antecedents of State Utility Regulation: The Chicago Gas Industry," in *The Regulated Economy: A Historical Approach to Political Economy*, ed. Claudia Goldin and Gary D. Libecap (Chicago: University of Chicago Press, 1994), 59.

9. Troesken, "Institutional Antecedents of State Utility Regulation," 61.

10. Troesken, 64.

11. Peoples Gas Light and Coke Company, *100 Years of Gas Service in Chicago* (Chicago: n.p., 1950), 12.

12. Troesken, "Institutional Antecedents of State Utility Regulation," 62.

13. Werner Troesken, "Regime Change and Corruption: A History of Public Utility Regulation," in *Corruption and Reform: Lessons from America's Economic*

*History*, ed. Edward L. Glaeser and Claudia Goldin (Cambridge, MA: National Bureau of Economic Research, 2006), 270.

14. Troesken, "Regime Change and Corruption," 262.

15. Gabriel Kolko, *The Triumph of Conservatism: A Reinterpretation of American History, 1900–1916* (New York: Free Press, 1962), 3.

16. Kolko, *Triumph of Conservatism*, 4.

17. Bradley, *Edison to Enron*, 119–20.

18. William D. Henderson and Richard D. Cudahy, "From Insull to Enron: Corporate (Re)Regulation After the Rise and Fall of Two Energy Icons," *Articles by Maurer Faculty*, paper 308, 2005, 49–50, https://pdfs.semanticscholar .org/1d74/73aebd21dbfeeae7dda3083fe43df62f47be.pdf.

19. Henderson and Cudahy, "From Insull to Enron," 57.

20. Federal Trade Commission, *Summary Report of the Federal Trade Commission to the Senate of the United States…on Economic, Financial, and Corporate Phases of Holding and Operating Companies of Electric and Gas Utilities*, vols. 68–69a (Washington, DC: Government Printing Office, 1934–1935).

21. Federal Trade Commission, *Summary Report*.

22. Gifford Pinchot, *The Power Monopoly: Its Make-Up and Its Menace* (Milford, CT: McFetridge, 1928), 6.

23. John N. Ingham, *Biographical Dictionary of American Business Leaders*, vol. 2 (Westport, CT: Greenwood, 1983), 948.

24. "Gas Merger," *Time*, March 31, 1930.

25. Stotz and Jamison, *History of the Gas Industry*, 387.

26. Stotz and Jamison, 348–49.

27. Glass, *Gas Boom in East Central Indiana*, 329.

28. Department of the Interior, United States Geological Survey, *Mineral Resources of the United States, Calendar Year 1901* (Washington, DC: Government Printing Office, 1902), 212.

29. Stotz and Jamison, *History of the Gas Industry*, 349.

30. Stotz and Jamison, 350.

31. Stotz and Jamison, 353.

32. Federal Trade Commission, *Report to the Senate on Public Utility Corporations*, Senate document no. 92, 70th Cong., 1st sess., part 82 (Washington, DC: Government Printing Office, 1935), 307. Taken from Christopher J. Castaneda, *Gas Pipelines and the Emergence of America's Regulatory State: A History of the Panhandle Eastern Corporation, 1928–1993* (Cambridge: Cambridge University Press, 1996), 15.

33. Castaneda, *Gas Pipelines*, 39.

34. George W. Niedringhaus, "Future of the Steel Industry in the Mississippi Valley: The Revolution Consequent on the Coking of Illinois Coal," *Executive's Magazine: The Business Journal of the Southwest*, June 5, 1922, 18.

35. Castaneda, *Gas Pipelines*, 39–40.

36. Castaneda, 21–22.

37. Castaneda, 27–28.

38. Castaneda, 37.

39. Browneller v. Natural Gas Pipeline Co., 8 N.W.2d 474 (Iowa 1943).

40. Chuck Knox of Knox Geological LLC, emails to author, August 22–24, 2016; Margaret Wynn, "Natural Gas in Indiana: An Exploited Resource," *Indiana Quarterly Magazine of History* 4, no. 1 (March 1908): 43.

41. John F. Kiefner and Cheryl J. Trench, *Oil Pipeline Characteristics and Risk Factors: Illustrations from the Decade of Construction* (Washington, DC: American Petroleum Institute, 2001), 13–16.

42. Karl T. Compton, "Biographical Memoir of Elihu Thomson," *National Academy of Sciences: Biographical Memoirs* 21, no. 4 (1939): 153.

43. Stotz and Jamison, *History of the Gas Industry*, 352.

44. Castaneda, *Gas Pipelines*, 37.

45. Castaneda, 41.

46. Castaneda, 44–45.

47. Castaneda, 35.

48. Castaneda, 45.

49. Castaneda, 51.

50. Castaneda, 54.

51. Castaneda, 53.

52. Castaneda, 64–65.

53. Castaneda, 63, 66–67.

54. Castaneda, 66–67.

55. Bradley, *Edison to Enron*, 194

56. Bradley, 199–201.

57. Bradley, 198.

58. Bradley, 202.

59. Castaneda, *Gas Pipelines*, 90.

60. John T. Smith II, *Cars, Energy, Nuclear Diplomacy and the Law: A Reflective Memoir of Three Generations* (Lanham, MD: Rowman & Littlefield, 2012), 8.

61. Castaneda, *Gas Pipelines*, 82–84.

62. Bobby D. Weaver, "Doherty, Henry Latham," *Encyclopedia of Oklahoma History and Culture*, https://www.okhistory.org/publications/enc/entry.php?entry=DO006.

63. Funding Universe, "CITGO Petroleum Corporation History," http://www.fundinguniverse.com/company-histories/citgo-petroleum-corporation-history/.

64. Charlotte Roberts, "House on Selby Property was Christy Payne's," Sarasota History Alive, http://www.sarasotahistoryalive.com/history/articles/house-on-selby-property-was-christy-payne-s/.

## CHAPTER 4: REGULATION AND REJUVENATION

1. Elizabeth Sanders, *The Regulation of Natural Gas* (Philadelphia: Temple University Press, 1981), 36.

2. Pinchot, *Power Monopoly*, 6.

3. Pinchot, 6.

4. Securities and Exchange Commission v. Electric Bond & Share Company, Southern District of New York, 18 F. Supp. 131 (1937).

5. The concept is known more formally as the Averch-Johnson effect.

6. Sanders, *Regulation of Natural Gas*, 36.

7. Sanders, 35.

8. Michael Hiltzik, *The New Deal: A Modern History* (New York: Free Press, 2011), 206–7.

9. Justin O'Brien, "Too Big to Fail or Too Hard to Remember: James M. Landis and Regulatory Design," Edmond J. Safra Center for Ethics, Harvard University, Cambridge, MA, October 27, 2014, http://ethics.harvard.edu/blog/too-big-fail-or-too-hard-remember-james-m-landis-and-regulatory-design.

10. O'Brien, "Too Big to Fail."

11. Richard H. K. Vietor, *Contrived Competition: Regulation and Deregulation in America* (Cambridge, MA: Belknap Press of Harvard University Press, 1994), 100.

12. *International Directory of Company Histories*, vol. 31 (Saint James Press, 2000), available at "CITGO Petroleum Corporation History," FundingUniverse, http://www.fundinguniverse.com/company-histories/citgo-petroleum-corporation-history/.

13. Sanders, *Regulation of Natural Gas*, 38.

14. Sanders, 49–50. "A pipeline" can actually be a collection of parallel pipelines running on the same right-of-way—many large pipelines today have four or more individual pipes, which are known as "loops."

15. Sanders, 51.

16. Herbert H. Hughes, *Minerals Yearbook 1939* (Washington, DC: Government Printing Office, 1939). Value of oil calculated as petroleum plus natural gasoline.

17. William A. Mogel and John P. Gregg, "Appropriateness of Imposing Common Carrier Status on Interstate Natural Gas Pipelines," *Energy Law Journal* 4, no. 2 (1983): 169.

18. Munn v. Illinois, 94 U.S. 113 (1876), 127.

19. Mogel and Gregg, "Common Carrier Status," 170n107.

20. Mogel and Gregg, 172.

21. Mogel and Gregg, 171.

22. *Natural Gas Act of 1938*, Public Law 75-688, *U.S. Statutes at Large 52* (1938).

23. James R. Sanders and Daniel L. Stufflebaum, *A Study of School without Schools: The Columbus, Ohio, Public Schools during the Natural Gas Shortage, Winter 1977* (Kalamazoo: Western Michigan University, 1977), http://files.eric.ed.gov/fulltext/ED151946.pdf. Columbus schools were closed February 4–25, 1977.

24. All quotes from the *Congressional Record*, August 14–19, 1937, via Sanders, *Regulation of Natural Gas*, 55–57.

25. Sanders, 39–40.

26. *Texas Almanac, 1978–1979*, 410.

27. Castaneda, *Gas Pipelines*, 18.

28. Jeff D. Makholm, *Theory of Relationship-Specific Investments, Long-Term Contracts and Gas Pipeline Development in the United States* (Boston: National Economic Research Associates, 2006), 24.

### CHAPTER 5: WARTIME PIPES AND THE POSTWAR BOOM

1. Bryan Burrough, *The Big Rich: The Rise and Fall of the Greatest Texas Oil Fortunes* (New York: Penguin, 2009), 155.

2. Texas Eastern Transmission Corporation (TETCO), *The Big Inch and Little Big Inch Pipelines: The Most Amazing Government-Industry Cooperation Ever Achieved* (Houston: n.p., 2000), 3–4.

3. "Q&A: The Man Who Succeeded Hitler," *BBC News*, May 5, 2005, http://news.bbc.co.uk/2/hi/europe/4514765.stm.

4. Stephen D. Nagiewicz, *Hidden History of Maritime New Jersey* (Charleston, SC: History Press, 2016).

5. TETCO, *Big Inch and Little Inch Pipelines*, 8.

6. "Norness," Uboat.net, http://uboat.net/allies/merchants/1248.html.

7. Burrough, *Big Rich*, 148.

8. TETC, *Big Inch and Little Inch Pipelines*, 9.

9. Christoper J. Castaneda and Joseph A. Pratt, *From Texas to the East: A Strategic History of Texas Eastern Corporation* (College Station: Texas A&M University Press, 1993), 33–34.

10. Castaneda and Pratt, *From Texas to the East*, 24.

11. Hiltzik, *New Deal*, 163.

12. Castaneda and Pratt, *From Texas to the East*, 19.

13. Castaneda and Pratt, 20.

14. Castaneda and Pratt, 22.

15. TETCO, *Big Inch and Little Inch Pipelines*.

16. Castaneda, *Regulated Enterprise*, 37.

17. Hughes, *Minerals Yearbook 1939*.

18. Bureau of Labor Statistics, "Technical Note: Labor Force, Employment, and Unemployment, 1929–1939: Estimating Methods," 51, https://www.bls.gov/opub/mlr/1948/article/pdf/labor-force-employment-and-unemployment-1929-39-estimating-methods.pdf.

19. Sanders, *Regulation of Natural Gas*, 50–51.

20. Sanders, 52–53.

21. Castaneda, *Regulated Enterprise*, 46.

22. Castaneda, 41–46.

23. Castaneda, 48–49.

24. Castaneda, 51.

25. Castaneda, 52.

26. Castaneda, 52.

27. Castaneda, 52–53.

28. Castaneda, 55–57.

29. Castaneda, 58–59.

30. Castaneda and Pratt, *From Texas to the East*, 35, 39.

31. Castaneda and Pratt, 40.

32. Castaneda and Pratt, 44–45.

33. Castaneda and Pratt, 48–49.

34. Castaneda and Pratt, 53.

35. Castaneda and Pratt, 64.

36. Castaneda, *Regulated Enterprise*, 123, 125–26.

37. Castaneda, 130.

38. Castaneda, 131. In fact, the FPC had issued a similar decision eighteen months earlier, in November 1946, to approve the Michigan-Wisconsin Pipeline over Panhandle Eastern's protests. Those two pipelines both served Detroit-area markets. However, Michigan-Wisconsin would also serve Wisconsin, which Panhandle did not.

39. Castaneda, 132–33.

40. Castaneda, 133–34.

41. Castaneda, 134.

42. Castaneda, 137.

43. Today, each of the three T's is able to deliver more than ten billion cubic feet on a peak winter day, as they have all built several parallel pipeline "loops" and operate large storage facilities.

44. Castaneda and Pratt, *From Texas to the East*, 62.

45. Castaneda, *Regulated Enterprise*, 150.

46. Castaneda, 152.

47. Castaneda, 153–54.

48. Castaneda, 155.

49. Castaneda, 156.

50. Castaneda, 157.

51. Castaneda, 162–64.

52. Stotz and Jamison, *History of the Gas Industry*, 320–22.

53. Orrin D. Kettelman Sr., "David H. Kettelman," *San Joaquin Historian* 7, no. 1 (January 1971): 1–3.

54. Stotz and Jamison, *History of the Gas Industry*, 315–16.

55. Frank Mangan, *The Pipeliners* (El Paso, TX: Guynes Press, 1977), 43.

56. Mangan, *Pipeliners*, 53.

57. Mangan, 85, 91.

58. Mangan, 117.

59. Mangan, 128–29.

60. Jacqueline Lang Weaver, *Unitization of Oil and Gas Fields in Texas: A Study of Legislative, Administrative, and Judicial Policies* (Washington, DC: RFF Press, 2011), 143.

61. David F. Prindle, *Petroleum Politics and the Texas Railroad Commission* (Austin: University of Texas Press, 1981), 61–63.

62. Prindle, *Petroleum Politics*, 64

63. Prindle, 64.

64. Weaver, *Unitization of Oil and Gas Fields in Texas*, 145.

## CHAPTER 6: REGULATION AND RUIN

1. *Natural Gas Act of 1938*, Public Law 75-688, *U.S. Statutes at Large 52* (1938).

2. *Congressional Record*, August 14–19, 1937, via Sanders, *Regulation of Natural Gas*, 56.

3. Alfred Long Scanlan, "Administrative Abnegation in the Face of Congressional Coercion: The Interstate Natural Gas Company Affair," *Notre Dame Law Review* 23, no. 2 (1948): 175.

4. Scanlan, "Administrative Abnegation," 176–77.

5. Scanlan, 175–76.

6. *Natural gas company* is the term for a company to whom the NGA applies.

7. Scanlan, "Administrative Abnegation," 179, 180.

8. Scanlan, 181.

9. Scanlan, 181.

10. FPC v. Hope Natural Gas, 320 U.S. 591 (1944), 594–96.

11. FPC v. Hope Natural Gas, 602.

12. FPC v. Hope Natural Gas, 605.

13. FPC v. Hope Natural Gas, 604.

14. FPC v. Hope Natural Gas, 614.

15. Leverett S. Lyon and Victor Abramson, *Government and Economic Life: Development and Current Issues of American Public Policy*, vol. 2 (Washington, DC: Brookings Institution, 1940), 692.

16. Lyon and Abramson, *Government and Economic Life*, 691.

17. Colorado Interstate v. FPC, 324 U.S. 581 (1945), 601.

18. Colorado Interstate v. FPC, 602–3.

19. Castaneda, *Regulated Enterprise*, 60.

20. FPC v. Hope, 629.

21. Colorado Interstate v. FPC, 612.

22. Burrough, *Big Rich*, 137.

23. Burrough, 135.

24. Scanlan, "Administrative Abnegation," 182.

25. Scanlan, 185.

26. Scanlan, 187.

27. Scanlan, 188–89.

28. Scanlan, 190–91.

29. Scanlan, 191–93.

30. Mr. Rizley on the floor of the House, Cong. Rec., 80th Cong., 1st Sess., March 10, 1947. See Scanlan, "Administrative Abnegation," 192n79.

31. Bradford Ross and Bernard A. Foster Jr., "Phillips and the Natural Gas Act," *Law and Contemporary Problems* 19 (Summer 1954): 387–88.

32. Interstate v. FPC, 331 U.S. 682 (1947), 690.

33. Interstate v. FPC, 692.

34. *Billings* was an FPC ruling substantially similar to Columbian Fuel, reaffirming the FPC's stance that it had no jurisdiction over arm's-length sales by independent  producers.

35. Richard H. K. Vietor, *Energy Policy in America Since 1945* (New York: Cambridge University Press, 1984), 76.

36. Ross and Foster, "Phillips and the Natural Gas Act," 393–94.

37. Sanders, *Regulation of Natural Gas*, 84.

38. Vietor, *Energy Policy*, 78.

39. "Natural Gas Amendments," *CQ Almanac* 1950, 6th ed., 598–602, Washington, DC, Congressional Quarterly, 1951, http://library.cqpress.com/cqalmanac/cqal50-1375676.

40. Harry S. Truman, *1950: Containing the Public Messages, Speeches, and Statements of the President, January 1 to December 31, 1950* (Washington, DC: Government Printing Office, 1951), 257.

41. Castaneda, *Gas Pipelines*, 92.

42. Castaneda, 113–14.

43. H. R. Rep. No. 709, 75th Cong., 1st Sess., 1–2, April 28, 1937.

44. Castaneda, *Gas Pipelines*, 112.

45. Castaneda, 115.

46. Castaneda, 111.

47. "ANR Pipeline Company," Funding Universe, http://www.fundinguniverse.com/company-histories/anr-pipeline-co-history/.

48. Phillips Petroleum Co., 7 FPC 983 (1948).

49. Edward T. James and Robert Livingston Schuyler, "Leland Olds," in *Dictionary of American Biography, Supplement 6: 1956–1960* (New York: American Council of Learned Societies, 1980).

50. Operations began on November 1, 1949.

51. Castaneda, *Gas Pipelines*, 146.

52. Castaneda, 147.

53. Wisconsin v. FPC, 205 F.2d 706 (D.C. Cir. 1953).

54. Castaneda, *Gas Pipelines*, 147.

55. Nicholas Johnson, "Producer Rate Regulation in Natural Gas Certification Proceedings: 'Catco' in Context," *Columbia Law Review* 62, no. 5 (May 1962): 784.

56. Linda Gugin and James E. St. Clair, *Sherman Minton: New Deal Senator, Cold War Justice* (Indianapolis: Indiana Historical Society, 1997), 87.

57. Truman, "Public Messages, Speeches, and Statements," 257–58.

58. Gugin and St. Clair, *Sherman Minton*, 179, 261.

59. This and later quotes from Phillips v. Wisconsin, 347 U.S. 672 (1954).

### CHAPTER 7: STOPPING A FREIGHT TRAIN

1. US Department of Energy, Energy Information Administration, http://www.eia.gov.

2. Burrough, *Big Rich*, 219.

3. Sanders, *Regulation of Natural Gas*, 99.

4. Dwight D. Eisenhower, "Veto of HR 6645," *Pittsburgh Post-Gazette*, February 18, 1956, 2. HR 6645 was also known as the Harris-Fulbright bill.

5. Sanders, *Regulation of Natural Gas*, 105.

6. Sanders, 98.

7. Paul W. MacAvoy and Robert S. Pindyck, *Price Controls and the Natural Gas Shortage* (Washington, DC: American Enterprise Institute for Public Policy Research, 1975), 12.

8. Kenneth L. Fisher, *100 Minds that Made the Market* (Hoboken, NJ: Wiley & Sons, 1993), 218.

9. James M. Landis, *Report on Regulatory Agencies to the President-Elect*, US Senate Committee on the Judiciary, 86th Cong., 2nd sess. (Washington, DC: Government Printing Office, 1960), 54.

10. J. Landis to J. F. Kennedy, February 20, 1961, Box 78, Presidential Office File, John F. Kennedy Library, Boston, Massachusetts.

11. Hope Natural Gas Co., 19 FPC 405 (1958).

12. Johnson, "Producer Rate Regulation," 789–93.

13. Atlantic Refining Company v. NY Public Service Commission, 360 U.S. 378 (1959).

14. Vietor, *Energy Policy*, 156.

15. DrillingInfo (by subscription only), https://www.enverus.com/; Energy Information Administration, Supply and Disposition of Petroleum and Other Liquids; *Minerals Yearbook* (1965), 313.

16. Permian Basin Area Rate Cases, 390 U.S. 747 (1968), 748. The rates given are for Permian Basin wells in Texas; Permian Basin wells in New Mexico received prices one cent lower, but excluded state taxes.

17. Permian Basin Area Rate Cases, 390 U.S. 747 (1968), 747–48.

18. Vietor, *Energy Policy*, 159.

19. Prindle, *Petroleum Politics*, 110.

20. Paul Burka, "Power Politics," *Texas Monthly*, May 1975, 84.

21. MacAvoy and Pindyck, *Price Controls*, 15.

22. Robert R. Nordhaus, "Producer Regulation and the Natural Gas Policy Act of 1978," *Natural Resources Journal* 19, no. 4 (Fall 1979): 835.

23. Affiliated producers had a much greater economic incentive to search for gas reserves than did independents.

24. Vietor, *Energy Policy*, 149.

25. Recall from chapter 7 the Texas Railroad Commission's Seeligson order from 1947, prohibiting flaring of associated gas from oil wells in the Seeligson Field, which became statewide policy thereafter.

26. *Seinfeld*, "The Muffin Tops," season 8, episode 21, NBC, May 8, 1997.

27. Vietor, *Energy Policy*, 154.

28. The point here is to evaluate how much oil and gas had been produced—that is, brought forth from the ground—by various dates. Only very minor amounts of oil were wasted, meaning that measured volumes correspond very closely with actual production. Trillions of cubic feet of gas were flared or vented, but data on these volumes simply do not exist in any uniform manner before 1936, and they were mostly never measured in the first place. Therefore, we cannot know exactly how much gas has been produced in the United States. An estimate based on disparate data sources and the author's interpretation of anecdotes suggests that something like 24 trillion cubic feet had been flared or vented by 1954, so that total cumulative production would have been 145 trillion cubic feet by that date. The 2016 number would then be 1,360 trillion cubic feet, with 55 trillion cubic feet of natural gas having been flared or vented in US history.

29. Castaneda, *Regulated Enterprise*, 41.

30. Castaneda, 123.

31. L. L. Waters, *Energy to Move* (Owensboro, KY: Texas Gas Transmission Company, 1985), 225.

32. Waters, *Energy to Move*, 226.

33. *Texas Almanac, 1978–1979*, 410.

34. Joseph A. Pratt, Tyler Priest, and Christopher J. Castaneda, *Offshore Pioneers: Brown & Root and the History of Offshore Oil and Gas* (Houston, TX: Gulf, 1997), 24.

35. United States v. Louisiana et al., 363 U.S. 1 (1960), II-125.

36. United States v. Louisiana et al., II–127.

37. Price Daniel, "Tidelands Controversy," *Handbook of Texas Online*, June 15, 2010, http://www.tshaonline.org/handbook/online/articles/mgt02.

38. "Submerged Lands Act." In *CQ Almanac* 1953, 9th ed., 09-388-09-396. Washington, DC, Congressional Quarterly, 1954.

39. Exxon Corporation, *Middle East Oil* (Washington, DC: Government

Printing Office, 1980), 7; Railroad Commission of Texas, "Crude Oil Production and Well Counts (Since 1935)," last update July 25, 2019, http://www.rrc.state .tx.us/oil-gas/research-and-statistics/production-data/historical-production-data/ crude-oil-production-and-well-counts-since-1935/; BP Statistical Review of World Energy, 2019, https://www.bp.com/en/global/corporate/energy-economics/statisti cal-review-of-world-energy.html.

40. US Department of Transportation, Bureau of Transportation Statistics, Historical Air Traffic Statistics, https://web.archive.org/web/20170702072400/ https://www.rita.dot.gov/bts/sites/rita.dot.gov.bts/files/subject_areas/airline_infor mation/air_carrier_traffic_statistics/airtraffic/annual/1954_1980.html.

41. Daniel Yergin, *The Prize: The Epic Quest for Oil, Money & Power* (New York: Free Press, 1991), 494; Energy Information Administration, Domestic crude oil first purchase price by area.

42. Yergin, *Prize*, 517.

43. Energy Information Administration, Supply and Disposition of Petroleum and Other Liquids.

44. Yergin, *Prize*, 473.

45. Yergin, 476. Allowances were eventually raised, but only after Standard of New Jersey and other majors increased the price they paid to Texas producers.

46. Robert Caro, *Master of the Senate: The Years of Lyndon Johnson* (New York: Random House, 2009), 413.

47. Yergin, *Prize*, 520.

48. *Texas Almanac, 1978–1979*; *BP Statistical Review of World Energy*, 2019.

## CHAPTER 8: THE END OF ABUNDANCE

1. L. C. Lawyer, Charles C. Bates, and Robert B. Rice, *Geophysics in the Affairs of Mankind* (Tulsa, OK: Society of Exploration Geophysics, 2001), 81.

2. Lawyer, Bates, and Rice, *Geophysics*, 83.

3. El Paso Natural Gas v. FPC, 281 F2d 567 (1960).

4. Vietor, *Contrived Competition*, 110–11.

5. *Minerals Yearbook*, 1970, 743.

6. *Minerals Yearbook*, 1970, 846–47.

7. Sanders, *Regulation of Natural Gas*, 154.

8. Vietor, *Energy Policy*, 159.

9. Vietor, 224.

10. Sanders, *Regulation of Natural Gas*, 127.

11. US Department of the Interior, Office of Economic Analysis, *Deregulation of Natural Gas Prices, Final Environmental Impact Statement* (Washington, DC: Government Printing Office, 1974), 1–9.

12. W. M. Jacobs to Lee White, December 16, 1968, reprinted in US Congress, House, *Natural Gas Act of 1971*, 139. From Vietor, *Energy Policy*, 160–61.

13. Sanders, *Regulation of Natural Gas*, 146–48.

14. Eileen Shanahan, "FPC Chief Urges Price Rise for Gas," *New York Times*, September 21, 1969, 67.

15. Vietor, *Contrived Competition*, 116.

16. The Advance Payment Program was contained in FPC Order 410.

17. Public Service Commission of New York v. FPC, 467 F2d 361 (D.C. Cir. 1972).

18. Elmer E. Staats, *The Advance Payment Program: An Uncontrolled Experiment* (Washington, DC: Government Printing Office, 1978), 1.

19. Order 428, 45 FPC (1971).

20. Vietor, *Energy Policy*, 279.

21. FPC v. Texaco, 417 U.S. 380 (1974).

22. Order 455, 48 FPC (1972).

23. Nordhaus, *Producer Regulation*, 838.

24. While the optional program, part of FPC Order 455, was never actually overturned in court, the FPC grew to dislike it, and abandoned it after approving about a dozen contracts utilizing the policy. See Vietor, *Energy Policy*, 281.

25. Vietor, 276.

26. Texas State Library and Archives Commission, "Hazardous Business: Industry, Regulation, and the Texas Railroad Commission," August 2011, https://www.tsl.texas.gov/exhibits/railroad/power/page1.html.

27. Texas Railroad Commission, "General Market Demand Order For All Oil Wells in the Various Districts and Fields In The State Of Texas," Docket no. 20-61, 640, March 16, 1972.

28. US Department of Energy, Energy Information Administration, http://www.eia.gov.

29. Yergin, *Prize*, 570.

30. Vietor, *Energy Policy*, 282.

31. Vietor, 290.

32. James Earl Carter Jr., *Public Papers of the Presidents of the United States, Jimmy Carter, 1977, Book I-January 20 to June 24, 1977*, Washington, DC: Government Printing Office, 1977, 21.

33. Erin L. Nissley, "Local History: 1977 Brought Frigid Temperatures, Natural Gas Shortage," *Times-Tribune* (Scranton, PA), September 13, 2015.

34. Sanders, *Regulation of Natural Gas*, 154.

35. This and all preceding quotes and figures in this subchapter taken from Paul Burka, "Power Politics," *Texas Monthly*, May 1975.

## CHAPTER 9: FROM SCARCITY TO SURPLUS

1. Henceforth, I will use FERC instead of FPC, even if I am referring to actions that were taken when the agency was still named the FPC, in order to avoid confusion.

2. Vietor, *Contrived Competition*, 146.

3. Sanders, *Regulation of Natural Gas*, 187.

4. Sanders, *Regulation of Natural Gas*, 187–88.

5. Sanders, 179–82.

6. William A. Mogel and William R. Mapes Jr., "Assessment of Incremental Pricing under the Natural Gas Policy Act," *Catholic University Law Review* 29, no. 4 (Summer 1980): 780–81.

7. US Department of Energy, Energy Information Administration, "An Analysis of Post-NGPA Interstate Pipeline Wellhead Purchases," DOE/EIA-0357, September 1982.

8. Vietor, *Contrived Competition*, 137.

9. Department of Energy, Energy Information Administration, "Crude Oil and Natural Gas Exploratory and Development Wells," http://www.eia.gov/dnav/pet/pet_crd_wellend_s1_m.htm.

10. Tennessee Gas Pipeline was launched as a means to monetize gas reserves near Corpus Christi that had been laying fallow for lack of a market. Transcontinental sourced gas from the Rio Grande Valley, about 150 miles south of Corpus Christi. Texas Eastern bought its gas from United (formerly EBASCO), which owned reserves in Louisiana. Meanwhile, expansions of El Paso into Arizona and California took gas from the Permian Basin in West Texas, many hundreds of miles from the other basins, while Panhandle Eastern, NGPL, and Colorado Interstate continued to sources from the massive Panhandle and Hugoton fields.

11. *Minerals Yearbook*, various years.

12. Vietor, *Energy Policy in America*, 299.

13. Castaneda, *Gas Pipelines*, 207, 215, 232.

14. US Department of Energy, Federal Energy Regulatory Commission, *LNG Trade: Past, Present, Future (?)* (Washington, DC: Government Printing Office, 1995), v, viii.

15. Vietor, *Contrived Competition*, 130.

16. The DOE told Algeria it would not consider a price of above $4 because that would result in a delivered price of upwards of $5. We use the $5 for an apples-to-apples comparison with domestic supplies. See Vietor, *Contrived Competition*, 130.

17. Castaneda, *Gas Pipelines*, 209.

18. US Office of Technology Assessment, *Description of LNG Technology and Import System* (Washington, DC: Government Printing Office, 1977), 27.

19. LNG imports rose again in the 2000s, peaking at above two billion cubic feet per day in 2006. History was bound to repeat itself, however, as the development of US shale gas made imports uneconomical, once again leading to large financial losses on the part of those who had signed LNG import contracts.

20. Vietor, *Contrived Competition*, 117.

21. Robert Hefner III, telephone interview with author, February 10, 2017.

22. Interviews with Richard Smead, who was at Tenneco during this period.

23. Author's analysis of DrillingInfo data (by subscription only), www.enverus .com.

24. Editorial Board, "Energetic Oklahoma," *Oklahoman* (Oklahoma City), April 2, 1975, 10.

25. US Department of Energy, *Economic Regulatory Administration, Division of Power Supply and Reliability, Compliance Problems of Small Utility Systems with the Powerplant and Industrial Fuel Use Act of 1978: Volume II—Appendices*, DOE/RG-0045 Vol 2 (Washington, DC: Government Printing Office, 1981), 68.

## CHAPTER 10: THE GAS BUBBLE AND THE END OF MERCHANT SERVICE

1. Interview with Bob Gibb, March 9, 2017, who led the team that renegotiated Tenneco's contracts in the late 1980s through the 1990s.

2. Technically, pipelines are still not regulated as common carriers, but as contract carriers, as they have discretion to deny service to certain shippers. However, this is mostly a technicality—for all intents and purposes interstate pipelines have operated as common carriers since the late 1980s.

3. Michael J. Hightower, "Penn Square: The Shopping Center Bank that Shook the World, Part 1–Boom," *Chronicles of Oklahoma* 90, no. 1 (Spring 2012): 75.

4. Covey Bean, "Oklahoma's Anadarko Basin: Where the High-Rollers Are," *Oklahoman* (Oklahoma City), October 25, 1981, 19.

5. Thomas J. Lueck, "Gas Slump in Oklahoma," *New York Times*, April 26, 1983.

6. Hightower, "Boom," 83.

7. Hightower, 84.

8. Hightower, 86.

9. Michael J. Hightower, "Penn Square: The Shopping Center Bank that Shook the World, Part 2–Bust," *Chronicles of Oklahoma* 90, no. 2 (Summer 2012): 206.

10. Mark Singer, *Funny Money* (New York: Knopf, 1985), 122.

11. Federal Deposit Insurance Corporation, "Managing the Crisis: The FDIC and RTC Experience," chapter 3, Penn Square Bank, n.a., 527–29. https://www .fdic.gov/bank/historical/managing/history2-03.pdf.

12. Singer, *Funny Money*, 156.

13. Singer, 132.

14. Robert Hefner III, telephone interview with author, February 10, 2017.

15. Hightower, "Bust," 214.

16. Hightower, 137.

17. Singer, *Funny Money*, 161.

18. Hightower, "Bust," 224.

19. Lueck, "Gas Slump in Oklahoma."

20. Robert A. Hefner III, *The Grand Energy Transition* (Hoboken, NJ: John Wiley & Sons, 2009), 120.

21. US Department of Commerce, National Oceanic and Atmospheric Administration, Pacific Marine Environmental Laboratory, "El Niño FAQs," https://www.pmel.noaa.gov/elnino/faq.

22. John Wyeth Griggs, "Restructuring the Natural Gas Industry: Order 436 and Other Regulatory Initiatives," *Energy Law Journal* 7, no. 1 (January 1986): 76–77.

23. Interview with Richard Smead, March 27, 2017.

24. Jeffery M. Petrash, "Long-Term Natural Gas Contracts: Dead, Dying, or Merely Resting?" *Energy Law Journal* 27, no. 2 (July 2006): 551.

25. Griggs, "Restructuring the Natural Gas Industry," 75.

26. Interview with Richard Smead, March 27, 2017.

27. Lueck, "Gas Slump in Oklahoma."

28. Preston Oil Company v. Transcontinental Gas Pipeline Corporation, No. 294491, Div. "L," 19th Judicial District Court, Parish of East Baton Rouge, LA (December 1986).

29. Mary Jo Nelson, "Congress Changing Gears on Oil Industry," *Oklahoman* (Oklahoma City), October 31, 1989, 22.

30. Stephen Bergstrom, interview with the author, November 14, 2017.

31. Maryland People's Counsel v. FERC, 768 F.2d 450, 455 (D.C. Cir. 1985).

32. Maryland People's Counsel v. FERC, 768 F.2d 450, 455 (D.C. Cir. 1985).

33. "FERC Responds to House Energy Committee Inquiry on Proposed Take-or-Pay Recovery Policy," *Foster Report*, April 9, 1987, 2; Michael J. Manning, Kenneth R. Carretta, and James P. White, "Gas Inventory Charges: Evolving Mechanisms for Allocating the Risks and Recovering the Costs of Maintaining Gas Supply," *Energy Law Journal* 11, no. 1 (January 1990): 7.

34. "Extent of Pipeline Take-or-Pay Problem Remains Focal Point of Gas Industry Disagreement during FERC Hearing on Order No. 500," *Foster Report*, April 14, 1988, 2.

35. David J. Teece, "The Uneasy Case for Mandatory Contract Carriage in the Natural Gas Industry," in *New Horizons in Natural Gas Deregulation*, ed. Jerry Ellig and Joseph P. Kalt (Westport, CT: Praeger, 1996), 63.

36. Manning, Carretta, and White, "Gas Inventory Charges," 7–8.

37. Manning, Carretta, and White, 8.

38. The amount of deficiency charge assigned to each customer was based on how much their consumption had dropped, relative to a base year. For example, if an LDC had purchased one hundred billion cubic feet of gas from the pipeline in 1982, but only fifty billion cubic feet in 1988, it would have a deficiency of fifty billion cubic feet. If the pipeline's total deficiency was two hundred billion cubic feet, then the LDC would be responsible for one-quarter of that pipeline's take-or-pay liability. If the pipeline's total liability was $100 million, and it was absorbing 50 percent, that that means that $50 million was left to pass on to its customers.

As such, the LDC would be responsible for paying one-quarter of $50 million, or $12.5 million.

39. "Applications for Rehearing of FERC Order No. 500 Reveal Industrywide Disaffection with Proposed Take-or-Pay Crediting and Cost Passthrough Mechanisms," *Foster Report*, September 10, 1987, 7.

40. Petrash, "Long-Term Contracts," 571.

41. Manning, Carretta, and White, "Gas Inventory Charges," 11.

42. Manning, Carretta, and White, 12.

43. Some would argue that, rather than repackaged minimum bills, gas inventory charges were more akin to insurance premiums. I believe that the difference is largely semantics.

### PART III: A (MOSTLY) FREE MARKET AND ITS DISCONTENTS, 1992–2020

1. Francis Fukuyama, "The End of History?" *National Interest* 16 (Summer 1989): 16.

### CHAPTER 11: GAS BECOMES A COMMODITY

1. John Barr, interview with author, June 10, 2017.

2. Bradley, *Edison to Enron*, 347.

3. US Department of Energy, Energy Information Administration, *Growth in Unbundled Natural Gas Transportation Services: 1982–1987* (Washington, DC: Government Printing Office, 1988).

4. Department of Energy, *Growth in Gas Transportation Services*.

5. Barr interview.

6. "Natural Gas Clearinghouse Arranges First Deal Entailing Short-Term Sale of 200,000 Mcf per Day to Four Northeast Users; Other Spot Market Services Discussed at Industry Conference," *Foster Report*, October 4, 1984, 5–6.

7. Stephen Bergstrom, interview with the author, November 14, 2017.

8. Bergstrom interview.

9. Greg Lander, interview with author, January 9, 2018.

10. Bradley, *Edison to Enron*, 225.

11. Bradley, 306.

12. Bradley, 308.

13. Bradley, 339.

14. Bradley, 340.

15. Robert D. Hershey Jr., "Scrambling for Profits in a Gas Glut," *New York Times*, December 30, 1984.

16. ANR is a two-pronged system that serves the Midwest markets from both the Texas/Oklahoma Panhandle area and the Gulf Coast. It began its life in 1945 as the Michigan-Wisconsin Pipeline, which caused the suit that eventually led to the *Phillips* decision to regulate wellhead prices.

17. Bethany McLean, *The Smartest Guys in the Room: The Amazing Rise and Scandalous Fall of Enron* (New York: Penguin, 2003), 11.

18. McLean, *Smartest Guys in the Room*, 12–13.

19. Robert L. Bradley Jr., *Enron Ascending: The Forgotten Years, 1984–1996* (Hoboken, NJ: Wiley & Sons, 2018), 215.

20. McLean, *Smartest Guys in the Room*, 30.

21. Bradley, *Enron Ascending*, 180–81, 234.

22. Bradley, 36.

23. Michelle Michot Foss, "Enron and the Energy Market Revolution," in *Enron, Corporate Fiascos, and their Implications*, ed. Nancy B. Rapoport and Bala G. Dharan (Los Angeles: Foundation, 2003), 5.

24. It is important, when talking to anyone in the oil or gas industry, to pronounce it *sah-BEAN*, not *SAY-bine*, or be immediately disregarded as a know-nothing.

25. Bradley, *Enron Ascending*, 260.

26. Jagjit Yadav, interview with author, March 1, 2018.

27. Jagjit Yadav, interview with author, March 1, 2018.

28. Greg Lander, interview with author, January 9, 2018.

29. Stewart Holmes, "The Development of Market Centers and Electronic Trading in Natural Gas Markets," Federal Energy Regulatory Commission, Office of Economic Policy, June 1999, https://www.ferc.gov/legal/maj-ord-reg/land-docs/mkt-ctrs.pdf; James Tobin, "Natural Gas Market Centers and Hubs: A 2003 Update," Department of Energy, Energy Information Administration, October 1, 2003, https://www.eia.gov/naturalgas/articles/mkthubs03index.php

30. Christopher A. Bartlett, "EnronOnline: Louise Kitchen, Intrapreneur," Harvard Business School Teaching Note 302–011, August 2001.

### CHAPTER 12: THE DASH FOR GAS

1. US Department of Energy, Energy Information Administration, *Natural Gas Productive Capacity for the Lower 48 States 1980–1995* (Washington, DC: Government Printing Office, 1994), 7.

2. I place quotation marks around the word *successful* because deregulation of these industries proceeded far from smoothly. The take-or-pay crisis for gas pipelines was caused by deregulation, and there are several interesting parallels between botched deregulation in the gas market and in the banking sector, where the savings and loan crisis cost US taxpayers about $100 billion. It suffices to say that 1980s-era deregulation in both industries was pro-cyclical, rather than countercyclical, and did not create a level playing field among market participants.

3. Linda Cohen, Martha Krebs, and B. B. Blevins, *Retrospective Report on California's Electricity Crisis* (Sacramento: California Council on Science and Technology, 2002), 13.

4. Cohen, Krebs, and Blevins, *Retrospective Report*, 18.

5. Abalone Alliance, "The Diablo Canyon Timeline," http://www.energy-net.org/01NUKE/DIABLO1.HTM.

6. Cohen, Krebs, and Blevins, *Retrospective Report*, 24.

7. Narda Zacchino, *California Comeback: How a 'Failed State' Became a Model for the Nation* (New York: Thomas Dunne Books, 2016), 80.

8. Timothy Egan, "Tapes Show Enron Arranged Plant Shutdown," *New York Times*, February 4, 2005.

9. McLean, *Smartest Guys in the Room*, 268–69.

10. Mark Gimein, "Who Turned the Lights Out?" *Fortune*, February 5, 2001.

11. Federal Energy Regulatory Commission (FERC), "Final Report on Price Manipulation in Western Markets: Fact-Finding Investigation of Potential Manipulation of Electric and Natural Gas Prices," Docket PA02-2-000, March 2003.

12. FERC, "Price Manipulation in Western Markets," I-2.

13. FERC, ES-6.

14. FERC, chapter 2.

15. Joskow, "California's Electricity Crisis," 1.

16. Joseph Kahn, "Administration Leaves Power Crisis in California's Hands," *New York Times*, January 23, 2001.

17. James Sweeney, *The California Electricity Crisis* (Stanford, CA: Hoover Institution, 2002), 185.

18. Sweeney, *California Electricity Crisis*, 232.

19. Sweeney, 136, 175. The average retail rate was about $125/MWh, of which $60 was dedicated to the amortization of transmission and distribution assets. Thus, $65/MWh of the total was used to refund the utility's wholesale purchases.

20. Joskow, "California's Electricity Crisis," 20.

21. Author's estimate derived from Sweeney and Joskow.

22. Energy Information Administration, Form 411.

23. David Chase, "Combined-Cycle Development Evolution and Future," http://physics.oregonstate.edu/~hetheriw/energy/topics/doc/elec/natgas/cc/combined%20cycle%20development%20evolution%20and%20future%20GER4206.pdf

24. National Petroleum Council, "Balancing Natural Gas Policy: Fueling the Demands of a Growing Economy," vol. 2, September 2003, https://www.npc.org/reports/ng.html;

US Department of Energy, Energy Information Administration, "Natural Gas Productivity for the Lower-48 States," May 2001.

25. Baker Hughes rig count.

26. Julie Gomez, "Natural Gas Overview," February 2001, slide 21. https://slideplayer.com/slide/11543387/.

27. Bob Fryklund, "North American Gas—Changing Mix," November 9, 2006, slide 6, https://slideplayer.com/slide/4567877/.

28. McLean, *Smartest Guys in the Room*, 282.

29. Nelson D. Schwartz, "Enron Fallout: Wide, but Not Deep. Enron's Collapse Has Hurt Big American Banks. But Not as Much as a Default by Argentina Would," *Fortune*, December 24, 2001.

30. Foss, "Enron and the Energy Market Revolution," 4.

## CHAPTER 13: SPECULATION ON A GALACTIC SCALE

1. Hedge Fund Marketing Association, "Hedge Fund Assets under Management Reach New High," March 28, 2017, http://www.hedgefundmarketing.org/hedge-fund-assets-under-management-reach-new-high/.

2. Jenny Anderson, "US Regulators Grow Alarmed over Hedge Fund Hotels," *International Herald Tribune*, January 1, 2007.

3. Barbara T. Dreyfuss, *Hedge Hogs* (New York: Random House, 2013), 168.

4. "The Case for Commodities as an Asset Class," Goldman, Sachs, slide 11, https://faculty.fuqua.duke.edu/~charvey/Teaching/BA453_2006/GSCI_Strategic_June_2004.ppt.

5. Dreyfuss, *Hedge Hogs*, 85.

6. In the long run, there are only three ways of enhancing returns above the market. The first is by having access to more information, which hedge funds always try to do, with variable success. The second is by increasing risk, but this cannot be said to be free—the added risk is a measurable cost. The last is by increasing portfolio diversity. Only diversification can enhance risk-adjusted returns, assuming perfect—or, at least, equal—information.

7. "Excessive Speculation in the Natural Gas Market," Staff Report with Additional Minority Staff Views, US Senate, Permanent Subcommittee on Investigations (2007), 57.

8. "Excessive Speculation," 57.

9. One lot is 10,000 million Btu, which is enough to heat 150 average US homes for a year.

10. "Excessive Speculation," 58.

11. US Department of Energy, "Comparing the Impacts of the 2005 and 2008 Hurricanes on U.S. Energy Infrastructure," February 2009, 11, https://www.oe.netl.doe.gov/docs/HurricaneComp0508r2.pdf.

12. "Excessive Speculation," 58.

13. "Excessive Speculation," 74.

14. Dreyfuss, *Hedge Hogs*, 146–47.

15. Dreyfuss, 173.

16. "Excessive Speculation," 62–63.

17. Dreyfuss, *Hedge Hogs*, 173.

18. Dreyfuss, 121.

19. Dreyfuss, 123–24.

20. "Excessive Speculation," 73.

21. Dreyfuss, *Hedge Hogs*, 180.

22. "Excessive Speculation," 98.

23. "Excessive Speculation," 81.

24. "Excessive Speculation," 85.

25. Dreyfuss, *Hedge Hogs*, 195. The fund that closed was MotherRock, led by

veteran gas trader Bo Collins. It had suffered major losses on both May 26, and the widening of the March/April spread on July 31 increased margin requirements beyond what MotherRock could pay.

26. "Excessive Speculation," 44.

27. "Excessive Speculation," 113.

28. "Excessive Speculation," 114.

29. Dreyfuss, *Hedge Hogs*, 229.

## CHAPTER 14: SHALE

1. Boucher, *History of Westmoreland County*, 550.

2. A third characteristic, which is actually a combination of many other characteristics, is referred to as *fraccability*. To be highly fraccable, the rock should be brittle, so as to better propagate fractures, with low clay content, as clay tends to absorb water and swell, rather than fracture.

3. The process by which the "impermeable" source rock allows hydrocarbons to migrate upward is still poorly understood. Suffice it to say, however, the large-scale migration occurs over geological time scales, and by the time the hydrocarbons have reached the more easily accessible reservoir rock, the "hard work" has been done.

4. In fact, we do frack conventional reservoirs, but with much less intensity. In the following pages, I will go over the differences between short fracks, which are essentially cleanup operations, and massive hydraulic fracks, which are meant to stimulate great volumes of rock.

5. D. N. Meehan and D. W. Bossert, "Experience with One Thousand Horizontal Wells" (paper presented at the Annual Technical Meeting of the Petroleum Society of Canada, Calgary, Alberta, June 7–9, 1995).

6. R. S. Parsons and R. W. Fincher, "Short-Radius Lateral Drilling: A Completion Alternative" (paper presented at the Society of Petroleum Engineers Eastern Regional Meeting, Columbus, OH, November 12–14, 1986).

7. J. M. Dees, T. G. Freet, and G. S. Hollabaugh, "Horizontal Well Stimulation Results in the Austin Chalk Formation, Pearsall Field, Texas" (paper presented at the Society of Petroleum Engineers Annual Technical Conference and Exhibition, New Orleans, LA, September 23–26, 1990).

8. Meehan and Bossert, "One Thousand Horizontal Wells," 1.

9. Gregory Zuckerman, *The Frackers* (New York: Penguin, 2013), 49.

10. Zuckerman, *Frackers*, 52, 57.

11. "Oryx Energy Company History," Company-Histories, http://www.company-histories.com/Oryx-Energy-Company-Company-History.html.

12. Darwin P. Roberts, "The Legal History of Federally Granted Railroad Rights-of-Way and the Myth of Congress's '1871 Shift,'" *University of Colorado Law Review* 82, no. 1 (2011): 88.

13. US Department of Energy, Energy Information Administration, *Drilling Sideways: A Review of Horizontal Well Technology and its Domestic Application*, DOE/EIA-TR-0565 (Washington, DC: Government Printing Office, 1993), 10.

14. "Oryx's Hauptfuhrer: Big Increase Due in U.S. Horizontal Drilling," *Oil & Gas Journal*, January 15, 1990, 28.

15. Meehan and Bossert, "One Thousand Horizontal Wells," 2.

16. Joseph W. Kutchin, *How Mitchell Energy & Development Corp. Got Its Start and How It Grew* (Irvine, CA: Universal Publishers: 2001), 14–15, 363–66.

17. Zuckerman, *Frackers*, 21.

18. Steward, *Barnett Shale Play*, 23–26.

19. "Mitchell Energy and Development Corporation Company History," Company-Histories, http://www.company-histories.com/Mitchell-Energy-and-Development-Corporation-Company-History.html.

20. H. G. Kozik and S. A. Holditch, "A Case History for Massive Hydraulic Fracturing the Cotton Valley Lime Matrix, Fallon and Personville Fields," *Journal of Petroleum Technology* 33, no. 2 (1981): 229.

21. Bill Grieser, Jimmie Hobbs, Jeff Hunter, and Jerry Ables, "The Rocket Science behind Water Frack Design" (paper presented at Society of Petroleum Engineers Production and Operations Symposium, Oklahoma City, OK, March 23–26, 2003).

22. Steward, *Barnett Shale Play*, 31.

23. Steward, 38, 44, 52.

24. Steward, 46.

25. Steward, 47.

26. Mitchell Energy's good fortune in receiving the highest price under the NGPA came after decades of frustration. It had received low, interstate prices for several decades while its neighbors, many of whom shunned interstate sales, realized much higher intrastate prices.

27. Steward, *Barnett Shale Play*, 90.

28. M. A. Mayerhofer, M. F. Richardson, R. N. Walker Jr., D. N. Meehan, M. W. Oehler, and R. R. Browning Jr., "Proppants? We Don't Need No Proppants" (paper presented at Society of Petroleum Engineers Annual Technical Conference and Exhibition, San Antonio, TX, October 5–8, 1997).

29. Steward, *Barnett Shale Play*, 113.

30. Steward, 114.

31. Steward, 125, 129.

32. Steward, 8.

33. Steward, 168–69.

34. Steward, 161.

35. Steward, 182.

## CHAPTER 15: THE AFTERMATH

1. Anne Hodges Morgan, "Kerr, Robert Samuel," *Encyclopedia of Oklahoma History and Culture*, https://www.okhistory.org/publications/enc/entry.php ?entry=KE011

2. Zuckerman, *Frackers*, 122–125.

3. Testimony of Federal Reserve Board Chairman Alan Greenspan to the Committee on Energy and Commerce, US House of Representatives, "Natural Gas Supply and Demand Issues," June 10, 2003.

4. Zuckerman, *Frackers*, 127.

5. "Calpine Corporate History," Funding Universe, http://www.fundinguni verse.com/company-histories/calpine-corporation-history/.

6. Zuckerman, *Frackers*, 144.

7. Zuckerman, 180–82.

8. Zuckerman, 197.

9. Zuckerman, 226, 241, 224.

10. David Brown, "Explorer of the Year: Zagorski Made His Mark With the Marcellus," *American Association of Petroleum Geologists Explorer* 34, no. 5 (2013): 33.

11. Jeffery Ventura, "Range's Path to Discovery and Commercialization of the Marcellus Shale—The Largest Producing Gas Field in the United States" (adapted from presentation of lecture presented at the AAPG Annual Convention and Exhibition, Pittsburgh, PA, May 20, 2013), http://www.searchanddiscovery.com/ documents/2013/110165ventura/ndx_ventura.pdf.

12. Range Resources, "Range Expands Barnett Shale Holdings and Provides Operations Update," December 10, 2007, http://ir.rangeresources.com/ news-releases/news-release-details/range-expands-barnett-shale-holdings-and-pro vides-operations.

13. Jonathan D. Silver, "The Marcellus Boom / Origins: The Story of a Professor, a Gas Driller and Wall Street," *Pittsburgh Post-Gazette*, March 20, 2011.

14. Terry Engelder, "Unconventional Natural Gas Reservoir Could Boost U.S. Supply," *Penn State News*, January 17, 2008.

15. Chesapeake Energy, "Chesapeake Announces Haynesville Shale Discovery and Seven Other New Unconventional Discoveries and Projects; Increases Capital Expenditures to Accelerate Development," March 24, 2008, https://www.businesswire.com/news/home/20080324005847/en/Chesapeake -Announces-Haynesville-Shale-Discovery-New-Unconventional.

16. Zuckerman, *Frackers*, 265.

17. Zuckerman, 246–47.

18. Zuckerman, 297.

19. Zuckerman, 313–14.

20. Jeff Goodell, "The Big Fracking Bubble: The Scam Behind Aubrey McClendon's Gas Boom," *Rolling Stone*, March 1, 2012.

21. Chesapeake Energy, "History," https://web.archive.org/web/200812090 41628/http://www.chk.com/About/Pages/History.aspx.

22. Zuckerman, *Frackers*, 261.

23. Zuckerman, 338.

24. Michael Brune, "The Sierra Club and Natural Gas," February 2, 2012, https://www.sierraclub.org/michael-brune/2012/02/sierra-club-and-natural-gas.

25. "Request for Rehearing of EarthReports, Inc. (dba Patuxent Riverkeeper); Potomac Riverkeeper, Inc.; Sierra Club; and Chesapeake Climate Action Network," Dominion Cove Point LNG, FERC CP13-113-000, October 15, 2014.

26. "Order Denying Rehearing and Stay," Dominion Cove Point LNG, FERC CP13-113-000, May 4, 2015, 10–11.

27. James Osborne, "Kelcy Warren on Growing up a Texan, the American Dream and How Oil Prices Still Keep Him Awake at Night (A Little)," *Dallas News*, December 2015, https://www.dallasnews.com/business/business/2015/12/02/kel cy-warren-on-growing-up-a-texan-the-american-dream-and-how-oil-prices-still -keep-him-awake-at-night-a-little.

28. "Order Issuing Certificates," FERC CP15-93, February 2, 2017, 2.

29. "Order Issuing Certificates," 2.

30. "Ohio EPA, Response to Rover Pipeline LLC's Request to Resume HDD Drilling," FERC CP15-93, September 7, 2017.

31. Craig Butler, Director Ohio EPA, "Violations from Inadvertent Returns by Rover Pipeline LLC Since September 8, 2017," FERC CP15-93, November 22, 2017.

32. Chris Sonneborn, Senior Vice President, Rover Pipeline LLC, "Response to January 24, 2018 Letter Tuscarawas Mainline B Horizontal Directional Drill," FERC CP15-93, January 28, 2018.

33. Chris Sonneborn, Senior Vice President, Rover Pipeline LLC, "Rover Pipeline LLC's Response to Ohio EPA Letter of February 16, 2018," FERC CP15-93, February 21, 2018.

34. West Virginia Department of Environmental Protection, Order 8813, March 5, 2018.

35. Christopher Helman, "How EOG Became One of America's Great Oil Companies," *Forbes*, July 24, 2013.

36. The Production Tax Credit (mostly for wind) and Investment Tax Credit (mostly for solar) gave developers of renewable resources subsidies by allowing them to reduce their tax bill.

37. FPC v. Hope, 629.

# BIBLIOGRAPHY

## GOVERNMENT SOURCES

Carter, James Earl, Jr. *Public Papers of the Presidents of the United States, Jimmy Carter, 1977. Book I-January 20 to June 24, 1977.* Washington, DC: Government Printing Office, 1977.

Federal Deposit Insurance Corporation. "Managing the Crisis: The FDIC and RTC Experience," chapter 3: Penn Square Bank, n.a. https://www.fdic.gov/bank/historical/managing/documents/history-consolidated.pdf.

Federal Energy Regulatory Commission. "Final Report on Price Manipulation in Western Markets: Fact-finding Investigation of Potential Manipulation of Electric and Natural Gas Prices," Docket PA02-2-000, March 2003.

Federal Trade Commission. *Summary Report of the Federal Trade Commission to the Senate of the United States . . . on Economic, Financial, and Corporate Phases of Holding and Operating Companies of Electric and Gas Utilities,* vols. 68–69a. Washington, DC: Government Printing Office, 1934–1935.

H.R. Rep. No. 709, 75th Cong., 1st sess, April 28, 1937.

Holmes, Stewart. "The Development of Market Centers and Electronic Trading in Natural Gas Markets." Federal Energy Regulatory Commission, Office of Economic Policy, June 1999. https://www.ferc.gov/legal/maj-ord-reg/land-docs/mkt-ctrs.pdf.

Kansas Geological Survey. "Public Information Circular." December 1996. http://www.kgs.ku.edu/Publications/pic5/pic5_3.html.

Landis, James M. *Report on Regulatory Agencies to the President-Elect.* U.S. Senate Committee on the Judiciary, 86th Cong., 2nd sess. Washington, DC: Government Printing Office, 1960.

Landis, James, to J. F. Kennedy, February 20, 1961, Box 78, Presidential Office File, John F. Kennedy Library, Boston, Massachusetts.

"Natural Gas Amendments." *CQ Almanac* 1950, 6th ed., 598–602. Washington, DC, Congressional Quarterly, 1951. https://library.cqpress.com/cqalmanac/document.php?id=cqal54-1359735.

Staats, Elmer E. *The Advance Payment Program: An Uncontrolled Experiment.* Washington, DC: Government Printing Office, 1978.

"Submerged Lands Act." In *CQ Almanac* 1953, 9th ed., 09–388–09–396. Washington, DC, Congressional Quarterly, 1954.

Texas Railroad Commission. "General Market Demand Order for All Oil Wells in the Various Districts and Fields in the State Of Texas." Docket No. 20–61, 640, March 16, 1972.

Truman, Harry S. 1950: *Containing the Public Messages, Speeches, and Statements of the President, January 1 to December 31, 1950.* Washington, DC: Government Printing Office, 1951.

US Department of Commerce, National Oceanic and Atmospheric Administration, Pacific Marine Environmental Laboratory. "El Niño Theme Page." https://www.pmel.noaa.gov/elnino/faq.

US Department of Energy. "Comparing the Impacts of the 2005 and 2008 Hurricanes on U.S. Energy Infrastructure." February 2009. https://www.oe.netl.doe.gov/docs/HurricaneComp0508r2.pdf.

US Department of Energy. Energy Information Administration. Washington, DC: Government Printing Office.

> *An Analysis of the Natural Gas Policy Act and Several Alternatives.* 1981–1983.
>> Preliminary Analysis: *Natural Gas Pipeline/Producer Contracts: A Preliminary Analysis.* DOE/EIA-0312. December 1981.
>> Part I: *Current State of the Natural Gas Market: An Analysis of the Natural Gas Policy Act and Several Alternatives.* DOE/EIA-0313. December 1981.
>> Part II: *Natural Gas Producer/Purchaser Contracts and Their Potential Impact on the Natural Gas Market.* DOE/EIA-0330. June 1982.
>> Part III: *An Analysis of Post-NGPA Interstate Pipeline Wellhead Purchases.* DOE/EIA-0357. September 1982.
> *Growth in Unbundled Natural Gas Transportation Services: 1982–1987.* DOE/EIA-0525. 1989.
> *Drilling Sideways: A Review of Horizontal Well Technology and Its Domestic Application.* DOE/EIA-TR-0565. 1993.
> *Natural Gas Productive Capacity for the Lower 48 States, 1980–1995.* DOE/EIA-0542. 1994.
> "Natural Gas Productivity for the Lower-48 States." May 2001. https://slideplayer.com/slide/7498530/.
> Tobin, James. "Natural Gas Market Centers and Hubs: A 2003 Update." October 2003. https://www.eia.gov/naturalgas/articles/mkthubs03index.php.

US Department of Energy, Economic Regulatory Administration, Division of Power Supply and Reliability. *Compliance Problems of Small Utility Systems with the Power Plant and Industrial Fuel Use Act of 1978: Volume II—Appendices.* DOE/RG-0045 Vol 2. Washington, DC: Government Printing Office, 1981.

US Department of Energy, Office of Fossil Energy. Quarterly Focus Report: *LNG Trade: Past, Present, Future (?).* Washington, DC: Government Printing Office, 1995.

US Department of Labor, Bureau of Labor Statistics. "Technical Note: Labor Force, Employment, and Unemployment, 1929–1939: Estimating Methods." https://www.bls.gov/opub/mlr/1948/article/pdf/labor-force-employment-and-unemployment-1929-39-estimating-methods.pdf.

US Department of the Interior, Office of Economic Analysis. *Deregulation of Natural Gas Prices, Final Environmental Impact Statement*. Washington, DC: Government Printing Office, 1974.

US Department of the Interior, United States Geological Survey. *Mineral Resources of the United States, Calendar Year 1901*. Washington, DC: Government Printing Office, 1902.

US Department of the Treasury, Bureau of Statistics. *Report on the Internal Commerce of the United States for the Fiscal Year 1881–82: Commerce and Navigation*. Washington, DC: Government Printing Office, 1884.

US Office of Technology Assessment. *Description of LNG Technology and Import System*. Washington, DC: Government Printing Office, 1977.

US Senate, Permanent Subcommittee on Investigations. *Excessive Speculation in the Natural Gas Market*. Washington, DC: Government Printing Office, 2007.

## PUBLIC LAW, COURT CASES, FPC AND FERC ORDERS AND INTERVENTIONS

Atlantic Refining Company v. NY Public Service Commission, 360 U.S. 378 (1959).

Browneller v. Natural Gas Pipeline Co., 8 N.W.2d 474 (Iowa 1943).

Colorado Interstate v. FPC, 324 U.S. 581 (1945).

El Paso Natural Gas v. FPC, 281 F2d 567 (1960).

FPC v. Hope Natural Gas, 320 U.S. 591 (1944).

FPC v. Texaco, 417 U.S. 380 (1974).

Hope Natural Gas Co., 19 FPC 405 (1958).

Interstate v. FPC, 331 U.S. 682 (1947).

Maryland People's Counsel v. FERC, 768 F.2d 450, 455 (D.C. Cir. 1985).

Munn v. Illinois, 94 U.S. 113 (1876).

*Natural Gas Act of 1938*, Public Law 75-688, *U.S. Statutes at Large 52* (1938).

Order 428, 45 FPC (1971).

Order 455, 48 FPC (1972).

Permian Basin Area Rate Cases, 390 U.S. 747 (1968).

Phillips Petroleum Co., 7 FPC 983 (1948).

Phillips v. Wisconsin, 347 U.S. 672 (1954).

Preston Oil Company v. Transcontinental Gas Pipeline Corporation, No. 294491, Div. "L," 19th Judicial District Court, Parish of East Baton Rouge, LA (December 1986).

Public Service Commission of New York v. FPC, 467 F2d 361 (D.C. Cir. 1972).

Securities and Exchange Commission v. Electric Bond & Share Company, Southern District of New York, 18 F. Supp. 131 (1937).

United States v. Louisiana et al., 363 U.S. 1 (1960).

Wisconsin v. FPC, 205 F.2d 706 (D.C. Cir. 1953).

Dominion Cove Point LNG:

"Request for Rehearing of EarthReports, Inc. (dba Patuxent Riverkeeper); Potomac Riverkeeper, Inc.; Sierra Club; and Chesapeake Climate Action Network," Dominion Cove Point LNG, FERC CP13-113-000, October 15, 2014.

"Order Denying Rehearing and Stay," Dominion Cove Point LNG, FERC CP13-113-000, May 4, 2015.

Rover Pipeline LLC:

"Order Issuing Certificates," FERC CP15-93, February 2, 2017, 2.

"Ohio EPA, Response to Rover Pipeline LLC's Request to Resume HDD Drilling," FERC CP15–93, September 7, 2017.

Craig Butler, Director Ohio EPA, "Violations from Inadvertent Returns by Rover Pipeline LLC since September 8, 2017," FERC CP15-93, November 22, 2017.

Chris Sonneborn, Senior Vice President, Rover Pipeline LLC, "Response to January 24, 2018 Letter Tuscarawas Mainline B Horizontal Directional Drill," FERC CP15-93, January 28, 2018.

Chris Sonneborn, Senior Vice President, Rover Pipeline LLC, "Rover Pipeline LLC's Response to Ohio EPA Letter of February 16, 2018," FERC CP15-93, February 21, 2018.

West Virginia Department of Environmental Protection, Order 8813, March 5, 2018.

### DATA

Baker Hughes. https://rigcount.bhge.com/.

Bloomberg. By subscription only. https://bloomberg.com.

BP Statistical Review of World Energy, 2019. https://www.bp.com/en/global/corporate/energy-economics/statistical-review-of-world-energy.html.

US Department of Energy, Energy Information Administration. http://www.eia.gov.

US Department of the Interior, U.S. Geological Survey. *Minerals Yearbook*. 1932–1990.

US Department of Transportation. Bureau of Transportation Statistics. Historical Air Traffic Statistics. https://web.archive.org/web/20170702072400/https://www.rita.dot.gov/bts/sites/rita.dot.gov.bts/files/subject_areas/airline_information/air_carrier_traffic_statistics/airtraffic/annual/1954_1980.html.

DrillingInfo. By subscription only. https://www.enverus.com/.

Railroad Commission of Texas. "Crude Oil Production and Well Counts (Since 1935)." Accessed June 21, 2019. http://www.rrc.state.tx.us/oil-gas/research-and-statistics/.

## PRIMARY AND SECONDARY SOURCES

Abalone Alliance. "The Diablo Canyon Timeline." http://www.energy-net.org/ 01NUKE/DIABLO1.HTM.

"Akron, Ohio, Takes Its First Step Towards Manufactured Gas." *American Gas Engineering Journal* 111 (1919): 123.

American Oil & Gas Historical Society. "First Texas Oil Boom." http://aoghs.org/ petroleum-pioneers/texas-oil-boom/.

Anderson, Jenny. "US Regulators Grow Alarmed over Hedge Fund Hotels." *International Herald Tribune*, January 1, 2007.

"ANR Pipeline Co. History." Funding Universe, http://www.fundinguniverse .com/company-histories/anr-pipeline-co-history/.

"Applications for Rehearing of FERC Order No. 500 Reveal Industrywide Disaffection with Proposed Take-or-Pay Crediting and Cost Passthrough Mechanisms." *Foster Report*, September 10, 1987.

Bartlett, Christopher A. "EnronOnline: Louise Kitchen, Intrapreneur." Harvard Business School Teaching Note 302-011, August 2001.

Bartlett, N. D. *Panhandle Petroleum*. Amarillo, TX: Miller National Corporation, 1982.

Bean, Covey. "Oklahoma's Anadarko Basin: Where the High-Rollers Are." *Oklahoman* (Oklahoma City), October 25, 1981, 19.

Boucher, John Newton. *History of Westmoreland County, Pennsylvania*, vol. 1. New York: Lewis Publishing Company, 1906.

Bradley Jr., Robert L. *Edison to Enron: Energy Markets and Political Strategies.* Hoboken, NJ: John Wiley & Sons, 2011.

Brookline Connection. "Short History of the Evolution of Coal Hill (Mount Washington)."http://www.brooklineconnection.com/history/Facts/CoalHill.html.

Brown, David. "Explorer of the Year: Zagorski Made His Mark with the Marcellus." *American Association of Petroleum Geologists Explorer* 34, no. 5 (2013): 32–38.

Brune, Michael. "The Sierra Club and Natural Gas." February 2, 2012. https:// www.sierraclub.org/michael-brune/2012/02/sierra-club-and-natural-gas.

Burka, Paul. "Power Politics." *Texas Monthly*. May 1975.

Burrough, Bryan. *The Big Rich: The Rise and Fall of the Greatest Texas Oil Fortunes.* New York: Penguin, 2009.

"Calpine Corporate History." Funding Universe, http://www.fundinguniverse .com/company-histories/calpine-corporation-history/.

Carnegie, Andrew. "The Natural Oil and Gas Wells of Western Pennsylvania." *MacMillan's Magazine* 51, November 1884–April 1885, 208–14.

Caro, Robert. *Master of the Senate: The Years of Lyndon Johnson*. New York: Random House, 2009.

"The Case for Commodities as an Asset Class." Goldman, Sachs & Co., June 2004, slide 11. https://faculty.fuqua.duke.edu/~charvey/Teaching/BA453_2006/GS CI_Strategic_June_2004.ppt.

Castaneda, Christopher J. *Invisible Fuel: Manufactured and Natural Gas in America, 1800–2000*. New York: Twayne, 1999.

Castaneda, Christopher J. *Regulated Enterprise: Natural Gas Pipelines and Northeast Markets, 1938–1954*. Columbus: Ohio State University Press, 1992.

Castaneda, Christopher J., and Clarence M. Smith. *Gas Pipelines and the Emergence of America's Regulatory State: A History of the Panhandle Eastern Corporation, 1928–1993*. Cambridge: Cambridge University Press, 1996.

Castaneda, Christopher J., and Joseph A. Pratt. *From Texas to the East: A Strategic History of Texas Eastern Corporation*. College Station: Texas A&M University Press, 1993.

Chase, David. "Combined-Cycle Development Evolution and Future." http://phys ics.oregonstate.edu/~hetheriw/energy/topics/doc/elec/natgas/cc/combined%20 cycle%20development%20evolution%20and%20future%20GER4206.pdf.

Chesapeake Energy. "Chesapeake Announces Haynesville Shale Discovery and Seven Other New Unconventional Discoveries and Projects; Increases Capital Expenditures to Accelerate Development." March 24, 2008. https://www.businesswire.com/news/home/20080324005847/en/Chesapeake-Announces-Haynesville-Shale-Discovery-New-Unconventional.

Chesapeake Energy. "History." https://web.archive.org/web/20081209041628/http://www.chk.com/About/Pages/History.aspx.

"CITGO Petroleum Corporation History." Funding Universe, http://www.funding universe.com/company-histories/citgo-petroleum-corporation-history/.

Cohen, Linda, Martha Krebs, and B. B. Blevins. *Retrospective Report on California's Electricity Crisis*. Sacramento: California Council on Science and Technology, 2002.

Compton, Karl T. "Biographical Memoir of Elihu Thomson." *National Academy of Sciences: Biographical Memoirs* 21, no. 4 (1939): 143–79.

Daniel, Price. "Tidelands Controversy." Texas State Historical Association. http://www.tshaonline.org/handbook/online/articles/mgt02.

Davidson, Cliff I. "Air Pollution in Pittsburgh: A Historical Perspective," *Journal of the Air Pollution Control Association* 29, no. 12 (2012): 1035–41.

Dees, J. M., T. G. Freet, and G. S. Hollabaugh. "Horizontal Well Stimulation Results in the Austin Chalk Formation, Pearsall Field, Texas." Paper presented at the Society of Petroleum Engineers Annual Technical Conference and Exhibition, New Orleans, LA, September 23–26, 1990.

Doane, Michael J., and Daniel F. Spulber. "Open Access and the Evolution of the Spot Market for US Natural Gas." *Journal of Law and Economics* 37, no. 2 (October 1984): 477–515.

Dreyfuss, Barbara T. *Hedge Hogs*. New York: Random House, 2013.

Egan, Timothy. "Tapes Show Enron Arranged Plant Shutdown." *New York Times*, February 4, 2005.

Eisenhower, Dwight D. "Veto of HR 6645." *Pittsburgh Post-Gazette*, February 18, 1956, 2.

"Energetic Oklahoma." *Oklahoman* (Oklahoma City), April 2, 1975, 10.

Engelder, Terry. "Unconventional Natural Gas Reservoir Could Boost U.S. Supply." *Penn State News*, January 17, 2008.

"Enron Fall 2001 Retention Bonus Payments." Accessed February 17, 2018. http://news.findlaw.com/hdocs/docs/enron/enron61702retenbonus.pdf.

"Extent of Pipeline Take-or-Pay Problem Remains Focal Point of Gas Industry Disagreement during FERC Hearing on Order No. 500." *Foster Report*, April 14, 1988.

Exxon Corporation. *Middle East Oil*. Washington, DC: Government Printing Office, 1980.

"FERC Responds to House Energy Committee Inquiry on Proposed Take-or-Pay Recovery Policy." *Foster Report*, April 9, 1987.

Fisher, Kenneth L. *100 Minds that Made the Market*. Hoboken, NJ: John Wiley & Sons, 1993.

Foss, Michelle Michot. "Enron and the Energy Market Revolution." In *Enron, Corporate Fiascos, and their Implications*, edited by Nancy B. Rapoport and G. Dharan Bala. Los Angeles: Foundation Press, 2003.

Fryklund, Bob. "North American Gas—Changing Mix." November 9, 2006, slide 6. https://slideplayer.com/slide/4567877/.

Fukuyama, Francis. "The End of History?" *National Interest* 16 (Summer 1989): 3–18.

"Gas Merger." *Time*, March 31, 1930.

"Gas Replacing Coal: The Great Change Which is to Take Place in Pittsburgh." *New York Times*, June 23, 1885.

Gimein, Mark. "Who Turned the Lights Out?" *Fortune*, February 5, 2001.

Glass, James A. "The Gas Boom in East Central Indiana." *Indiana Magazine of History* 96, no. 4 (December 2000): 313–35.

Gomez, Julie. "Natural Gas Overview." February 2001, slide 21. https://slideplayer.com/slide/11543387/.

Goodell, Jeff. "The Big Fracking Bubble: The Scam Behind Aubrey McClendon's Gas Boom." *Rolling Stone*, March 1, 2012.

Greenspan, Alan. Testimony of Federal Reserve Board Chairman Greenspan to the Committee on Energy and Commerce, US House of Representatives, "Natural Gas Supply and Demand Issues," June 10, 2003.

Grieser, Bill, Jimmie Hobbs, Jeff Hunter, and Jerry Ables. "The Rocket Science behind Water Frac Design." Paper presented at Society of Petroleum Engineers Production and Operations Symposium, Oklahoma City, OK, March 23–26, 2003.

Griggs, John Wyeth. "Restructuring the Natural Gas Industry: Order 436 and Other Regulatory Initiatives." *Energy Law Journal* 7, no. 1 (January 1986): 71–99.

Gugin, Linda, and James E. St. Clair. *Sherman Minton: New Deal Senator, Cold War Justice.* Indianapolis: Indiana Historical Society, 1997.

Hedge Fund Marketing Association. "Hedge Fund Assets under Management Reach New High." March 28, 2017. http://www.hedgefundmarketing.org/hedge-fund-assets-under-management-reach-new-high/.

Hefner III, Robert A. *The Grand Energy Transition.* Hoboken, NJ: John Wiley & Sons, 2009.

Henderson, William D., and Richard D. Cudahy. "From Insull to Enron: Corporate (Re)Regulation after the Rise and Fall of Two Energy Icons." Articles by Maurer Faculty, Paper 308 (2005): 36–110.

Hightower, Michael J. "Penn Square: The Shopping Center Bank that Shook the World, Part 1–Boom." *Chronicles of Oklahoma* 90, no. 1 (Spring 2012): 68–99.

Hightower, Michael J. "Penn Square: The Shopping Center Bank that Shook the World, Part 2–Bust." *Chronicles of Oklahoma* 90, no. 2 (Summer 2012): 204–36.

Hiltzik, Michael. *The New Deal: A Modern History.* New York: Free Press, 2011.

Hoover, Michael. "The Whiskey Rebellion." US Department of the Treasury, Alcohol and Tobacco Tax and Trade Bureau. Accessed March 15, 2017. https://www.ttb.gov/public_info/whisky_rebellion.shtml.

Hope Natural Gas. *From Gas Lights to New Energy Heights.* N.p: Consolidated Natural Gas, 1998.

Ingham, John N. *Biographical Dictionary of American Business Leaders*, vol. 2. Westport, CT: Greenwood, 1983.

James, Edward T., and Robert Livingston Schuyler. "Leland Olds." In *Dictionary of American Biography, Supplement 6: 1956–1960.* New York: American Council of Learned Societies, 1980.

Jensen Associates. *U.S. Open Access Gas Pipeline Transportation: A Model for Europe?* Boston: n.p., 1990.

Johnson, Nicholas. "Producer Rate Regulation in Natural Gas Certification Proceedings: 'Catco' in Context." *Columbia Law Review* 62, no. 5 (May 1962): 773–820.

Joskow, Paul L. "California's Electricity Crisis." Working Paper, September 2001, tables 3–4. https://economics.yale.edu/sites/default/files/files/Workshops-Seminars/Industrial-Organization/joskow1-011115.pdf.

Kettelman, Sr., Orrin D. "David H. Kettelman." *San Joaquin Historian* 7, no. 1 (January 1971): 1–3.

Kiefner, John F., and Cheryl J. Trench. *Oil Pipeline Characteristics and Risk Factors: Illustrations from the Decade of Construction.* Washington, DC: American Petroleum Institute, 2001.

Kitchen, Louise. "Enron Net Works." June 12, 2000. https://www.slideserve.com/lyle-bates/enron-net-works-june-12-2000.

Kolko, Gabriel. *The Triumph of Conservatism: A Reinterpretation of American History, 1900–1916*. New York: Free Press, 1962.

Kozik, H. G., and S. A. Holditch. "A Case History for Massive Hydraulic Fracturing the Cotton Valley Lime Matrix, Fallon and Personville Fields." *Journal of Petroleum Technology* 33, no. 2 (1981): 229–45.

Kutchin, Joseph W. *How Mitchell Energy & Development Corp. Got Its Start and How It Grew*. Irvine, CA: Universal Publishers, 2001.

Lane, Franklin K. Speech to Natural Gas Conservation Conference on January 15, 1919, published in "Plan to Save Natural Gas," *Gas Record* 17, no. 2 (1920): 63–65.

Lawyer, L. C., Charles C. Bates, and Robert B. Rice. *Geophysics in the Affairs of Mankind*. Tulsa, OK: Society of Exploration Geophysics, 2001.

Lueck, Thomas J. "Gas Slump in Oklahoma." *New York Times*, April 26, 1983.

Lyon, Leverett S., and Victor Abramson. *Government and Economic Life: Development and Current Issues of American Public Policy*, vol. 2. Washington, DC: Brookings Institution, 1940.

MacAvoy, Paul W., and Robert S. Pindyck. *Price Controls and the Natural Gas Shortage*. Washington, DC: American Enterprise Institute for Public Policy Research, 1975.

Makholm, Jeff D. *Theory of Relationship-Specific Investments, Long-Term Contracts and Gas Pipeline Development in the United States*. Boston: National Economic Research Associates, 2006.

Mangan, Frank. *The Pipeliners*. El Paso, TX: Guynes Press, 1977.

Manning, Michael J., Kenneth R. Carretta, and James P. White. "Gas Inventory Charges: Evolving Mechanisms for Allocating the Risks and Recovering the Costs of Maintaining Gas Supply." *Energy Law Journal* 11, no. 1 (January 1990): 1–35.

Mayerhofer, M. A., M. F. Richardson, R. N. Walker Jr., D. N. Meehan, M. W. Oehler, and R. R. Browning Jr. "Proppants? We Don't Need No Proppants." Paper presented at Society of Petroleum Engineers Annual Technical Conference and Exhibition, San Antonio, TX, October 5–8, 1997.

Mayo, Travis. "Monroe Was World's Natural Gas Capital." *News-Star*, http://www.thenewsstar.com/story/news/local/2016/04/08/monroe-worlds-natural-gas-capital/82700550/.

McCullough, Hax, and Mary Brignano. *The Vision and Will to Succeed: A Centennial History of the Peoples Natural Gas Company*. N.p.: Peoples Natural Gas, 1985.

McLean, Bethany. *The Smartest Guys in the Room: The Amazing Rise and Scandalous Fall of Enron*. New York: Penguin, 2003.

Meehan, D. N., and D. W. Bossert. "Experience with One Thousand Horizontal Wells." Paper presented at the Annual Technical Meeting of the Petroleum Society of Canada, Calgary, Alberta, June 7–9, 1995.

"Mitchell Energy and Development Corporation Company History." Company-Histories, http://www.company-histories.com/Mitchell-Energy-and-Development-Corporation-Company-History.html.

Mogel, William A., and John P. Gregg. "Appropriateness of Imposing Common Carrier Status on Interstate Natural Gas Pipelines." *Energy Law Journal* 4, no. 2 (1983): 155–87.

Mogel, William A., and William R. Mapes Jr. "Assessment of Incremental Pricing under the Natural Gas Policy Act." *Catholic University Law Review* 29, no. 4 (Summer 1980): 763–98.

Moran, Richard. *Executioner's Current: Thomas Edison, George Westinghouse, and the Invention of the Electric Chair.* New York: Knopf Doubleday, 2007.

Morgan, Anne Hodges. "Kerr, Robert Samuel." *Encyclopedia of Oklahoma History and Culture.* https://www.okhistory.org/publications/enc/entry.php?entry=KE011.

Mott, Reginald Arthur. *The Triumphs of Coke: To Commemorate the Jubilee of the Coke Oven Managers' Association, 1915–1965.* N.p.: Coke Oven Managers' Association, 1965.

Nagiewicz, Stephen D. *Hidden History of Maritime New Jersey.* Charleston, SC: History Press, 2016.

National Petroleum Council. "Balancing Natural Gas Policy: Fueling the Demands of a Growing Economy." Vol. 2. September 2003. https://www.npc.org/reports/ng.html.

"Natural Gas Clearinghouse Arranges First Deal Entailing Short-Term Sale of 200,000 Mcf per Day to Four Northeast Users; Other Spot Market Services Discussed at Industry Conference." *Foster Report,* October 4, 1984, 5–6.

Nehring, Richard. *The Discovery of Significant Oil and Gas Fields in the United States.* Santa Monica, CA: RAND Corporation, 1981.

Nelson, Mary Jo. "Congress Changing Gears on Oil Industry." *Oklahoman* (Oklahoma City), October 31, 1989, 22.

Niedringhaus, George W. "Future of the Steel Industry in the Mississippi Valley: The Revolution Consequent on the Coking of Illinois Coal." *Executive's Magazine: The Business Journal of the Southwest,* June 5, 1922, 18, 32, 36.

Nissley, Erin L. "Local History: 1977 Brought Frigid Temperatures, Natural Gas Shortage." *Times-Tribune* (Scranton, PA), September 13, 2015.

Nordhaus, Robert R. "Producer Regulation and the Natural Gas Policy Act of 1978." *Natural Resources Journal* 19, no. 4 (Fall 1979): 829–57.

"Norness." Uboat.net. http://uboat.net/allies/merchants/1248.html.

O'Brien, Justin. "Too Big To Fail or Too Hard To Remember: James M. Landis and Regulatory Design." Edmond J. Safra Center for Ethics, Harvard University, October 27, 2014. http://ethics.harvard.edu/blog/too-big-fail-or-too-hard-remember-james-m-landis-and-regulatory-design.

O'Connor, John J., Jr. *The Economic Cost of the Smoke Nuisance to Pittsburgh*. Pittsburgh: University of Pittsburgh, 1913.

"Oryx Energy Company History." Company-Histories, http://www.company-histories.com/Oryx-Energy-Company-Company-History.html.

"Oryx's Hauptfuhrer: Big Increase Due in U.S. Horizontal Drilling." *Oil & Gas Journal*, January 15, 1990, 28.

Osborne, James. "Kelcy Warren on Growing up a Texan, the American Dream and How Oil Prices Still Keep Him Awake at Night (A Little)." *Dallas Morning News*, December 2015. https://www.dallasnews.com/business/business/2015/12/02/kelcy-warren-on-growing-up-a-texan-the-american-dream-and-how-oil-prices-still-keep-him-awake-at-night-a-little.

Parsons, R. S., and R. W. Fincher. "Short-Radius Lateral Drilling: A Completion Alternative." Paper presented at the Society of Petroleum Engineers Eastern Regional Meeting, Columbus, OH, November 12–14, 1986.

Peoples Gas Light and Coke Company. *100 Years of Gas Service in Chicago*. Chicago: n.p., 1950.

Petrash, Jeffery M. "Long-Term Natural Gas Contracts: Dead, Dying, or Merely Resting?" *Energy Law Journal* 27, no. 2 (July 2006): 545–82.

Pinchot, Gifford. *The Power Monopoly: Its Make-Up and Its Menace*. Milford, CT: n.p., 1928.

Pratt, Joseph A., Tyler Priest, and Christopher J. Castaneda. *Offshore Pioneers: Brown & Root and the History of Offshore Oil and Gas*. Houston, TX: Gulf Publishing, 1997.

Prindle, David F. *Petroleum Politics and the Texas Railroad Commission*. Austin: University of Texas Press, 1981.

"Q&A: The Man Who Succeeded Hitler," *BBC News*, May 5, 2005. http://news.bbc.co.uk/2/hi/europe/4514765.stm.

Range Resources. "Range Expands Barnett Shale Holdings and Provides Operations Update." December 10, 2007. http://ir.rangeresources.com/news-releases/news-release-details/range-expands-barnett-shale-holdings-and-provides-operations.

Roberts, Charlotte. "House on Selby Property was Christy Payne's." Sarasota History Alive. http://www.sarasotahistoryalive.com/history/articles/house-on-selby-property-was-christy-payne-s/.

Roberts, Darwin P. "The Legal History of Federally Granted Railroad Rights-of-Way and the Myth of Congress's '1871 Shift.'" *University of Colorado Law Review* 82, no. 1 (2011): 85–166.

Ross, Bradford, and Bernard A. Foster, Jr. "Phillips and the Natural Gas Act." *Law and Contemporary Problems* 19 (Summer 1954): 382–412.

San Joaquin Valley Geology. "Famous Gushers of the World." http://www.sjvgeology.org/history/gushers_world.html.

Sanders, Elizabeth. *The Regulation of Natural Gas*. Philadelphia: Temple University Press, 1981.

Sanders, James R., and Daniel L. Stufflebaum. *A Study of School Without Schools: The Columbus, Ohio, Public Schools During the Natural Gas Shortage, Winter 1977*. Kalamazoo: Western Michigan University, 1977.

Scanlan, Alfred Long. "Administrative Abnegation in the Face of Congressional Coercion: The Interstate Natural Gas Company Affair." *Notre Dame Law Review* 23, no. 2 (1948): 173–207.

Schofield, Robert E. "The Science Education of an Enlightened Entrepreneur: Charles Willson Peale and His Philadelphia Museum, 1784–1827." *American Studies* 30, No. 2 (Fall 1989): 21-40.

Schwartz, Nelson D. "Enron Fallout: Wide, But Not Deep Enron's Collapse Has Hurt Big American Banks. But Not as Much as a Default by Argentina Would." *Fortune*, December 24, 2001.

Shanahan, Eileen. "FPC Chief Urges Price Rise for Gas." *New York Times*, September 21, 1969, 67.

Silver, Jonathan D. "The Marcellus Boom / Origins: The Story of a Professor, a Gas Driller and Wall Street." *Pittsburgh Post-Gazette*, March 20, 2011.

Singer, Mark. *Funny Money*. New York: Knopf, 1985.

Smith, John T., II. *Cars, Energy, Nuclear Diplomacy and The Law: A Reflective Memoir of Three Generations*. Lanham, MD: Rowman & Littlefield, 2012.

Steward, Dan. *The Barnett Shale Play: Phoenix for the Fort Worth Basin*. Fort Worth: Fort Worth Geological Society and the North Texas Geological Society, 2007.

Stotz, Louis, and Alexander Jamison. *History of the Gas Industry*. New York: Stettiner Bros, 1938.

Sweeney, James. *The California Electricity Crisis*. Stanford, CA: Hoover Institution Press, 2002.

Tankersley, G. J. *The Story of Consolidated Natural Gas Company: Innovation, Ingenuity, and Accomplishment*. New York: Newcomen Society in North America, 1980.

Teece, David J. "The Uneasy Case for Mandatory Contract Carriage in the Natural Gas Industry." In *New Horizons in Natural Gas Deregulation*, edited by Jerry Ellig and Joseph P. Kalt. Westport, CT: Praeger, 1996.

Texas Eastern Transmission Corporation. *The Big Inch and Little Big Inch Pipelines: The Most Amazing Government-Industry Cooperation Ever Achieved*. Houston: n.p., 2000.

"Texas Minerals." *Texas Almanac* 1978–1979.

Texas State Library and Archives Commission. "Hazardous Business: Industry, Regulation, and the Texas Railroad Commission." August 2011. https://www .tsl.texas.gov/exhibits/railroad/power/page1.html.

Troesken, Werner. "Regime Change and Corruption. A History of Public Utility Regulation." In *Corruption and Reform: Lessons from America's Economic*

*History*, edited by Edward L. Glaeser and Claudia Goldin, 259–81. Cambridge, MA: National Bureau of Economic Research, 2006.

Troesken, Werner. "The Institutional Antecedents of State Utility Regulation: The Chicago Gas Industry." In *The Regulated Economy: A Historical Approach to Political Economy*, edited by Claudia Goldin and Gary D. Libecap, 55–80. Chicago: University of Chicago Press, 1994.

Ventura, Jeffrey. "Range's Path to Discovery and Commercialization of the Marcellus Shale—The Largest Producing Gas Field in the United States." Adapted from presentation of lecture presented at the AAPG Annual Convention and Exhibition, Pittsburgh, PA, May 20, 2013. http://www.searchanddiscovery .com/documents/2013/110165ventura/ndx_ventura.pdf.

Vietor, Richard H. K. *Contrived Competition: Regulation and Deregulation in America*. Cambridge, MA: Belknap Press of Harvard University Press, 1994.

Vietor, Richard H. K. *Energy Policy in America since 1945*. New York: Cambridge University Press, 1984.

Waples, David A. *The Natural Gas Industry in Appalachia*. Jefferson, NC: McFarland, 2012.

Waters, L.L. *Energy to Move*. Owensboro: Texas Gas Transmission Company, 1985.

Weaver, Bobby D. "Doherty, Henry Latham." *Encyclopedia of Oklahoma History and Culture*. https://www.okhistory.org/publications/enc/entry.php?entry =DO006.

Weaver, Jacqueline Lang. *Unitization of Oil and Gas Fields in Texas: A Study of Legislative, Administrative, and Judicial Policies*. Washington, DC: Resources for the Future Press, 2011.

Western Pennsylvania Conservancy and Mount Washington Community Development Corporation. "Master Implementation Plan for the Mount Washington 'Emerald Link.'" November 2005. http://waterlandlife.org/assets/Section_ II_word.pdf.

Williamson, Harold F., Ralph L. Andreano, and Carmen Menezes. "The American Petroleum Industry." In *Output, Employment, and Productivity in the United States after 1800*, edited by Dorothy Brady, 349–404. Cambridge, MA: National Bureau of Economic Research, 1966.

Wynn, Margaret. "Natural Gas in Indiana: An Exploited Resource." *Indiana Quarterly Magazine of History* 4, no. 1 (March 1908): 31–45.

Yergin, Daniel. *The Prize: The Epic Quest for Oil, Money & Power*. New York: Free Press, 1991.

Zacchino, Narda. *California Comeback: How a 'Failed State' Became a Model for the Nation*. New York: Thomas Dunne Books, 2016.

Zuckerman, Gregory. *The Frackers*. New York: Penguin, 2013.

# INDEX